T0205703

Advances in Intelligent Systems and Computing

Volume 382

Series editor

Janusz Kacprzyk, Polish Academy of Sciences, Warsaw, Poland
e-mail: kacprzyk@ibspan.waw.pl

About this Series

The series "Advances in Intelligent Systems and Computing" contains publications on theory, applications, and design methods of Intelligent Systems and Intelligent Computing. Virtually all disciplines such as engineering, natural sciences, computer and information science, ICT, economics, business, e-commerce, environment, healthcare, life science are covered. The list of topics spans all the areas of modern intelligent systems and computing.

The publications within "Advances in Intelligent Systems and Computing" are primarily textbooks and proceedings of important conferences, symposia and congresses. They cover significant recent developments in the field, both of a foundational and applicable character. An important characteristic feature of the series is the short publication time and world-wide distribution. This permits a rapid and broad dissemination of research results.

Advisory Board

More information about this series at http://www.springer.com/series/11156

Joong Hoon Kim · Zong Woo Geem
Editors

Harmony Search Algorithm

Proceedings of the 2nd International
Conference on Harmony Search Algorithm
(ICHSA 2015)

Springer

Editors
Joong Hoon Kim
Civil, Environmental and Architectural
  Engineering
Korea University
Seoul
Korea, Republic of (South Korea)

Zong Woo Geem
Energy IT
Gachon University
Seongnam-Si
Korea, Republic of (South Korea)

ISSN 2194-5357   ISSN 2194-5365   (electronic)
Advances in Intelligent Systems and Computing
ISBN 978-3-662-47925-4   ISBN 978-3-662-47926-1   (eBook)
DOI 10.1007/978-3-662-47926-1

Library of Congress Control Number: 2015946091

Springer Heidelberg New York Dordrecht London

Printed on acid-free paper

Springer-Verlag GmbH Berlin Heidelberg is part of Springer Science+Business Media
(www.springer.com)

Preface

Since the Harmony Search Algorithm (HSA) was first introduced in 2001, it has drawn a world-wide attention which comes to the citations of over 2,000 citations in the domain of not only soft computing but also engineering optimization. The first International Conference on HSA (ICHSA) was held in February of 2014 out of an acknowledged need to bring together researchers who study on and with HSA.

The second ICHSA, which will be held in Korea University in Seoul, South Korea from 19 to 21 August, 2015, is intended to be an international forum for researchers in the area of developing, design, variants, and hybrid methods of HSA. The Conference created an excellent opportunity to introduce various aspects of HSA to junior researchers as well as professors in various disciplines. Topics ranged from the most recent variants of HSA to the application to new engineering problems.

This proceedings contain many interesting papers each of which has a lot of potential to be extended to a journal paper in high-quality journals. The papers can be divided into seven groups: various aspects of optimization algorithms, large scale applications of HSA, recent variants of HSA, other nature-inspired algorithms, related areas and computational intelligence, optimization in civil engineering, and multi-objectives variants of HSA. We hope that you gain a deeper insight into HSA and other optimization algorithms and enjoy reflecting and discussing the proceedings with your colleagues.

In the first part, the papers on various aspects of optimization algorithms are included. In the following chapters, large scale applications and recent variants of HSA, other nature-inspired algorithms, and computational intelligence are included. Finally, multi-objectives variants of HSA are introduced after optimization in civil engineering.

The editors would like to express our deep gratitude to the ICHSA 2015 patron, keynote speakers, members of International Steering Committee and International Scientific Committee, and Reviewers. The Conference would not have been successful without their support. We are also grateful to Springer and its team for their work in the publication of this proceedings.

Finally, financial support from National Research Foundation and Korea University was absolutely essential to the Conference. Our sincere appreciation goes to the two sponsors of ICHSA 2015.

August 2015   Joong Hoon Kim
                                                                                     LOC Chair

Contents

Part I
Various Aspects of Optimization Algorithms

Investigating the Convergence Characteristics of Harmony Search

Joong Hoon Kim, Ho Min Lee and Do Guen Yoo

Abstract Harmony Search optimization algorithm has become popular in many fields of engineering research and practice during the last decade. This paper introduces three major rules of the algorithm: harmony memory considering (HMC) rule, random selecting (RS) rule, and pitch adjusting (PA) rule, and shows the effect of each rule on the algorithm performance. Application of example benchmark function proves that each rule has its own role in the exploration and exploitation processes of the search. Good balance between the two processes is very important, and the PA rule can be a key factor for the balance if used intelligently.

Keywords Harmony search · Convergence · Exploration and exploitation · Harmony memory considering · Pitch adjustment

1 Introduction

Optimization is the process of selecting the best element from some sets of available alternatives under certain constraints. In each iteration of the optimization process, choosing the values from within an allowed set is done systematically until the minimum or maximum result is reached or when the stopping criterion is met [1]. Meta-heuristic algorithms are well known approximate algorithms which can solve optimization problems with satisfying results [2, 3]. The Harmony Search (HS) algorithm [4, 5] is one of the most recently developed optimization algorithm and at a same time, it is one the most efficient algorithm in the field of combinatorial

J.H. Kim(✉) · H.M. Lee
School of Civil, Environmental and Architectural Engineering,
Korea University, Seoul 136-713, South Korea
e-mail: {jaykim,dlgh86}@korea.ac.kr

D.G. Yoo
Research Center for Disaster Prevention Science and Technology,
Korea University, Seoul 136-713, South Korea
e-mail: godqhr425@naver.com

© Springer-Verlag Berlin Heidelberg 2016
J.H. Kim and Z.W. Geem (eds.), *Harmony Search Algorithm,*
Advances in Intelligent Systems and Computing 382,
DOI: 10.1007/978-3-662-47926-1_1

optimization [6]. The HS algorithm can be conceptualized from a musical performance process involving searching for a best harmony. In the HS algorithm, random
selecting (RS) rule, harmony memory considering (HMC) rule, and pitch adjusting
(PA) rule are used for generation of new solution, and then adopts two parameters of
harmony memory considering rate (HMCR) and pitch adjustment rate (PAR), which
mean a selection probability of one of the processes. In addition, harmony memory
size (HMS) representing the size of memory space (harmony memory, HM) and
band width (BW) meaning the adjustment width during the pitch adjustment are
used as the parameters. Recently, the HS algorithm's three rules were analyzed by
using various parameters for applications of continuous benchmark functions by
Ahangaran and Ramesani [7]. In this study, six benchmark functions have varied
characteristics (e.g., continuous, discrete and mixed discrete functions) were used
for analysis of the effect of each rule on the algorithm performance.

2   Three Rules of Harmony Search Algoritm

2.1   Random Selecting (RS) Rule

In the RS operation, the values of decision variables are generated randomly in the
boundary condition with probability of (1-HMCR). The RS rule is one of the exploration (global search) parts in the optimization process. The role of the RS rule
is inducement to escaping from local optima for new solution by using sketchy
search with whole solution domain for each dicision variable.

2.2   Harmony Memory Considering (HMC) Rule

The HMC rule selects the solution value for each decision variable from the memory space (HM) of the HS algorithm. The probability of selecting HMC rule is
HMCR and it can have a value between 0 and 1. In general, in the cases with between 0.70 and 0.95 of HMCR produce good results. The HMC rule is exploitation (local search) part in the optimization process of the HS algorithm.

2.3   Pitch Adjusting (PA) Rule

After finish the HMC operation, the PA operation can be selected with probability
of PAR. In the PA operation, a selected solution value of decision variable from
HMC operation is adjusted with upper or lower value. The parameter PAR can
have a value between 0 and 1 and it is usually set between 0.01 and 0.30. The PA
rule has composite role in the HS algorithm. It is an exploration part for escaping
from local optima, and it is also an exploitation part in the optimization process
for finding exact optimal point by using fine tuning of decision variables.

3   Methodology

In this study, the HS algorithm with various parameter combinations was applied
for solving six unconstrained benchmark functions widely examined in the

literature (two continuous benchmark functions, two descrete bechmark functions, and two mixed descrete benchmark functions). The optimization task was carried out using 30 individual runs for problems.

Table 1 Benchmark Functions (BFs)

BF 1 (continuous) : Six-hump camel back function

$$\text{Minimize } f(x) = \left(4 - 2.1x_1^2 + x_1^4 / 3\right)x_1^2 + x_1 x_2 + (-4 + 4x_2^2)x_2^2$$

$-3 \leq x_1 \leq 3, -2 \leq x_2 \leq 2$, $\min f(x) = -1.0316$

BF 2 (continuous) : Goldstein price's function

$$\text{Minimize } f(x) = \left[1 + \left(x_1 + x_2 + 1\right)^2 \left(19 - 14x_1 + 3x_1^2 - 14x_2 + 6 + x_1 x_2 + 3x_2^2\right)\right]$$

$$\times \left[30 + \left(2x_1 - 3x_2\right)^2 \left(18 - 32x_1 + 12x_1^2 + 48x_2 - 36x_1 x_2 + 27x_2^2\right)\right]$$

$-2 \leq x_i \leq 2, i \in \{1, 2\}$, $\min f(x) = 3$

BF 3 (discrete) : Gear function

$$\text{Minimize } f(x) = \left(\frac{1}{6.931} - \frac{x_1 x_2}{x_3 x_4}\right)$$

$12 \leq x_i \leq 60$ (integer variables), $i \in \{1, 2, 3, 4\}$, $\min f(x) = 0$

BF 4 (discrete) : Simpleton-25 function

$$\text{Minimize } f(x) = -\sum_{i=1}^{n} x_i$$

$0 \leq x_i \leq 10$ (integer variables), $i \in \{1, 2, \cdots, n\}$, $n = 25$, $\min f(x) = -250$

BF 5 (mixed discrete) : Mixed Griewank function

$$\text{Minimize } f(x) = (1 / 4000)\sum_{i=1}^{n} x_i^2 - \prod_{i=1}^{n} \cos(x_i / \sqrt{i}) + 1$$

$-600 \leq x_i \leq 600$ (continuous variables), $i \in \{1, 2, 3, 4\}$,

$-600 \leq x_i \leq 600$ (integer variables), $i \in \{5, 6, 7, 8\}$,

$n = 8$, $\min f(x) = 0$

BF 6 (mixed discrete) : Mixed Ackley function

$$\text{Minimize } f(x) = -a \exp\left(-b\sqrt{(1/n)\sum_{i=1}^{n} x_i^2}\right) - \exp\left((1/n)\sum_{i=1}^{n} \cos(cx_i)\right) + a + \exp(1)$$

$-32 \leq x_i \leq 32$ (continuous variables), $i \in \{1, 2, 3, 4\}$,

$-32 \leq x_i \leq 32$ (integer variables), $i \in \{5, 6, 7, 8\}$,

$a = 20, b = 0.2, c = 2\pi, n = 8$, $\min f(x) = 0$

Table 2 Applied Parameters

Cases	HMS	HMCR	PAR	BW	NFEs
Case 1		1.0			
Case 2		0.8			
Case 3		0.5	0.2		
Case 4	10	0.2			
Case 5	(for BFs 1-3),	0.0			20,000
Case 6			1.0	0.01	
Case 7	30		0.8		
Case 8	(for BFs 4-6)	0.8	0.5		
Case 9			0.2		
Case 10			0.0		

Tables 1 and 2 show the definitions and specifications of benchmark functions and applied parameter combinations of HS algorithm in this study respectively. In this case study, HMS of 10 and 30 were applied to bechmark functions respectively in consideration of the number of decision variables in each function. HMCR and PAR were applied differently in each case as shown in Table 3. The total number of function evaluations (NFEs) was fixed value 20,000 and also BW is fixed value 0.01.

Case 1 has HMC and PA rules, Cases 2-4, 7-9 have RS, HMC and PA rulse in accordance with HMCR and PAR, Case 5 only has RS rule, Case 6 has RS and PA rules, and Case 9 has RS and HMC rules respectively. Cases 2 and 9 are same case as a default parameter combination for the comparison criterion of Cases 1-5 and Cases 6-9 respectively.

4   Results and Discussions

Table 3 and Figures 1-3 show the analysis results from appications of HS algorithm with various combinations of parameters for becnmark functions (Figure 1 for BFs 1, 2, Figure 2 for BFs 3, 4, and Figure 3 for BFs 5, 6).

Table 3 Analysis Results Comparison

Cases	BF 1 Avg. error	BF 2 Avg. error	BF 3 Avg. error	BF 4 Avg. error	BF 5 Avg. error	BF 6 Avg. error
Case 1	1.09E-01	1.67E+01	9.19E-05	1.67E+00	1.10E+01	1.03E+01
Case 2	0.00E+00	7.88E-06	2.36E-05	0.00E+00	2.77E-01	2.18E-01
Case 3	0.00E+00	4.33E-05	3.03E-05	2.52E+01	2.34E+00	5.31E+00
Case 4	4.53E-06	6.73E-04	4.19E-05	5.55E+01	1.60E+01	1.34E+01
Case 5	3.23E-01	1.76E+01	6.71E-05	7.47E+01	9.38E+01	1.84E+01
Case 6	6.33E-07	2.76E-04	5.43E-05	1.29E+01	7.77E-01	2.80E+00
Case 7	0.00E+00	9.43E-06	1.85E-05	7.57E+00	7.47E-01	2.82E+00
Case 8	0.00E+00	4.76E-06	1.74E-05	9.60E+00	4.97E-01	1.64E+00
Case 9	0.00E+00	7.88E-06	2.36E-05	0.00E+00	2.77E-01	2.18E-01
Case 10	1.95E-04	7.71E-03	1.31E-04	8.67E-01	4.04E-01	1.22E+00

In most benchmark functions, the combined cases with three rules of the HS algorithm (Cases 2-4, 7-9) showed better efficiency than combined cases with two rules (Cases 1, 6 and 10) and cases with one rule (Cass 5). This results mean the importance of each rule in the HS algorithm and each rule has own role in the optimization process of HS algorithm. Meanwhile, the cases with the value of HMCR above 0.5 and the cases with the value of PAR below 0.5 showed better results of average error stably than other cases.

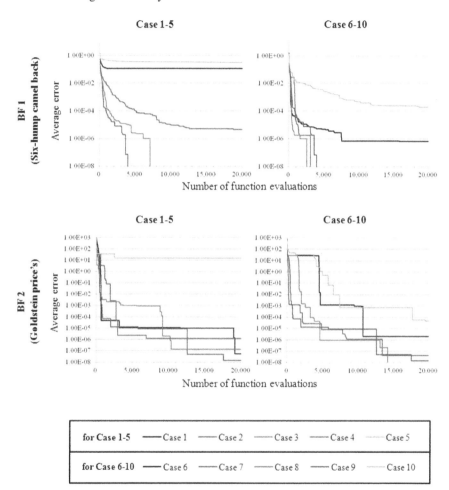

Fig. 1 Average Error Results for BFs 1, 2 (continuous functions)

Average error results of benchmark functions with parameter combination Case 1, only includes RS rule, showed the effect of the RS rule in early stages is far more than the final iterations. Therfore, we can conclude that in early stages of optimization process RS and PA rules work together as an exploration part, and during optimization progresses the influences of HMC and PA rules are increased gradually for exploitation.

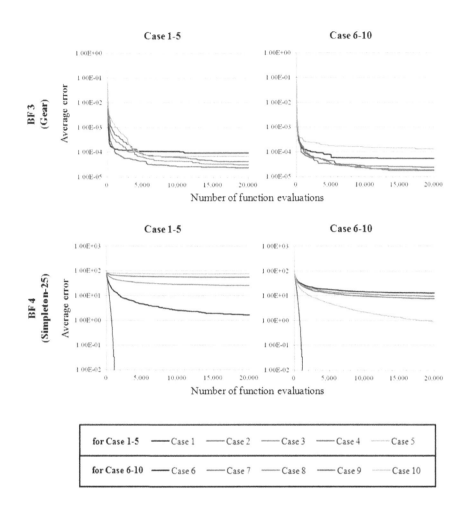

Fig. 2 Average Error Results for BFs 3, 4 (discrete functions)

Meanwhile, for the optimization results of BF 4 (Figure 2), Case 1 without RS rule and Case 10 without HMC rule produced second ranking results among Cases 1-5 and Cases 6-10 respectivley. The reason is characteristics of BF 4. This function has 25 dicision variables, more dicision variables than other benchmark functions, however BF 4 does not include local optima. Moreover BF 4 is discrete problem with 10 possible solutions for each dicision variable. Therefor we should consider the problem characteristics in the optimization when we apply optimization algorithms for particular problem.

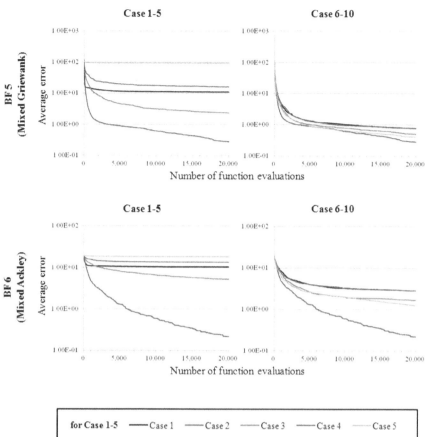

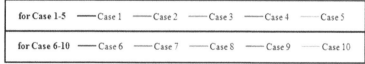

Fig. 3 Average Error Results for BFs 5, 6 (mixed discrete functions)

In this study the varied combinations of parameters HMCR and PAR were applied to benchmark functions to compare the convergence characteristics. However, the HS algorithm includes two more parameters HMS and BW, also important parameters in optimization process. Therfore the anlysis of results by considering various combinations of HMS and BW should be studied on our future research.

5   Conclusion

Optimization is the process of selecting the best solution among available alternatives under certain constraints. Meta-heuristic algorithms are well known approximate algorithms which can solve optimization problems with satisfying results and have own rules for finding best solution. The Harmony Search (HS) algorithm is one of the most recently developed optimization algorithm, and it has three rules in the optimization. The optimization process of HS algorithm includes three operation rules, harmony memory considering (HMC) rule, random selecting (RS) rule, and pitch adjusting (PA) rule. In this study, six benchmark functions have varied characteristics were selected for analysis of the effect of each rule on the algorithm performance.

Applications of benchmark functions prove that each rule has its own role in the exploration and exploitation processes of the optimization. In addition, the selection of suitable parameter combination with considering characteristics of object problem is essential for using optimization algorithms. In early stages of optimization process RS and PA rules have a leading role for exploration, and as optimization progresses the roles of HMC and PA rules are important for exploitation. Good balance between exploration and exploitation is very important for every optimization algorithm, and the intelligent use of PA rule can be a key factor for the balance in the HS algorithm.

Acknowledgement This work was supported by the National Research Foundation of Korean (NRF) grant funded by the Korean government (MSIP) (NRF-2013R1A2A1A01013886).

References

1. Moh'd Alia, O., Mandava, R.: The variants of the harmony search algorithm: an overview. Artificial Intelligence Review **36**(1), 49–68 (2011)
2. Blum, C., Roli, A.: Metaheuristics in combinatorial optimization: Overview and conceptual comparison. ACM Computing Surveys (CSUR) **35**(3), 268–308 (2003)
3. Blum, C., Roli, A.: Hybrid metaheuristics: an introduction. In: Blum, C., Aguilera, M.J.B., Roli, A., Sampels, M. (eds.) Hybrid Metaheuristics, vol. 114, pp. 1–30. Springer, Heidelberg (2008)
4. Geem, Z.W., Kim, J.H., Loganathan, G.V.: A new heuristic optimization algorithm: harmony search. Simulation **76**(2), 60–68 (2001)
5. Kim, J.H., Geem, Z.W., Kim, E.S.: Parameter estimation of the nonlinear Muskingum model using harmony search. Journal of the American Water Resources Association **37**(5), 1131–1138 (2001)
6. Geem, Z.W.: Music-inspired Harmony Search Algorithm: Theory and Applications. Springer (2009)
7. Ahangaran, M., Ramesani, P.: Harmony search algorithm: strengths and weaknesses. Journal of Computer Engineering and Information Technology (2013)

Performance Measures of Metaheuristic Algorithms

Joong Hoon Kim, Ho Min Lee, Donghwi Jung and Ali Sadollah

Abstract Generally speaking, it is not fully understood why and how metaheuristic algorithms work very well under what conditions. It is the intention of this paper to clarify the performance characteristics of some of popular algorithms depending on the fitness landscape of specific problems. This study shows the performance of each considered algorithm on the fitness landscapes with different problem characteristics. The conclusions made in this study can be served as guidance on selecting algorithms to the problem of interest.

Keywords Fitness landscape · Metaheuristic algorithms · Nature-Inspired algorithms · Optimization · Performance measures

1 Introduction

Numerous optimization algorithms have been proposed to tackle a number of problems that cannot be solved analytically. Generally, a newly developed algorithm is compared with a set of existing algorithms with respect to their performances on a set of well-known benchmark functions. The development is considered as a success if the new algorithm outperforms the existing algorithms considered. However, conventional benchmark test problems have a limited range of fitness landscape structure (e.g., the number and height of big valley), which makes it difficult to investigate the performance of newly developed algorithm on

J.H. Kim(✉) · H.M. Lee
School of Civil, Environmental and Architectural Engineering,
Korea University, Seoul 136-713, South Korea
e-mail: {jaykim,dlgh86}@korea.ac.kr

D. Jung · A. Sadollah
Research Center for Disaster Prevention Science and Technology,
Korea University, Seoul 136-713, South Korea
e-mail: donghwiku@gmail.com, sadollah@korea.ac.kr

© Springer-Verlag Berlin Heidelberg 2016
J.H. Kim and Z.W. Geem (eds.), *Harmony Search Algorithm,*
Advances in Intelligent Systems and Computing 382,
DOI: 10.1007/978-3-662-47926-1_2

the landscape with specific geometric property [1-2]. Therefore, previous studies provided little guidance for practitioners on selecting the best-suitable algorithm to the problem of interest [3-4].

Recently, a fitness landscape generator proposed by Gallagher and Yuan [5] has drawn attention in the study of various nature-inspired algorithms. The proposed landscape generator is used to generate optimization solution surfaces for continuous, boundary-constrained optimization problems and parameterized by a small number of parameters each of which controls a particular geometric feature of the generating landscapes. Therefore, by using the generator, a number of fitness landscapes of various geometric features can be generated and used for the full investigation of relative strengths and weaknesses of algorithms. General guidance on the algorithm selection can be extracted from the results of the investigations.

This paper compared the performances of eight optimization algorithms using fitness landscapes generated by a modified Gaussian fitness landscape generator originally proposed in Gallagher and Yuan [5]. Eight algorithms are compared with respect to their expected performance and the performance variation (performance reliability). Radar plots of several algorithms were drawn and compared to indicate the level of the two performance measures.

2   Methodology

The following sections describe the selected eight algorithms, methodologies for test problem generation, and performance measures and its visualizations.

2.1   Algorithm Selection

In this study, total of eight optimization algorithms are compared with respect to their performances on generated fitness landscapes. Eight algorithms are listed as follows: random search (RS) as a comparison target, simulated annealing (SA) [6], particle swarm optimization (PSO) [7], water cycle algorithm (WCA) [8], genetic algorithms (GAs) [9], differential evolution (DE) [10], harmony search (HS) [11, 12], and cuckoo search (CS) [13]. Most algorithms were inspired by nature phenomena or animal behavior and their fundamental optimization mechanisms are based on generating new solutions while adopting different strategies for the task.

RS keeps randomly generating new solutions within the allowable range until stopping criteria are met. SA, inspired by annealing process in metallurgy, moves the current state (solution) of a material to some neighboring state with a probability that depends on two energy states and a temperature parameter. PSO simulates social behavior of organisms in a bird flock or fish school in which particles in a swarm (solutions in a population) are guided by their own best position as well as the entire swarm's best known position. WCA mimics the river network generation process where streams are considered as candidate solutions. GAs has gained inspiration from natural adaptive behaviors, i.e., "the survival of the fittest". In DE, a new solution is generated by combining three existing randomly selected solutions from the population. HS was inspired by the musical ensemble and contains a solution storage function called harmony memory. CS was inspired by the obligate brood parasitism of some cuckoo species and their various strategies for choosing the nest to lay their eggs.

For more details on the algorithm, please refer to the references supplied above.

2.2 Test Problem Generation

To test and compare a newly developed metaheuristic algorithm, several well-known benchmark problems (e.g., Ackley and Rosenbrock functions) have been used [14-17]. In this study, however, a set of fitness landscapes was generated using a Gaussian fitness landscape generator proposed in Gallagher and Yuan [5] and used for testing the reported algorithms. In the generator, a set of n-dimensional Gaussian functions are combined to generate a n-dimensional fitness landscape where "the value of a point is given by the maximum value of any of the Gaussian components at that point" [5].

There are several advantages of using such fitness landscape generators compared to using classical benchmark problems [3]. First, the structure of test problems can be easily tunable by a user by altering a small number of parameters. Therefore, general conclusions on the performance of an algorithm can be made by relating its performance to the fitness landscapes in the specific structures. Finally, a large number of fitness landscapes in similar structure can be generated and used to increase the reliability of comparison results. The generated landscape provides a platform for consistent comparison of the eight algorithms listed in Section 2.1.

We considered two new parameters in the Gallagher and Yuan's Gaussian landscape generator to additionally manipulate the structure of big valley and the range of optimums. The modified generator has six input parameters: n, m, ul, r, w, and d. n indicates the dimensionality of the generated landscape, while m sets the number of local optimum. ul defines the rectangular boundary of the solution space. r indicates the ratio between the fitness values of the best possible local optimum and the global optimum. w is an identical component in the covariance matrix of the Gaussian functions and controls the orientation and shape of each valley in the landscape. Finally, d defines the boundary of the centers of Gaussian components.

Total of twenty-four landscapes were generated using the default parameters (bold numbers in Table 1) of four dimensions ($n = 4$), three local optimums ($m = 3$), the Euclidean distance between the upper and lower limits of 20 ($-10 \leq \begin{bmatrix} x_1 \\ x_2 \end{bmatrix} \leq 10$ for the 2-D problem), average ratio of local optimum to global optimum of 0.3, $w = 0.03$, and $d = 0.6$, with only changing a single parameter's value for each landscape.

Table 1 Parameters in the modified Gaussian fitness landscape generator

Parameter	Values used
n (dimensionality)	[2, **4**, 6, 8]
m (number of local optima)	[0, **3**, 6, 9]
ul (interval span of side constraints)	[10, **20**, 30, 40]
r (ratio of local optima)	[0.1, **0.3**, 0.6, 0.9]
w (valley structure coefficient)	[0.01, **0.03**, 0.06, 0.09]
d (peak density ratio)	[0.4, **0.6**, 0.8, 1.0]

Fig. 1 shows the 2-D fitness landscape generated using the maximum parameters ($m = 9$, $ul = 40$, $r = 0.9$, $w = 0.09$, and $d = 1.0$). Therefore, there exist ten peaks that include nine local optimums and one global optimum. As entered inputs, the heights of nine local optimums were lower than 0.9, while global maximum is 1.0. The range of two decision variables varies from -20 to 20, while the height of peaks is bounded within ±20 (i.e., ±1.0 x 20).

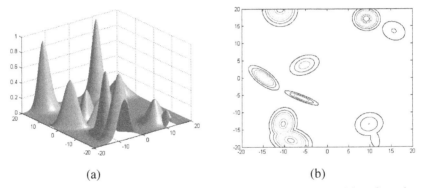

(a)   (b)

Fig. 1 A 2-D landscape generated by the Gaussians landscape generator: (a) surface plot, and (b) contour plot

2.3 Performance Measures and Their Visualizations

We ran each algorithm for 20 times on each landscape in the twenty-four landscapes each of which represents each particular characteristic structure. Stochastic natures of the algorithms result in the different optimal solution from each optimization. In this study, therefore, we compared the expected performance and reliability of each algorithm to changing landscape structures, in the form of a radar plot as shown in Fig. 2. The former is measured by the averaged fitness distance (error) of final solutions from the known global optimum. On the other hand, the reliability of each algorithm is quantified by the standard deviation (SD) of the average error. Therefore, more robust algorithm results in smaller standard deviations of the error.

A radar plot is in the form of hexagon where six axes connect the center and corners of hexagon. The values of performance measures decrease from the center to the corners of hexagon. Algorithm's performance on a particular characteristic structure is represented by positioning each corner of the colored hexagon. Therefore, an algorithm with larger surface area has better performance.

3 Optimization Results

For each of the twenty optimization trials, independent initial population is randomly generated. The maximum number of function evaluations (NFEs) was set as 2,500 and consistently used as a stopping criterion for each reported algorithm.

Fig. 2 shows radar plots indicating average error (blue areas in Fig. 2) and the standard deviation (SD) of the average error (red areas in Fig. 2) for each of eight algorithms with respect to different landscape features. Values close to the corners of the hexagon indicate a smaller value of the average error and the SD in the blue and red areas, respectively. Therefore, a robust algorithm has a large surface area.

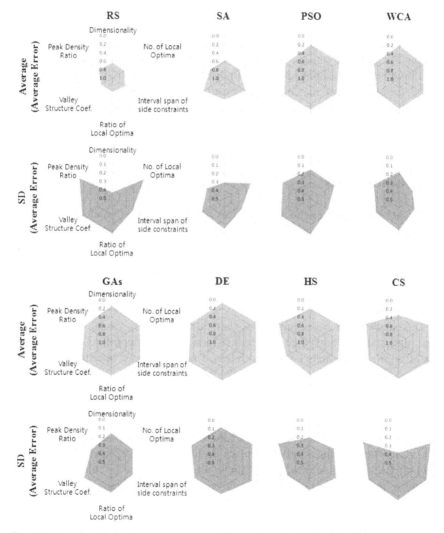

Fig. 2 Radar plots indicating average error (blue surfaces) and the SD of the average error (red surfaces) for each of eight algorithms with respect to different landscape features

The largest surface area was observed from DE. The average errors in DE are very close to zero for all fitness landscapes, also showing robust performances for SD (standard deviations are close to zero). PSO, GAs, and HS have shown overall good performances and outperformed the rest of the algorithms. HS was especially good in the fitness landscape with wide-spreading local optimums. The performance of PSO and WCA were variable in the wide fitness landscapes compared to its performances in other landscapes. CS performed poorly at the high dimensional landscapes (its performance variation was also large). The worst algorithm with the smallest surface area in the blue and red radar plots was RS. Its average errors are around 0.6 regardless of the landscape features. SA with having the largest values of average error and standard deviation has been placed in one before the last ranking after RS.

In this paper, all reported algorithms do not use the derivative information for finding the global solution. Therefore, by altering the ratio of local optima, we should witness no meaningful change in their performances. As can be seen from Fig. 2, for all ratio parameters, RS demonstrates similar performances as it selects new solutions randomly from the entire search space. Algorithms with possessing features of strong global search such as DE and GAs show better performances to avoid being stuck in local optima.

4   Conclusions

This paper has compared the performances of eight optimization algorithms using fitness landscapes generated by a modified Gaussian fitness landscape generator. The modified generator can produce a fitness landscape with particular geometric features. The eight algorithms, namely as RS, SA, PSO, WCA, GAs, DE, HS, and CS, are compared with respect to their expected performances and performance variations (performance reliability). A radar plot was drawn to indicate the level of the two performance measures.

This study has several limitations that future research should be addressed. First, this study has compared the original version of the algorithms, while a number of improved versions have been released in the last two decades. Therefore, the most effective improved versions should be selected for each algorithm and compared to investigate the impact of the improvement on the original algorithm performance. Second, in order to provide full guidance for selecting an algorithm, more algorithms including recently developed algorithms need to be included for having comprehensive comparison. Finally, the optimization results presented in this study were obtained under fixed value for number of function evaluations (NFEs). Because the allowed NFEs limits the performance of some algorithms, therefore, the sensitivity analyses on different NFEs (i.e., higher NFEs) will be performed to examine the radar plots and efficiency of the algorithms.

Acknowledgement This work was supported by the National Research Foundation of Korean (NRF) grant funded by the Korean government (MSIP) (NRF-2013R1A2A1A01013886).

References

1. Barr, R., Golden, B., Kelly, J.: Designing and reporting on computational experiments with heuristic methods. Journal of Heuristics **1**(1), 9–32 (1995)
2. Merz, P.: Fitness landscape analysis and memetic algorithms for the quadratic assignment problem. Evolutionary Computation, IEEE **4**(4), 337–352 (2000)
3. Yuan, B., Gallagher, M.: On building a principled framework for evaluating and testing evolutionary algorithms: a continuous landscape generator. In: Proceedings of the 2003 Congress on Evolutionary Computation, pp. 451–458. IEEE, Canberra (2003)
4. Yuan, B., Gallagher, M.: Statistical racing techniques for improved empirical evaluation of evolutionary algorithms. In: Yao, X., Burke, E.K., Lozano, J.A., Smith, J., Merelo-Guervós, J.J., Bullinaria, J.A., Rowe, J.E., Tiňo, P., Kabán, A., Schwefel, H.-P. (eds.) PPSN 2004. LNCS, vol. 3242, pp. 172–181. Springer, Heidelberg (2004)
5. Gallagher, M., Yuan, B.: A General-purpose tunable landscape generator. IEEE Transactions on Evolutionary Computation (2006) (to appear)
6. Kirkpatrick, S., Gelatt, C.D., Vecchi, M.P.: Optimization by Simulated Annealing. Science **220**(4598), 671–680 (1983)
7. Eberhart, R.C., Kennedy, J.: A new optimizer using particle swarm theory. In: Proceedings of the Sixth International Symposium on Micro Machine and Human Science, pp. 39–43 (1995)
8. Eskandar, H., Sadollah, A., Bahreininejad, A., Hamdi, M.: Water cycle algorithm–A novel metaheuristic optimization method for solving constrained engineering optimization problems. Computers & Structures **110**, 151–166 (2012)
9. Golberg, D.E.: Genetic algorithms in search, optimization, and machine learning. Addion Wesley (1989)
10. Storn, R., Price, K.: Differential evolution–a simple and efficient heuristic for global optimization over continuous spaces. Journal of Global Optimization **11**(4), 341–359 (1997)
11. Geem, Z.W., Kim, J.H., Loganathan, G.V.: A new heuristic optimization algorithm: harmony search. Simulation **76**(2), 60–68 (2001)
12. Kim, J.H., Geem, Z.W., Kim, E.S.: Parameter estimation of the nonlinear Muskingum model using harmony search. Journal of the American Water Resources Association **37**(5), 1131–1138 (2001)
13. Yang, X.S., Deb, S.: Cuckoo search via Lévy flights. In: World Congress on Nature & Biologically Inspired Computing, NaBIC 2009, pp. 210–214. IEEE (2009)
14. Bäck, T.: Evolutionary algorithms in theory and practice: evolution strategies, evolutionary programming, genetic algorithms. Oxford university press (1996)
15. Digalakis, J.G., Margaritis, K.G.: An experimental study of benchmarking functions for genetic algorithms. International Journal of Computer Mathematics **79**(4), 403–416 (2002)
16. Im, S.S., Yoo, D.G., Kim, J.H.: Smallest-small-world cellular harmony search for optimization of unconstrained benchmark problems. Journal of Applied Mathematics (2013)
17. Milad, A.: Harmony search algorithm: strengths and weaknesses. Journal of Computer Engineering and Information Technology (2013)

Harmony Search Algorithm with Ensemble of Surrogate Models

Krithikaa Mohanarangam and Rammohan Mallipeddi

Abstract Recently, Harmony Search Algorithm (HSA) is gaining prominence in solving real-world optimization problems. Like most of the evolutionary algorithms, finding optimal solution to a given numerical problem using HSA involves several evaluations of the original function and is prohibitively expensive. This problem can be resolved by amalgamating HSA with surrogate models that approximate the output behavior of complex systems based on a limited set of computational expensive simulations. Though, the use of surrogate models can reduce the original functional evaluations, the optimization based on the surrogate model can lead to erroneous results. In addition, the computational effort needed to build a surrogate model to better approximate the actual function can be an overhead. In this paper, we present a novel method in which HSA is integrated with an ensemble of low quality surrogate models. The proposed algorithm is referred to as HSAES and is tested on a set of 10 bound-constrained problems and is compared with conventional HSA.

Keywords Harmony search algorithm · Surrogate modeling · Ensemble · Global optimization · Polynomial regression model

1 Introduction

Harmony Search Algorithm (HSA) is a meta-heuristic algorithm that emulates the music improvisation process by musicians. In other words, finding rapport between pitches to achieve a better state of harmony in music and searching for optimality in optimization process using HSA is much alike. This comprehensible nature of HSA has led to its application in many optimization problems including

K. Mohanarangam · R. Mallipeddi(✉)
College of IT Engineering, Kyungpook National University,
1370 Sanyuk-Dong, Daegu 702-701, South Korea
e-mail: {krithikaamohan,mallipeddi.ram}@gmail.com

© Springer-Verlag Berlin Heidelberg 2016
J.H. Kim and Z.W. Geem (eds.), *Harmony Search Algorithm,*
Advances in Intelligent Systems and Computing 382,
DOI: 10.1007/978-3-662-47926-1_3

19

the traveling salesperson problem [1], the layout of pipe networks [1,2], pipe capacity design in water supply networks [3,4], hydrologic model parameter calibrations [3,4], cofferdam drainage pipe design [5], and optimal school bus routings [1].

Surrogate models are used to mimic the complex behavior of the underlying simulation model by using the evolution history of the population members and thus help in reducing original fitness evaluations [6-8]. Like most of the evolutionary algorithms, finding optimal solution to a given numerical problem using HSA is expensive. To overcome this, HSA can be integrated with surrogate model. Though, the use of surrogate model can reduce the original functional evaluations, the use of approximate model during the evolution can lead to erroneous results. Surrogate models can be built using a variety of techniques, such as, radial basis functions [7], artificial neural networks, Kriging models [8], support vector machines, splines, and polynomial approximation models.

In this paper, we tried to reduce the number of actual evaluations of the function to be optimized and widen the probability of getting optimal solution by commingling HSA with ensemble of low quality surrogate models. In other words, we use an ensemble of low quality surrogates instead of an expensive surrogate model to estimate the quality of newly developed harmony vectors. The proposed algorithm is referred as Harmony Search Algorithm with ensemble of surrogates (HSAES). In the current work, to build the low quality surrogate models we employed polynomial regression (PR) technique which is one of the best methods to solve problems with fewer dimensions, uni or low-modality, and where data are inexpensive to procure [12]. In PR, polynomials are used as an approximation function for its austere simplicity. We employ an ensemble comprising of cubic and quadratic polynomial models to accelerate convergence of Harmony search (HS). The efficiency of the proposed algorithm is tested on a set of 10 bound-constrained problems.

The reminder of this paper is organized as follows: Section 2 presents a literature survey on HSA and surrogate modeling using polynomial regression. Section 3 presents the proposed HSA with ensemble of surrogates. Section 4 presents the experimental results and discussions, while Section 5 concludes the paper.

2   Literature Review

2.1   *Harmony Search Algorithm*

Harmony Search is homologous with music improvisation phenomenon where musicians improvise the pitch of their instruments by searching for a perfect state of harmony. These prominent characteristics of HS are used to distinguish it from other metaheuristics [1]: 1) all the existing solution vectors are contemplated while generating a new vector, and (2) each decision variable are independently considered and examined singly in a solution vector. Standard HSA is presented in Table 1.

Table 1 Standard Harmony Search Algorithm

STEP 1: Initialize the HM with HMS randomly generated solutions. Set generation count $G = 0$.
STEP 2: WHILE stopping criterion is not satisfied
        /*Generate a new solution*/
        FOR each decision variable **DO**
            IF $rand_1$ < HMCR
                Pick the value from one of the solutions in HM
                IF $rand_2$ < PAR
                    Pertub the value picked   */*New solution generated*/*
                END IF
            END IF
        END FOR

        IF new solution better than the worst solution in HM (in terms of fitness)
        Replace the worst solution in HM with new solution
        END IF
        Increment the generation count $G = G + 1$
STEP 3: END WHILE

Musicians, notes, harmonies, and harmony memory are the key concepts of HSA [9]. Each musician is correlated with a decision variable; pitch range in the musical instrument's pertain to the alphabet of the decision variable; the musical harmony extemporized at a certain period corresponds to a solution vector at a given iteration; and Objective function to measure the fitness of solutions corresponding to the optimization problem is represented by audience's aesthetic impression [10]. A New Harmony vector is produced every time by a set of new optimization parameters in the HM, which provides an estimation of optimal solution.

Harmony search has captured much attention and gained a wide range of acceptance in various engineering applications. Based on the requirements of the problem different researchers have proposed improvements related to HS and can be classified as [11]: (1) HS improvement by appropriate parameters setting; and (2) improvement of HS by hybridizing with other metaheuristic algorithms.

2.2 Surrogate Modeling

Surrogate model bear's semblance of underlying simulation model. These low-end models are used to replace computationally extortionate original function evaluations. The integration of surrogate models with harmony search is used to solve expensive optimization problems. The surrogate model helps in acquiring a better harmony in every generation with minimum number of samples. Many approximation models are used to construct surrogate model in engineering design optimization. Polynomial Regression (PR), Artificial Neural Network (ANN), Radial Basis Function (RBF), and Gaussian Process (GP) are the most eminent and commonly used techniques [13], [14], [15]. Here we use polynomial regression model to productize surrogate models.

Polynomial models are useful to approximate functions which has complex nonlinear relationship. It is the Taylor series expansion of the unknown function. In general, polynomial regression is used to fit complex nonlinear relationship model to the data. These models are very popular as they are comprehensible and have well known properties, easy mapping of raw data to a polynomial model and more importantly computationally inexpensive. A nonlinear phenomenon of a dependent variable `$y'$ can be modeled in terms of an independent variable `β' as an nth degree polynomial as below in (1).

$$y = \beta_0 + \beta_1 x + \beta_2 x^2 + \cdots + \beta_n x^n + \varepsilon \tag{1}$$

Where, `$n'$ is the order of the polynomial, `β_n' is the coefficient of the corresponding order term and `ε' is unobserved random error of independent variable. For instance, polynomial regression with one variable of order 2 (quadratic form) and 3 (cubic form) takes the following forms as shown in (2) and (2) respectively [16].

$$y = \beta_0 + \beta_1 x + \beta_2 x^2 + \varepsilon \tag{2}$$

$$y = \beta_0 + \beta_1 x + \beta_2 x^2 + \beta_3 x^3 + \varepsilon \tag{3}$$

3   Harmony Search with Ensemble of Surrogate Models

During the evolution process, the process of evaluating the newly produced harmony vector to obtain information regarding its fitness in comparison with the members in the harmony memory is a computationally expensive task at least in real-world engineering applications. Therefore, to reduce the number of actual function evaluations that are computationally expensive it would helpful to know if the produced harmony vector is capable of replacing the worst vector in the harmony memory without actually evaluating. In other words, if we have some idea that the produced harmony vector can replace the worst in the memory by some means prior to the actual evaluation then the actual number of computationally expensive function evaluations can be reduced.

In literature, researchers employed surrogate models to reduce the actual number of function evaluations. However, most of the researchers try to build a surrogate model that is good approximation of the function that is being optimized. To build a surrogate model that is as good as the actual function is tedious and computationally expensive since it requires large number of samples.

In this paper, we propose to reduce the actual number of function evaluations using an ensemble of low quality surrogate models. The term "low quality" means that the surrogate models are built with minimum number of samples that are produced during the evolution. Since the surrogate models are of low quality, we use an ensemble of surrogate models so that the probability of classifying the harmony vector as the one that can replace the worst one in the memory is high. In each generation, every original evaluation is preceded by an ensemble of surrogate function evaluations. In other words, if the produced harmony vector is said to be better than the worst one in the memory based on the surrogate function evaluations in the ensemble then the solution is evaluated using the actual function.

This process culminates towards a better search direction with lesser number of original functions. In each generation, the available members in the memory are used to construct surrogate model. Due to this reason, there is no need for an extra memory as compared to some of the surrogate assisted models [7], where the population members corresponding to the earlier generations are stored.

The outline of the proposed HSAES is presented in Table 2. HM is initialized and the sample points are randomly generated. A new solution vector is generated which is computed using the surrogate models in the ensemble. Based on the

Table 2 Outline of proposed HSAES

STEP 1: Initialize the HM with HMS randomly generated solutions. Set generation count $G = 0$.

STEP 2: WHILE stopping criterion is not satisfied

   Build Surrogate Models using vectors in HM

   /*Generate a new solution*/

   FOR each decision variable **DO**

      IF $rand_1 <$ HMCR

      Pick the value from one of the solutions in HM

         IF $rand_2 <$ PAR

            Perturb the value picked   /*New solution generated*/

         END IF

      END IF

   END FOR

   Evaluate the produced harmony vector using the Surrogate Models

   IF $\min(f_{s,new}$ for all s$) < f_{worst}$

      Evaluate the harmony vector using the actual objective function

   END IF

   IF $f_{new} < f_{worst}$

      Replace the worst solution in HM with new solution

   END IF

   Increment the generation count $G = G + 1$

STEP 3: END WHILE

---

surrogate model evaluation if the new vector is better the worst member in the memory the vector is evaluated using the actual objective function. Based on the actual objective function evaluation, if the new solution is better than the worst one stored in HM, the algorithm will progress to the next step of replacing the worst solution with the new vector.

4 Optimization Results and Discussions

In this section, we try to evaluate and compare the performance of the proposed algorithm with standard HSA. In addition, to highlight the need of an ensemble of surrogate models we also compare the performance of the proposed HSAES with HSA with one surrogate model. To evaluate the performance of the algorithms, we have used the 10D version of the test problems presented in Appendix. The maximum number of function evaluations used is 1000 for all the instances of the algorithms.

For all the algorithm instances, the parameters of the HSA such as HM, HMCR and PR are set to be 10, 0.8 and 0.4 respectively. HSAES contains an ensemble of two low quality surrogate models built using polynomial regression with quadratic and cubic orders. The results of the algorithms on 10D are summarized in terms of min, max, median, mean and standard deviation (std.) values in Tables 3. The best results in terms of mean and standard deviation values are highlighted in Table 3. In Table 3, HSA_{cub} and HSA_{quad} refers to HAS algorithm employing PR with cubic and quadratic order polynomials to construct the surrogate models which help in the evolution process. In other words, HSA_{cub} and HSA_{quad} use single surrogate models unlike HSAES that employs both the surrogate models.

Table 3 The Performance results of different algorithms on 10D test functions

		Min	Max	Median	Mean	Std
f_1	HS	4.4647E+01	6.6156E+02	1.8907E+02	2.3107E+02	1.3483E+02
	HSA_{cub}	3.6776E+01	5.4381E+01	1.5344E+01	1.7964E+01	1.2257E+01
	HSA_{quad}	2.003E − 01	3.8555E +01	1.1265E+01	1.1617E+01	6.9185E+00
	HSAES	3.8743E−01	2.2534E+01	5.0317E+00	6.2099E+00	4.9005E+00
f_2	HS	9.6735E+04	6.5062E+06	1.0492E+06	1.3961E+06	1.3355E+06
	HSA_{cub}	4.5932E+02	2.7894E+05	1.6532E+04	3.8901E+04	6.1430E+04
	HSA_{quad}	6.8194E+02	1.9865E+05	9.6770E+03	2.2292E+04	3.9583E+04
	HSAES	1.3775E+02	2.7143E+04	4.1957E+03	5.4966E+03	5.4625E+03
f_3	HS	3.7362E+00	9.4201E +00	6.3699E +00	6.4346E +00	1.2157E +00
	HSA_{cub}	1.8654E+00	5.1958E +00	3.0999E +00	3.1470E +00	6.6822E−01
	HSA_{quad}	1.4994E +00	4.2557E +00	2.7754E +00	2.8271E +00	6.3901E−01
	HSAES	1.1756E +00	3.2535E +00	2.6739E +00	2.6624E+00	4.5920E−01

Table 3 (*Continued*)

		Min	Max	Median	Mean	Std
f_4	HS	1.7727 E+00	4.8065 E+00	2.8202 E+00	3.0056 E+00	9.1251E−01
	HSA$_{cub}$	5.464E−01	1.3229 E+00	9.128E−01	9.0540E−01	2.3072E−01
	HSA$_{quad}$	3.665E−01	1.7997 E+00	8.529E−01	9.0540E−01	3.5151E−01
	HSAES	1.7532−01	1.0278 E+00	5.586E−01	6.0753E−01	2.5593E−01
f_5	HS	7.2332 E+00	2.0013E+01	1.1766 E+01	1.2392E+01	3.1575 E+00
	HSA$_{cub}$	6.417E−01	9.4587 E+00	4.1175 E+00	4.3264E+00	2.0383 E+00
	HSA$_{quad}$	8.943E−01	1.0481E+01	3.7188 E+00	4.1445E+00	2.4647 E+00
	HSAES	1.300E−01	9.9796 E+00	3.0367 E+00	3.2702E+00	1.9014 E+00
f_6	HS	2.1928E+03	1.3800E+04	5.1439E+03	6.1099E+03	3.0689E+03
	HSA$_{cub}$	2.5650E+02	1.5322E+03	4.9863E+02	6.0313E+02	3.1741E+02
	HSA$_{quad}$	1.1797E+02	1.3190E+03	3.5110E+02	4.3391E+02	2.6380E+02
	HSAES	1.1402E+02	7.8640E+02	2.5545E+02	3.1436E+02	1.9264E+02
f_7	HS	8.5306E+02	5.5028E+03	2.5342E+03	2.7770E+03	1.2712E+03
	HSA$_{cub}$	4.0174E+02	5.0331E+03	1.2463E+03	1.4409E+03	8.9298E+02
	HSA$_{quad}$	4.5663E+02	4.9231E+03	2.0745E+03	2.1011E+03	1.0593E+03
	HSAES	4.3069E+02	4.5642+03	1.6212E+03	1.8822E+03	1.0699E+03
f_8	HS	1.7435E+04	2.9157E+06	5.7169E+05	8.2204E+05	8.0983E+05
	HSA$_{cub}$	5.1778E+03	3.8892E+05	2.7811E+04	6.6735E+04	8.6896E+04
	HSA$_{quad}$	3.0105E+03	8.2352E+06	2.3642E+04	3.2960E+05	1.4951E+06
	HSAES	7.5127E+02	5.8096E+05	1.3654E+04	3.7971E+04	1.0507E+05
f_9	HS	1.9940 E+00	4.1886 E+00	2.8960 E+00	2.9852E+00	6.082E−01
	HSA$_{cub}$	8.602E−01	2.3624 E+00	1.7644 E+00	1.7035E+00	3.9382E−01
	HSA$_{quad}$	6.755E−01	2.3913 E+00	1.4969 E+00	1.4734E+00	3.5853E−01
	HSAES	5.184E−01	2.2300 E+00	1.1924 E+00	1.2482E+00	3.8363E−01
f_{10}	HS	1.5336 E+00	4.0344 E+00	2.5480 E+00	2.5453E+00	6.668E−01
	HSA$_{cub}$	2.148E−01	1.4271 E+00	7.991E−01	8.2200E−01	3.2551E−01
	HSA$_{quad}$	2.129E−01	1.0598 E+00	6.709E−01	6.671E−01	2.1902E−01
	HSAES	1.771E−01	1.3471 E+00	3.856E−01	5.058E−01	2.6732E−01

From the results in Table 3, it can be observed that HSA algorithms assisted with surrogate models HSA$_{cub}$, HSA$_{quad}$ and HSAES show improved performance compared to standard HS on all the problems. The performance of HSA algorithm is accepted to improve compared to the standard HSA since the algorithms evaluate the newly generated vectors using surrogate model(s) to check in advance if it necessary to evaluate using the actual function which is expensive.

In addition, it can also be observed that HSAES that uses ensemble of surrogate models perform better than HSA_{cub} and HSA_{quad}. From the results it can be observed that HSAES is better than HSA_{cub} and HSA_{quad} in terms of mean performance on all the cases. However, in some cases (f_1, f_2 and f_8) the improvement is statistically significant and is verified using statistical t-test. The improved performance of HSAES can be attributed the ensemble of surrogate models which can better information regarding quality of the newly generated vector compared to a single surrogate model. In other words, since the surrogate models employed are of low quality the use of ensemble can provide strong evidence compared to a single surrogate model regarding the quality of the newly generated harmony vector. Then the algorithm can decide to evaluate the generated vector using the actual function evaluation or not.

Based on the experimental results, we can observe that low quality surrogate models generated using the solution vectors present the harmony memory alone can be good enough to improve the performance of HSA algorithm within a limited number of function evaluations.

5   Conclusions

In this paper, we propose to improve the convergence speed of Harmony Search Algorithm (HSA) by reducing the number of actual evaluations that is being optimized using an ensemble of low quality surrogate models (HSAES). In the proposed algorithm, the surrogate models are generated using the solution vectors in the harmony memory and no addition memory is used. The performance of HSAES is compared with standard HSA and HSA with single surrogate model on 10D versions of 10 bound constrained optimization problems present in the literature to demonstrate the advantage of the proposed algorithm. Based on the results, we would like to apply the proposed algorithm to solve computational expensive optimization problems in the future.

Acknowledgement This research was supported by BK21 plus, Kyungpook National University, Daegu, South Korea.

References

1. Geem, Z.W., Kim, J.H., Loganathan, G.V.: A new heuristic optimization algorithm: harmony search. Simulation **76**(2), 60–68 (2001). doi:10.1177/003754970107600201
2. Geem, Z.W., Kim, J.H., Loganathan, G.V.: Harmony search optimization: application to pipe network design. Int. J. Model. Simul. **22**(2), 125–133 (2002)
3. Geem, Z.W., Tseng, C.L.: Engineering Applications of Harmony Search. In: GECCO Late Breaking Papers, pp. 169–173, July 2002
4. Geem, Z.W., Tseng, C.L.: New Methodology, Harmony Search, its Robustness. In: GECCO Late Breaking Papers, pp. 174–178, July 2002

5. Paik, K.R., Jeong, J.H., Kim, J.H.: Use of a harmony search for optimal design of coffer dam drainage pipes. J. KSCE 21(2-B), 119–128 (2001)
6. Jin, Y.: A comprehensive survey of fitness approximation in evolutionary computation. Soft Computing 9(1), 3–12 (2005)
7. Zhang, J., Sanderson, A.C.: DE-AEC: a differential evolution algorithm based on adaptive evolution control. In: IEEE Congress on Evolutionary Computation, CEC 2007, pp. 3824–3830, September 2007
8. Díaz-Manríquez, A., Toscano-Pulido, G., Gómez-Flores, W.: On the selection of surrogate models in evolutionary optimization algorithms. In: 2011 IEEE Congress on Evolutionary Computation (CEC), pp. 2155–2162, June 2011
9. Diao, R., Shen, Q.: Feature selection with harmony search. IEEE Transactions on Systems, Man, and Cybernetics, Part B: Cybernetics 42(6), 1509–1523 (2012)
10. Manjarres, D., Landa-Torres, I., Gil-Lopez, S., Del Ser, J., Bilbao, M.N., Salcedo-Sanz, S., Geem, Z.W.: A survey on applications of the harmony search algorithm. Eng. Appl. Artif. Intell. 26(8), 1818–1831 (2013)
11. Moh'd Alia, O., Mandava, R.: The variants of the harmony search algorithm: an overview. Artif. Intell. 36(1), 49–68 (2011)
12. Forrester, A.I., Keane, A.J.: Recent advances in surrogate-based optimization. Prog. Aerosp. Sci. 45(1), 50–79 (2009)
13. Giunta, A.A., Watson, L.T., Koehler, J.: A comparison of approximation modeling techniques: polynomial versus interpolating models. AIAA paper, 98–4758 (1998)
14. Daberkow, D.D., Mavris, D.N.: New approaches to conceptual and preliminary aircraft design: A comparative assessment of a neural network formulation and a response surface methodology (1998)
15. Jin, R., Chen, W., Simpson, T.W.: Comparative studies of metamodelling techniques under multiple modelling criteria. Struct. Multidiscip. Optim. 23(1), 1–13 (2001)
16. Quinn, G.P., Keough, M.J.: Experimental design and data analysis for biologists. Cambridge University Press (2002)

Appendix

1) Sphere function :

$$f_1(x) = \sum_{i=1}^{D} x_i^2$$

2) Generalized Rosenbrock's function

$$f_2(x) = \sum_{i=1}^{D-1} (100(x_i^2 - x_{i+1})^2 + (x_i - 1)^2)$$

3) Ackley function

$$f_3(x) = -20\exp(-0.2\sqrt{\frac{1}{D}\sum_{i=1}^{D} x_i^2}) - \exp(\frac{1}{D}\sum_{i=1}^{D}\cos(2\pi x_i)) + 20 + e$$

4) Griewank's function

$$f_4(x) = \sum_{i=1}^{D} \frac{x_i^2}{4000} - \prod_{i=1}^{D} \cos(\frac{x_i}{\sqrt{i}}) + 1$$

5) Rastrigin's function

$$f_5(x) = \sum_{i=1}^{D} (x_i^2 - 10\cos(2\pi x_i) + 10)$$

6) Non-continuous Rastrigin's function

$$f_6(x) = \sum_{i=1}^{D} (y_i^2 - 10\cos(2\pi y_i) + 10)$$

$$y_i = \begin{cases} x_i & |x_i| < 1/2 \\ round(2x_i)/2 & |x_i| >= 1/2 \end{cases} \quad i = 1, 2, .., D$$

7) Shifted Schwefel's Problem 1.2 with Noise in Fitness

$$f_9(x) = \left(\sum_{i=1}^{D} (\sum_{j=1}^{i} x_j)^2 \right) (1 + 0.4|N(0,1)|)$$

8) High Conditioned Elliptic Function

$$f_{10}(x) = \sum_{i=1}^{D} (10^6)^{\frac{i-1}{D-1}} x_i^2$$

9) Weierstrass's Function

$$f_{11}(x) = \sum_{i=1}^{D} (\sum_{k=0}^{k_{max}} [a^k \cos(2\pi b^k (x_i + 0.5))]) - D\sum_{k=0}^{k_{max}} [a^k \cos(2\pi b^k .0.5)]$$
$$a = 0.5, \ b = 3, \ k_{max} = 20$$

10) Schwefel's Problem 2.22

$$f_{12}(x) = \sum_{i=1}^{D} |x_i| + \prod_{i=1}^{D} |x_i|$$

Precision Motion Control Method Based on Artificial Bee Colony Algorithm

Jinxiang Pian, Dan Wang, Yue Zhou, Jinxin Liu and Yuanwei Qi

Abstract The parameters of traditional PID controller cannot varies with load and environment for the precision motion control system. In this paper, an efficient scheme for proportional-integral-derivation (PID) controller using bee colony algorithm is applied to precision motion control system. The simulation results show that the feasibility of bee colony PID control algorithm in precision motion field. Furthermore, the bee colony PID control algorithm make the precision motion control system has faster response speed, high positioning accuracy, and its parameters can optimize automatically.

Keywords Precision motion control · PID controller · Bee colony algorithm

1 Introduction

As human beings continue to explore in the field of micro-scale, precision motion positioning technology has become one of the key technologies in the 21st century. The precision motion positioning technology is widely used in ultra-precision machining and ultra-precision measurement. Its various technical indicators have

J. Pian · D. Wang · Y. Zhou · Y. Qi
Faculty of Information and Control Engineering, Shenyang Jianzhu University,
Shenyang, People's Republic of China
e-mail: {Pianjx,wangdan1575,qqyw2000}@163.com

Y. Zhou
SOU College of Engineering Science and Technology, Shanghai Ocean University,
Shanghai, People's Republic of China
e-mail: zhouhappy2899@163.com

J. Liu(✉)
Key Laboratory of Networked Control System, Chinese Academy of Sciences,
CAS, Shenyang, People's Republic of China
e-mail: liujinxin@sia.cn

© Springer-Verlag Berlin Heidelberg 2016
J.H. Kim and Z.W. Geem (eds.), *Harmony Search Algorithm,*
Advances in Intelligent Systems and Computing 382,
DOI: 10.1007/978-3-662-47926-1_4

29

become one of the important indicators that measure high-tech development level of the nation [1,2]. At present, nPoint company of United States and Physik Instrument (PI) company of Germany are famous precision motion control system manufacturers in the foreign. Harbin Core Tomorrow Science & Technology Co., Ltd. has produced relatively mature nanoscale precision positioning platforms in the domestic.

MCS is used as the main chip for precision motion control systems in the early. Its execution speed is slow and control cycle is long, so it is difficult to meet the requirements of fast and accurate positioning of ultra-precision positioning system. In order to improve the positioning performance of the platform, DSP that is used as the master controller chip is becoming popular [3].Texas instruments of United States, Free scale, Motorola, Intel, IBM, ARM, Hitachi, Sun, MIPS are main companies for producing DSP chips. In the process of DSP application, it is particularly important to select the appropriate control strategy. Due to simple structure and high reliability of PID control, it has become the most common control strategy in precision motion control system. How to select controller parameters is the core of PID controller. Adaptive control method of the micro displacement platform is proposed in the reference [4], which improves the performance of traditional PID controller and the displacement accuracy of platform. Neural network adaptive PID control method of micro displacement platform is put forward on the basis of neural network theory in the reference [5] and [6]. It has both the advantages of high reliability and good robustness of traditional PID controller and embodies the characteristics of self-learning and self-organization of neural network, so the tracking error of system can be reduced effectively [5,6]. The fuzzy PID algorithm is applied to nano positioning stage in the reference [7]. The designed fuzzy controller dynamically adjusts the incremental of PID parameters, which improves the dynamic performance of micro displacement platform system effectively and can make the positioning accuracy of platform be up to 10nm under certain disturbances [7]. The identification method of micro displacement platform is proposed based on genetic algorithm in the reference [8], and can improve the performance of the identification structure of neural network. Neural network adaptive PID control algorithm based on genetic algorithm can compensate the displacement error effectively and further improve the precision of output of micro displacement platform [8].

Artificial bee colony (ABC) algorithm proposed by simulating the behavior of the honey is a kind of swarm intelligence optimization algorithm. It has a potential parallelism and can make many points search simultaneously. When dealing with optimization problem, the position of food source represents feasible solution of optimization problem and the food source with the largest yield rate represents optimal solution.

The parameters of traditional PID controller cannot change with the environment and load. ABC PID controller of precision motion positioning platform is designed by combining ABC algorithm with traditional PID control, and the effectiveness of its can be proved by experiments. The experimental results show that ABC PID compared with the traditional PID control and adaptive PID control can make precision positioning platform acquire better static and dynamic performance. When the environment changes, ABC PID can also achieve online real-time high precision control for the platform.

2   The Description of the Micro-displacement Platform Motion Process

Micro-displacement system is mainly composed of three parts: the control system, the executing agency and the detection mechanism. Its structure is shown in Fig. 1.

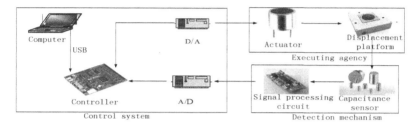

Fig. 1 Micro-displacement system structure

The control system of micro-displacement platform includes computer and controller. The controller utilizes the PID control algorithm and other advanced control algorithm to control the displacement of the platform, and it sends commands to implement positioning control. The executive mechanism drives micro-displacement platform to locate. The detection includes displacement sensors and hardware circuit. It achieves closed-loop control through detecting current position of the platform and giving feedback to the control system [9,10]. The control system of micro-displacement platform selects DSP as the core of the controller, and USB as an interface between the PC and the controller.

3   PID Control Method Based on ABC Algorithm

3.1   Objective Function

In order to improve system performance and reduce positioning errors, the performance index that is used to evaluate the performance of the optimum PID controller is a weighted combination, which includes the *ISE*, three times settling time, and the maximum overshoot in percent for the step response as follow.

$$J = 10(ISE) + 3(t_s) + \sigma \qquad (1)$$

The fitness function is expressed as:

$$fit_i = \begin{cases} \dfrac{1}{1+J_i} & J_i \geq 0 \\ 1+|J_i| & J_i < 0 \end{cases} \tag{2}$$

where $ISE = \int_0^\infty e^2(t)dt$. $e(t) = y(t) - r(t)$ is steady-state error of the system. J_i is the objective function of solution S_i . The system requires that the objective function is minimum, therefore the fitness function is maximum.

The physical meaning of the above-mentioned objective function is that PID control parameters are determined by the ABC PID control optimization for given points. PID control parameters sent to the traditional PID controller make the controller have a good step response output, which reflected in the precision motion positioning platform is a control point with the minimum error, at the fastest speed to approach the given point.

3.2   PID Control Algorithm Based on ABC Algorithm

At the first step of the algorithm, the artificial employed bees are randomly scattered in the search domain producing SN initial solutions $S_i(i = 1, 2 \cdots SN)$. Here SN represents the number of food sources. It is notable that any of these solutions $S_i(i = 1, 2 \cdots SN)$ is a 3-dimensional vector representing 3 design parameters constructing the PID controller of precision motion control system. The ABC generates a candidate food position from the old one by:

$$S_{ij}' = S_{ij} + rand(-1,1) \times (S_{ij} - S_{kj}) \tag{3}$$

where $k \in \{1, 2, \cdots, SN\}$, $j \in \{1, 2, 3\}$, $k \neq i$, S_{ij}' is a new solution comes from food source S_{ij} and its neighbor S_{kj} . Therefore, the search comes close to optimum solution in the search space.

An employed bee starts neighborhood search firstly depending on the local information and tests the nectar amount of new sources replace the previous one if better than previous position, otherwise keep the position of previous one. This method is called greedy selection mechanism.

After the search process is completed by all employed bees, an onlooker bee evaluates the nectar information taken from all employed bees and then it chooses a food source by using roulette wheel. It is expressed as:

$$p_i = \frac{fit_i}{\sum\limits_{i=1}^{SN} fit_i} \qquad (4)$$

After the search process is completed by all onlooker bees, if a solution cannot be improved by employed or onlooker bees after certain iterations called *limit*, then the solution is abandoned and the bee becomes a scout. In that case, the scout bee searches randomly for a new solution within the search space as follow:

$$S_{ij} = S_{min,j} + rand(0,1) \times (S_{max,j} - S_{min,j}) \qquad (5)$$

where $j \in \{1,2,3\}$. $S_{min,j}$, $S_{max,j}$ are lower and upper boundary of the abandoned source respectively [11-13].

The steps of the ABC algorithm are given below.

1) Initialization of the parameters SN, t and *limit*.
2) In the search space, randomly generated SN food sources.
3) According to the formula (3), the employed bees attempt to find new food source in the neighbor of the previous one. They selected by greedy selection mechanism.
4) Calculate fit_i for food source by using (2). Onlooker bees choose a food source by using (4).
5) The onlooker bees attempt to find new food source in the neighbor of the previous one. They selected by greedy selection mechanism.
6) The number of mining of same food source is determined. If the number is larger than *limit*, the employed bee becomes scout bee. The scout bee searches randomly for a new solution within the search space by using (5). Otherwise, the previous food source continue to be exploited.
7) Recording to S_{i1}, S_{i2} and S_{i3} representing the parameters of PID controller.
8) The number of iteration is determined. If the number of iteration is larger than t, output S_{i1}、S_{i2}、S_{i3} as control parameter for PID controller Otherwise, return to step 4).

Basic flow chart of ABC PID algorithm is shown in fig. 2.

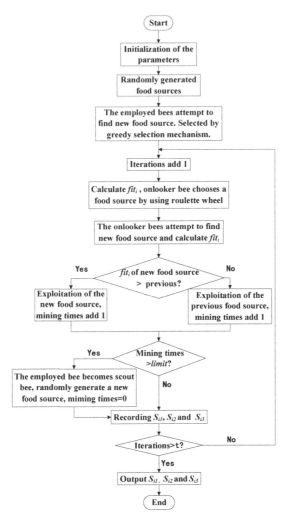

Fig. 2 Basic flow chart of ABC PID algorithm

4 Simulation

In order to study the effectively of the ABC PID algorithm in the control system of precision motion. We designed a simulation experiment. The experimental results of ABC PID algorithm were compared with the traditional PID algorithm and the adaptive PID algorithm [14]. SN =20, t=40, $limit$=20 were set as the Initialization of the parameters.

The change curves of settling time, steady-state error, overshoot and integral square error in iterative process for ABC PID algorithm are shown in fig. 3. The change curves of objective function and fitness function are shown in fig. 4. The change curves of k_p, k_i and k_d are shown in fig. 5.

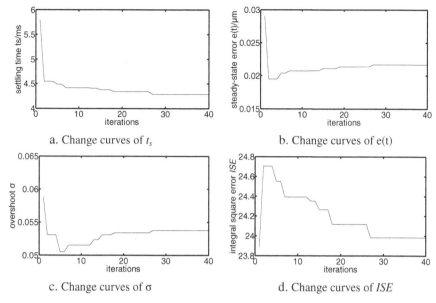

a. Change curves of t_s b. Change curves of e(t)

c. Change curves of σ   d. Change curves of ISE

Fig. 3 Change curves of t_s, e(t), σ and ISE

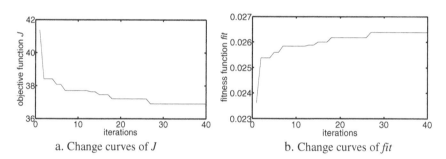

a. Change curves of J b. Change curves of fit

Fig. 4 Change curves of J and fit

From the figures above, these parameters are unchanged after 27 iterations. e(t)=0.0217μm, t_s=4.2913ms, σ%=5.37%. The objective function is determined by these three parameters. J=36.9073, fit=0.0264. The fitness function is the maximum value in the iterative process. Therefore, select the current k_p, k_i and k_d as the optimized control parameters of the ABC PID algorithm, where k_p=0.0407, k_i=0.0227, k_d=0.0201.

Simulation of step response curve for control system of precision motion is carried out with the optimal k_p, k_i and k_d parameters above. When the system is stable, we add interference to it. The experimental results of ABC PID algorithm were compared with the traditional PID algorithm and the adaptive PID algorithm proposed by reference [14]. The simulation results are shown in Fig. 6. The performance is shown in Tab.1.

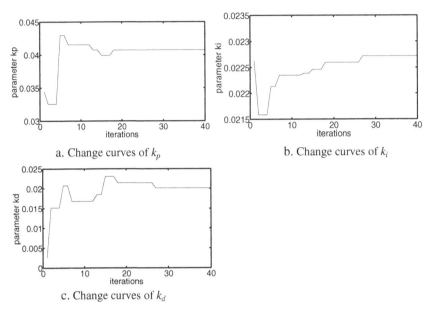

a. Change curves of k_p b. Change curves of k_i

c. Change curves of k_d

Fig. 5 Change curves of k_p, k_i and k_d

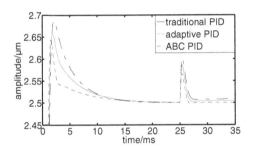

Fig. 6 Step response of traditional PID, adaptive PID and ABC PID

Table 1 PID parameters and performance indicators

	k_p	k_i	k_d	$e(t)$	t_s	$\sigma\%$	J	fit
traditional PID	0.0326	0.0216	0.0152	0.0472	6.7142	7.63	46.7524	0.0214
adaptive PID	0.0351	0.0220	0.0176	0.0325	6.1361	6.41	43.0257	0.0232
ABC PID	0.0407	0.0227	0.0201	0.0217	4.2913	5.37	36.9073	0.0264

The experimental results show that the step response of ABC PID algorithm are better than the traditional PID algorithm and the adaptive PID algorithm, which has a smaller overshoot, settling time and steady-state error. The ABC PID algorithm makes the precision motion control system has faster response speed and higher positioning accuracy. When the time is 25ms, we add interference to the system. The ABC PID can automatically adjust the parameters and make the system return to optimal working condition. When the external environment or load changes, if you still want precision motion control system with fast, accurate position performance, you need to change the control parameters of the controller. The traditional method is to adjust by using experience, but the ABC PID algorithm can realize automatic optimization.

5 Conclusions

This paper used the parallelism of ABC algorithm and combined it with the traditional PID controller to realize automatic optimization of PID controller parameters, and applied it in the precision motion control system. The simulation results proved the effectiveness of the algorithm. The designed precision motion control system has a more favorable static and dynamic performance. The certain reference value for the construction of nanoscale positioning control system.

Acknowledgement This work was supported by the national science and technology support project of the Ministry of Science and Technology of the People's Republic of China under project No. 2011BAK15B09 and the Natural Science Foundation of the People's Republic of China under project No. 61440004.

References

1. Chen, M.Y., Liu, J.S.: High-precision motion control for a linear permanent magnet iron core synchronous motor drive in position platform. IEEE T. Ind. Inform. (2014). doi:10.1109/TII.2013.2247044
2. Liu, C.F., Du, Z.T.: Immune control for permanent magnet synchronous linear servo system of CNC machine. J. Shenyang Uni. Tech. (2015). doi:10.7688/j.issn.1000-1646.2015.01.01
3. Chen, M.F., Zou, P., Han, G., Hou, B.M.: Research on NC auxiliary ultra-precision positioning work platform. J. Northeastern Uni. (Natural Science) **34**, 875–879 (2013)
4. Liaw, H.C., Oetom, D., Shirinzadeh, B., Alici, G.: Adaptive control strategy for micro/nano manipulation systems. In: 9th International Conference on Intelligent Autonomous Systems, pp. 375-382 (2006)
5. Zhang, J.L., Liu, Y., Guo, Y.Q., Liu, J.N.: Research on ultra-precision nanopositioning stage. J. Mech. Eng. (2011). doi:10.3901/JME.2011.09.187
6. Zhang, D., Zhang, C.J., Wei, Q., Tian, Y.B.: Modeling and control of piezo-stage using neural networks. Optics and Precision Engineering (2012). doi:10.3788/OPE.20122003.0587
7. Liu, J.Y., Yin, W.S., Zhu, Y.: Application of adaptive fuzzy PID controller to nanoscale precision motion stage system. Control Engineering of China **18**, 254–257 (2011)
8. Wei, Q., Zhang, Y.L., Hao, H.J., Lu, W.J.: Control of STM micro-displacement stage based on genetic neural network. Microfabrication Technology, 10–14 (2006)

9. Zhou, Y., Su, Y.Q., Pian, J.X., Yao, X.X.: Design and implementation of multifunctional monitoring platform in nano-displacement positioning control system. Manufacturing Automation (2013). doi:10.3969/j.issn.1009-0134.2013.11.09
10. Zhou, Y., Yao, X.X., Pian, J.X., Sun, L.J.: Precision motion monitoring software design and development. Control Engineering of China 21, 1–5 (2014)
11. Li, G.C., Yang, P., Wang, S.E.: Study on optimal method for PID parameter based on artificial bee colony algorithm. In: 2014 International Conference on Mechatronics, Materials and Manufacturing (2014). doi:10.4028/www.scientific.net/AMM.624.454
12. Zhong, W.Q., Zhang, Y.B., Li, X.Y., Zheng, C.S.: Hole machining path planning optimization based on dynamic tabu artificial bee colony algorithm. Research Journal of Applied Sciences, Engineering and Technology 5, 1454–1460 (2013)
13. Bi, X.J., Wang, Y.J.: A modified artificial bee colony algorithm and its application. Journal of Harbin Engineering University 33, 117–123 (2012)
14. Xiao, S.L., Li, Y.M., Liu, J.G.: A model reference adaptive PID control for electromagnetic actuated micro-positioning stage. In: 8th IEEE International Conference on Automation Science and Engineering, pp. 97–102 (2012)

A Scatter Search Hybrid Algorithm for Resource Availability Cost Problem

Hexia Meng, Bing Wang, Yabing Nie, Xuedong Xia and Xianxia Zhang

Abstract This paper discusses the resource availability cost problem (RACP) with the objective of minimizing the total cost of the unlimited renewable resources by a prespecified project deadline. A tabued scatter search (TSS) algorithm is developed to solve the RACP. The deadline constraint is handled in coding. A tabu search module is embedded in the framework of scatter search. A computational experiment was conducted and the computational results show that the proposed TSS hybrid algorithm is effective and advantageous for the RACP.

Keywords RACP · Scatter search · Tabu search

1 Introduction

The resource-constrained project scheduling problem (RCPSP) is one of the main branches of project scheduling, which is known to be NP-hard [1]. It involves minimizing the completion time or time cost of the project subject to precedence relations and the limited resources. The RCPSP has been studied extensively and several effective exact [2] and heuristic [3,4] algorithms have been proposed. Among these algorithms, tabu search has been proved to be an efficient algorithm to solve the RCPSP [5,6].

This paper discusses the resource availability cost problem (RACP), which aims to minimize the total cost of the unlimited renewable resources required to complete the project by a prespecified project deadline. It was introduced by Möhring [7] as an NP-hard problem. The RACP is a variant of the RCPSP [8], but it differs from the RCPSP in the sense that the time for completing the project is limited and the resources are unlimited at a non-decreasing discrete cost function. Because RACP and RCPSP are closely related and the RCPSP has been well

H. Meng · B. Wang(✉) · Y. Nie · X. Xia · X. Zhang
School of Mechatronic Engineering and Automation, Shanghai University,
Shanghai 200072, China
e-mail: susanbwang@shu.edu.cn

© Springer-Verlag Berlin Heidelberg 2016
J.H. Kim and Z.W. Geem (eds.), *Harmony Search Algorithm,*
Advances in Intelligent Systems and Computing 382,
DOI: 10.1007/978-3-662-47926-1_5

39

studied, it is natural to think of solving the RACP by iteratively solving a serial of RCPSPs. We call it the indirect way. The literature on solution methods for the RACP is relatively scarce.

Möhring [7] proposed an exact procedure based on graph theoretical algorithms, Demeulemeester [9] proposed an exact cutting plane procedure, and Rangaswamy [10] proposed a branch-and-bound for the RACP. Rodrigues and Yamashita [11] developed a hybrid method in which an initial feasible solution is found heuristically, and proposed new bounds for the branching scheme. Drexl and Kimms [12] proposed two lower bounds for the RACP using Lagrangean relaxation and column generation techniques. Shadrokh and Kianfar [13] presented a genetic algorithm to solve the RACP, where tardiness is permitted with defined penalty. Yamashita et al. [14] used scatter search to solve the RACP, and computational results show that scatter search is capable of providing high-quality solutions of the RACP in reasonable computational time. Further, Yamashita et al. [15] adopted scatter search to solve the RACP with uncertain activity durations and created robust optimization models.

However, Qi et al. [16] pointed that the process of solving the RACP in indirect way is very complicated and inefficient. He presented two methods to improve the efficiency of solving the RACP, one of which is directly solving the RACP instead of transforming it to the RCPSP. We shall refer to this way as the direct way. In the RACP, the completion time and total cost of the project are determined by the start time of each activity. Therefore, determining the start time of each activity is the key to the direct way to solve the RACP. Qi et al. [16] predigested the process of solving the RACP by using the start times to code the schedule. Then he proposed a pseudo particle swarm optimization (PPSO) to make the process of looking for the best solution efficiently. Ranjbar et al. [17] developed two metaheuristics, path relinking and genetic algorithm to tackle the RACP. The problem is represented as a precedence feasible priority list and converted to a real schedule directly using an available schedule generation scheme.

The direct way can simplify the process of solving the RACP by avoiding transforming the RACP to the RCPSP. In this paper, we study the solution method for the RACP based on the direct way. Firstly, we use the start time of each activity to code the schedule and handle the deadline constraint by coding. Then, a tabued scatter search (TSS) algorithm combined with scatter search and tabu search is developed to solve the RACP.

The remainder of this paper is organized as follows. The RACP is described in Section 2. Section 3 describes the proposed TSS algorithm and its components for the RACP. Section 4 shows the computational results on the benchmark problem instances. The conclusions of this study are given in Section 5.

2   The Resource Availability Cost Problem

The RACP can be stated as follows. A project consists of $n + 2$ activities subject to finish-start precedence relations $(i, j) \in H$, where activities 0 and $n + 1$ are dummy activities that indicate the start and finish of the project respectively.

For each activity i, it is started at time st_i and requires r_{ik} units of renewable resource type k $(k = 1,...,m)$ in every time unit of its deterministic and non-preemptive duration d_i. Let D represent the project deadline, which is prespecified and the project must be completed before the deadline. Let A_t denote the set of activities in progress during the time interval $(t-1,t]$ and a_k represent the availability of resource type k. The RACP can be conceptually modeled as follows:

$$\min \sum_k C_k(a_k) \tag{1}$$

$$\text{s.t.} \quad st_i + d_i \le st_j \quad \forall (i,j) \in H \tag{2}$$

$$st_0 = 0 \tag{3}$$

$$st_{n+1} + d_{n+1} \le D \tag{4}$$

$$\sum_{i \in A_t} r_{ik} \le a_k \quad k = 1,...,m, \, t = 1,...,D \tag{5}$$

The objective function (1) is to minimize the total resource cost of the project, wherein $C_k(a_k)$ denotes a discrete non-decreasing cost function associated with the availability a_k of resource type k. Constraint (2) takes the precedence relations among the activities into account, where activity i immediately precedes activity j. Constraint (3) denotes that the project should be started at time zero and constraint (4) indicates that the project must be completed before the deadline. Constraint (5) shows that the renewable resource constraints are satisfied.

From the model, it can be seen that both the resource availability values a_k $(k = 1,...,m)$ and the start times st_i $(i = 1,...,n)$ of the activities are variables, and the decision variables of the RACP are a_k $(k = 1,...,m)$. Indeed, the value of a_k depends on the start time of each activity, i.e. $a_k = \max_{t=1,...,D} \sum_{i \in A_t} r_{ik}$. Different start times of activities may result in different values of a_k. This is also the main basis of the direct way to solve the RACP. The objective of the RACP is to find a precedence feasible schedule, such that the project can be completed before the deadline and the total resource cost is minimized.

3   TSS for RACP

In this section, a tabued scatter search (TSS) algorithm combined with scatter search and tabu search is developed to solve the RACP based on the direct way. We use the start time of each activity to code the schedule, so that the RACP can be solved directly. The specific coding and the proposed TSS algorithm will be introduced in the following.

3.1   Coding

To solve the RACP directly, we adopt the direct coding method. The start time of each activity is used to code the schedule. Let s denote a solution of the RACP, then $s = (st_1,...,st_n)$. In $s = (st_1,...,st_n)$, each component $st_i(i = 1,...,n)$ is the start time of activity $i(i = 1,...,n)$. For the RACP, if the start time of each activity $st_i(i = 1,...,n)$ in s satisfies the precedence relations and the completion time of the project does not exceed the deadline D, then s is a feasible solution of the RACP.

> **Step 1.** Calculate the earliest start time est_i of each activity
> from $i = 1$ to n;
> **Step 2.** Let $lst_{n+1} = D$;
> **Step 3.** Calculate the latest start time lst_i of each activity
> from $i = n$ to 1;

Fig. 1 Procedure to obtain $[est_i, lst_i]$

For a given solution s, the completion time of the project can be determined directly. To satisfy the deadline constraint of the project, we handle the deadline constraint in coding in this paper. Let est_i and lst_i denote the earliest and latest start time of activity i respectively, then $est_i \leq st_i \leq lst_i$. Interval $[est_i, lst_i]$ is called the effective value interval of st_i by us. It contains all possible values of st_i, and the value of st_i can only be selected from $[est_i, lst_i]$ when generating new solutions. The procedure to obtain $[est_i, lst_i]$ is shown in Fig. 1. The earliest start time est_i of each activity from $i = 1$ to n is calculated according to the critical path method (CPM) [18] in step 1. The latest start time of activity $n + 1$ is initialized with the deadline D, i.e. $lst_{n+1} = D$ in step 2. Finally, the latest start time lst_i of each activity from $i = n$ to 1 is calculated according to CPM in step 3.

By setting effective value interval $[est_i, lst_i]$ for st_i, the deadline constraint of the project is transformed to the restrictions of the start times of the activities. However, for $s = (st_1,...,st_n)$, $st_i \in [est_i, lst_i](i = 1,...,n)$ cannot guarantee that s is feasible due to that the precedence relations may not be satisfied. For the solution which is infeasible, check up the start time of each activity successively and update the start times which do not satisfy the precedence relations. For a solution s which satisfies the precedence relations, a theorem is given as follows.

Theorem 1. If solution s satisfies the precedence relations, then the completion time *makespan* of the project must not exceed the deadline D, i.e. $makespan \leq D$.

Proof. Because s satisfies the precedence relations, then $makespan = st_{n+1}$. As $st_{n+1} \leq lst_{n+1}$ and $lst_{n+1} = D$, then $st_{n+1} \leq D$, i.e. $makespan \leq D$.

Theorem 1 gives the relationship between the precedence relations and the deadline. If **s** satisfies the precedence relations, it must satisfy the deadline constraint and thus it is a feasible solution of the RACP. The completion time and total cost of the project can be determined directly without transforming the RACP to RCPSP.

3.2   TSS Algorithm

Scatter search (SS) is an evolutionary method based on heuristic proposed by Glover [19]. Glover [20] further identified a template for SS. Martí et al. [21] provided the fundamental concepts and principles of SS. According to the template [20], SS contains five systematic methods, i.e. diversification generation method, improvement method, reference set update method, subset generation method and solution combination method. The implementation of SS is based on these methods. SS considers both intensification and diversification in search process, thus it has strong global search capability. SS has been successfully applied to solve the RACP [14,15].

Tabu search (TS) was initially proposed by Glover [22] and has been applied to many combinatorial optimization problems. It can avoid falling into local optimum by introducing tabu list and has proved to have strong search ability. However, TS depends on the initial solution strongly and the search process is serial. In project scheduling, the application of TS focuses on the RCPSP.

To solve the RACP more efficiently, we develop a tabued scatter search (TSS) algorithm combined with SS with TS in this paper. A tabu search module (denoted as M-TS) is embedded in the framework of SS. The framework of TSS algorithm is presented in Fig. 2.

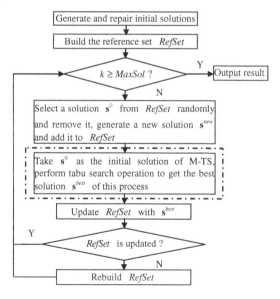

Fig. 2 Framework of TSS algorithm

As shown in Fig. 2, TSS starts with generating and repairing initial solutions. Then select b solutions from the initial solutions to build the reference set $RefSet$, including b_1 high-quality solutions and b_2 diverse solutions. If the stopping condition is not satisfied, i.e. the number of evaluated solutions k does not exceed the maximum number of solution evaluations $MaxSol$, select a solution $\mathbf{s}^0$ from $RefSet$ randomly and remove it. To ensure the size of $RefSet$, generate a new solution $\mathbf{s}^{new}$ and add it to $RefSet$. After that, M-TS is carried out with the initial solution $\mathbf{s}^0$ and the best solution of this process $\mathbf{s}^{best}$ is got. Then update $RefSet$ with $\mathbf{s}^{best}$. If $RefSet$ is not updated, rebuild it. Finally, output the optimization result.

From Fig. 2, we can see that TSS takes SS as the main framework and replaces subset generation and solution combination with M-TS. Compared with SS [14], TSS differs in initial solutions, M-TS and reference set update.

3.2.1 Initial Solutions

Generate $PSize$ initial solutions with the diversification generation method [14]. For the start time st_i of activity i, divide interval $[est_i, lst_i]$ in g sub-intervals. Let $M[i][h], i=1,...,n, h=1,...,g$ denote a frequency matrix. Each sub-interval is selected with probability inversely proportional to its frequency in M, and then a value for st_i is randomly generated in this interval. The procedure to generate diverse solutions can refer to [14]. If the new generated solution is infeasible, repair it. The quality of a solution is evaluated by objective function (1) and the diversity of a solution is evaluated by the minimum distance from the solution to the reference set [14].

3.2.2 M-TS

In TSS, M-TS is introduced to replace the subset generation and solution combination of SS. It takes $\mathbf{s}^0$ from $RefSet$ as initial solution and the best solution $\mathbf{s}^{best}$ obtained by M-TS is used to update $RefSet$. The flowchart of M-TS is shown in Fig. 3.

(1) Initial solution of M-TS
TS depends on the initial solution strongly. To improve the search quality and efficiency of M-TS, the initial solution $\mathbf{s}^0$ of M-TS is selected randomly from $RefSet$. Because $RefSet$ contains both high-quality and diverse solutions, taking the solution in $RefSet$ as the initial solution of M-TS can ensure the intensification and diversification of TSS algorithm.

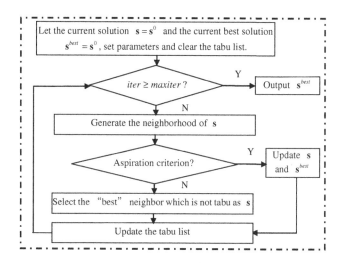

Fig. 3 Flowchart of M-TS

To avoid the search process trapping into circulation, remove the selected solution s^0 from *RefSet*. Then generate a new feasible solution s^{new} according to diversification generation method. If s^{new} is better than the worst solution in *RefSet* ether in quality or diversity, add it to *RefSet*. Otherwise, generate a new solution again until a new solution is added to *RefSet*.

(2) Neighborhood structure
Neighborhood structure is an important element for TS. In this paper, a kind of neighborhood structure is constructed based on the coding. For a current solution $s = (st_1, ..., st_n)$, the procedure of constructing neighborhood is presented in Fig. 4.

Step 1. Select *NSize* activities randomly from the current solution s ;
Step 2. for each selected activity i ͨ
        perform the move operation on s and get a new neighbor solution s^c ;
           if s^c is infeasible then
               repair s^c ;
           end if
        add the neighbor solution s^c into the neighborhood $N(s)$.
      end for

Fig. 4 Procedure of constructing neighborhood

As shown in Fig. 4, a neighbor solution is got by performing move operation on the selected activities of the current solution, and neighborhood $N(s)$ includes

NSize neighbor solutions. The move operation means that generate a random value *random* from $[est_i, lst_i]$ which is not equal to $st_i(\mathbf{s})$ and then let $st_i(\mathbf{s}) = random$. If the neighbor solution is infeasible, repair it and then add it into $N(\mathbf{s})$. The value of *NSize* will be determined experimentally.

(3) Tabu element and move selection
Because the neighborhood is constructed based on move operations on activities, the tabu elements are the selected activities. When a neighbor solution is selected as the new current solution, the corresponding moved activity is put into the tabu list.

The move selection rule is to select the move that is non-tabu that has the lowest cost or satisfies the aspiration criterion. When all of the possible moves are tabu and none of them can satisfy the aspiration criterion, select any solution randomly as the current solution.

In each iteration of TSS, M-TS stops when the number of continuous search steps k exceeds the maximum number of search steps *maxiter*. The best solution $\mathbf{s}^{best}$ is obtained and further used to update *RefSet*. The value of *maxiter* and the tabu list length L will be determined in experiments.

3.2.3   Reference Set Update

The reference set plays an important role in SS. The reference set update method is used to build and maintain a *RefSet*. In SS [14], *RefSet* is updated by the solutions produced by subset generation and solution combination. In each iteration, several new solutions are used to update *RefSet*. In TSS, *RefSet* is updated by the $\mathbf{s}^{best}$ obtained by M-TS. As M-TS is serial, only one solution is obtained to update *RefSet* in each iteration, which is quite different with that in SS [14]. M-TS takes the solution in *RefSet* as the initial solution and constructs an effective neighborhood for the current solution, and the best solution of the search process is saved as $\mathbf{s}^{best}$. Then $\mathbf{s}^{best}$ is further used to update *RefSet*, which improves the quality of the solution used to update *RefSet* and reduces the number of updates. The performance of TSS will be tested in the following experiment.

4   Optimization Results and Discussions

This section tests the effectiveness of the TSS algorithm proposed in this paper. The computational experiments consist of two parts. In the first part, we tuned the TSS algorithm. In the second part, the effectiveness of TSS for the RACP was tested by comparing TSS with SS [14]. All algorithms will be performed for 10 runs for each instance and the best solution among 10 runs is taken as the result. All procedures were coded in C++ language under the Microsoft Visual Studio 2012 programming environment. All computational experiments were performed on a PC Pentium G630 2.7 GHz CPU and 2.0 GB RAM.

4.1   Generation of Problem Instances

The RACP and RCPSP share common data, thus it is easy to adapt existing instances of the RCPSP to the RACP. Progen [23] is an instance generator for RCPSP, which was used to generate the project scheduling problem library (PSPLIB) involving 30, 60, 90 and 120 activities and four resource types. In this paper, the tested instances are directly obtained from the PSPLIB and there are three important parameters as follows:

- Resource factor (RF): reflects the density of the different resource types needed by an activity and takes the values 0.25, 0.5, 0.75 and 1.0.
- Network complexity (NC): reflects the average number of immediate successors of an activity and takes the values 1.5, 1.8 and 2.1.
- Deadline factor (DF): reflects the deadline of the project such that $D = DF \cdot \max_{i=1,\dots,n} eft_i$. In this paper, DF is fixed at 1.2 according to [14].

Each combination of n, m, RF, NC, DF gives one instance, resulting in a total $4 \times 1 \times 4 \times 3 \times 1 = 48$ instances. For TSS, the parameters are set as below:

The number of initial solutions $PSize = 50$, the size of the reference set $b = 10$, the number of high-quality solutions $b_1 = 7$, the number of diverse solutions $b_2 = 3$ and $g = 4$. Stopping condition $MaxSol = \beta \cdot m \cdot n^2$, where β is a coefficient used to adjust the value of $MaxSol$. When n and m are determined, the value of $MaxSol$ is determined by β. β is a positive integer and its value will be tuned in the experiment. For each instance, the costs $C_k (k = 1,\dots,4)$ are drawn from a uniform distribution $U[1,10]$ randomly. To make comparison easier, $C_k (k = 1,\dots,4)$ are fixed at 2,8,5,1 in this paper. For the M-TS of TSS, we set the tabu list length L experimentally for each instance as follows:

$$L = [5 \times 1.2^{n/30}] \qquad (6)$$

The coefficient 5 was obtained by tuning and the value of L is decided by n. The neighborhood size $NSize$ is set to 3,4,5,6 for $n = 30,60,90,120$ respectively and the maximum number of search steps $maxiter$ is set to 20 experimentally.

4.2   Determining the Termination Condition

This section is designed to determine the termination condition $MaxSol$ by testing the performance of TSS under different $MaxSol$. For the 12 instances with 30 activities, $MaxSol = \beta \cdot 4 \cdot 30^2$. β is set to 5, 10, 15 and 25, the corresponding values of $MaxSol$ and the obtained best results under different values of $MaxSol$ are presented in Table 1. In Table 1, $Cost$ represents the objective value of the solution, and CPU represents the CPU time consumed by TSS. The symbol "*" marks the best obtained $Cost$ which cannot be improved with the increment of $MaxSol$.

From Table 1, we can see that TSS performed quite differently in 12 instances. Different instances got the obtained best solution under different values of $MaxSol$. For example, Instance 2 and 6 got their best solutions under $MaxSol = 5 \cdot 4 \cdot 30^2$, and Instance 5 got the best solution under $MaxSol = 15 \cdot 4 \cdot 30^2$. However, there are 7 instances (Instance 1, 3, 4, 9, 10, 11, 12) in 12 instances got their best solutions under $MaxSol = 10 \cdot 4 \cdot 30^2$ with small time cost. For instances with 60, 90 and 120 activities, TSS can also get better solutions in reasonable time under $MaxSol = 10 \cdot m \cdot n^2$. Therefore, $MaxSol = 10 \cdot m \cdot n^2$ can achieve a compromise between solution quality and CPU time, and it would be adopted in the TSS and SS in the following.

Table 1 The performance of TSS under different $MaxSol$

Instances	$5 \cdot 4 \cdot 30^2$		$10 \cdot 4 \cdot 30^2$		$15 \cdot 4 \cdot 30^2$		$25 \cdot 4 \cdot 30^2$	
	Cost	CPU	Cost	CPU	Cost	CPU	Cost	CPU
1	156	0.9	149*	1.7				
2	227*	1.0						
3	297	1.0	294*	1.9				
4	466	0.7	453*	1.4				
5	194	1.0	194	1.7	189*	2.6		
6	231*	1.1						
7	238	1.1	236	2.1	236	3.1	226*	5.0
8	321	1.0	308	2.0	308	2.9	303*	4.7
9	132	1.1	130*	2.2				
10	269	1.0	266*	1.8				
11	294	1.0	283*	2.7				
12	325	1.0	314*	1.8				

4.3   Comparing TSS with SS

To investigate the effectiveness of TSS, we compared it with SS [14]. For SS, the parameters $PSize$, b, b_1, b_2, $MaxSol$ and g take the same values with that in TSS. Fig. 5 and Fig. 6 respectively show the comparisons on solution quality and computational efficiency between TSS and SS.

In Fig. 5, $Ave.\ Cost$ denotes the average cost of the 12 instances with the same activity number n. For $n = 30, 60, 90, 120$, the $Ave.\ Cost$ obtained by TSS is smaller than that obtained by SS, demonstrating that TSS outperforms SS on solution quality. In Fig. 6, $Ave.\ CPU$ denotes the average CPU time of the 12 instances with the same activity number n. For $n = 30, 60, 90, 120$, the $Ave.\ CPU$ consumed by TSS is larger than that consumed by SS, which demonstrates that TSS is worse than SS on computational efficiency. From the above analysis, it can be seen that the proposed TSS algorithm is able to obtain better solutions by spending relatively large CPU time. Thus the TSS algorithm proposed in this paper is an effective algorithm to solve the RACP.

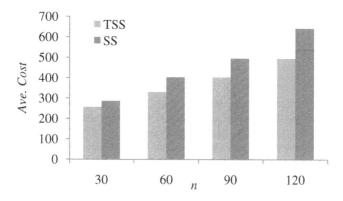

Fig. 5 Comparison on solution quality

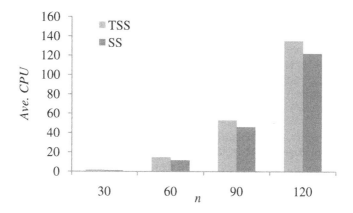

Fig. 6 Comparison on computational efficiency

5   Conclusions

In this paper, we have presented the TSS algorithm combined with SS and TS for solving the RACP. To solve the RACP directly, we use the start times of activities to code the schedule. The deadline constraint is handled in coding by setting an effective value interval for the start time of each activity. In computational experiments, TSS is compared with SS and the results show that TSS can obtain better solutions by spending relatively larger CPU time than SS, and it is an efficient method for the RACP.

For further research, TSS is recommended to solve other project scheduling problems. The RACP could be extended to the multi-mode RACP and the RACP with uncertainty, which is more practical.

Acknowledgement This work is partly supported by National Natural Science Foundation of China # 61273182.

References

1. Blazewicz, J., Lenstra, J.K., Rinnooy Kan, A.H.G.: Scheduling subject to resource constraints: Classification and complexity. Discrete Appl. Math. **5**, 13–24 (1983)
2. Herroelen, W., De Reyck, B., Demeulemeester, E.: Resource-constrained project scheduling: A survey of recent developments. Comput. Oper. Res. **25**, 279–302 (1998)
3. Kolisch, R., Hartmann, S.: Heuristic algorithms for the resource-constrained project scheduling problem: classification and computational analysis. In: Weglarz, J. (ed.) Project Scheduling: Recent Models, Algorithms, and Applications, pp. 147–178. Kluwer Academic Publishers (1998)
4. Kolisch, R., Hartmann, S.: Experimental investigation of heuristics for resource-constrained project scheduling: An update. Eur. J. Oper. Res **174**, 23–37 (2006)
5. Al-Fawzan, M.A., Haouari, M.: A bi-objective model for robust resource-constrained project scheduling. Int. J. Prod. Econ. **96**, 175–187 (2005)
6. Lambrechts, O., Demeulemeester, E., Herroelen, W.: A tabu search procedure for developing robust predictive project schedules. Int. J. Prod. Econ. **111**, 493–508 (2008)
7. Möhring, R.H.: Minimizing costs of resource requirements in project networks subject to a fix completion time. Oper. Res. **32**, 89–120 (1984)
8. Van Peteghem, V., Vanhoucke, M.: An artificial immune system algorithm for the resource availability cost problem. Flexible Int. J. Flexible Manuf. Syst. **25**, 122–144 (2013)
9. Demeulemeester, E.: Minimizing resource availability costs in time-limited project networks. Manage. Sci. **41**, 1590–1598 (1995)
10. Rangaswamy, B.: Multiple Resource Planning and Allocation in Resource-Constrained Project Networks. University of Colorado, Colorado (1998)
11. Rodrigues, S.B., Yamashita, D.S.: An exact algorithm for minimizing resource availability costs in project scheduling. Eur. J. Oper. Res. **206**, 562–568 (2010)
12. Drexl, A., Kimms, A.: Optimization guided lower and upper bounds for the resource investment problem. J. Oper. Res. Soc. **52**, 340–351 (2001)
13. Shadrokh, S., Kianfar, F.: A genetic algorithm for resource investment project scheduling problem, tardiness permitted with penalty. Eur. J. Oper. Res. **181**, 86–101 (2007)
14. Yamashita, D.S., Armentano, V.A., Laguna, M.: Scatter search for project scheduling with resource availability cost. Eur. J. Oper. Res. **169**, 623–637 (2006)
15. Yamashita, D.S., Armentano, V.A., Laguna, M.: Robust optimization models for project scheduling with resource availability cost. J. Sched. **12**, 67–76 (2007)
16. Qi, J.J., Guo, B., Lei, H.T., Zhang, T.: Solving resource availability cost problem in project scheduling by pseudo particle swarm optimization. J. Syst. Eng. Electron. **25**, 69–76 (2014)
17. Ranjbar, M., Kianfar, F., Shadrokh, S.: Solving the resource availability cost problem in project scheduling by path relinking and genetic algorithm. Appl. Math. Comput. **196**, 879–888 (2008)
18. Shaffer, L.R., Ritter, J.B., Meyer, W.L.: The critical-path method. McGraw-Hill, New York (1965)

19. Glover, F.: Heuristics for integer programming using surrogate constraints. Decision Sci. **8**, 156–166 (1977)
20. Glover, F.: A template for scatter search and path relinking. In: Hao, J.-K., Lutton, E., Ronald, E., Schoenauer, M., Snyers, D. (eds.) AE 1997. LNCS, vol. 1363, pp. 1–51. Springer, Heidelberg (1998)
21. Martí, R., Laguna, M., Glover, F.: Principles of Scatter Search. Eur. J. Oper. Res. **169**, 359–372 (2006)
22. Glover, F.: Future paths for integer programming and links to artificial intelligence. Comput. Oper. Res. **13**, 533–549 (1986)
23. Kolisch, R., Sprecher, A., Drexl, A.: Characterization and generation of a general class of resource-constrained project scheduling problems. Manage. Sci. **41**, 1693–1703 (1995)

A Study of Harmony Search Algorithms: Exploration and Convergence Ability

Anupam Yadav, Neha Yadav and Joong Hoon Kim

Abstract Harmony Search Algorithm (HSA) has shown to be simple, efficient and strong optimization algorithm. The exploration ability of any optimization algorithm is one of the key points. In this article a new methodology is proposed to measure the exploration ability of the HS algorithm. To understand the searching ability potential exploration range for HS algorithm is designed. Four HS variants are selected and their searching ability is tested based on the choice of improvised harmony. An empirical analysis of the proposed method is tested along with the justification of theoretical findings and experimental results.

Keywords Harmony search · Exploration · Convergence

1 Introduction

Harmony Search algorithm is one of the nature inspired optimization algorithms which is inspired from music improvisation process. Geem et. al. [1] proposed the very first idea of Harmony Search. It was an outstanding idea to design optimization algorithms which is inspired from the tuning of music over music

A. Yadav(✉)
Department of Sciences and Humanities National Institute of Technology Uttarakhand,
Srinagar 246-174, Uttarakhand, India
e-mail: anupam@nituk.ac.in, anupuam@gmail.com

N. Yadav · J.H. Kim
School of Civil, Environmental and Architectural Engineering,
Korea University, Seoul 136-713, South Korea
e-mail: nehayad441@yahoo.co.in, jaykim@korea.ac.kr

© Springer-Verlag Berlin Heidelberg 2016
J.H. Kim and Z.W. Geem (eds.), *Harmony Search Algorithm*,
Advances in Intelligent Systems and Computing 382,
DOI: 10.1007/978-3-662-47926-1_6

instruments. Many developments of the HS algorithm have been recorded in literature from the very first inception of this novel method. Mahadavi et. al. [2] designed an improved version of HS algorithm for solving optimization problems, later on may more versions of HS algorithm has been proposed to improve the performance of the algorithm on real life applications as well as benchmark problems. Few of them are, Self adaptive HSA [3,4] , Global best Harmony search [5], Adaptive binary HSA [6] and Novel global HSA [7]. Some of the hybridization based HS algorithms are also recorded for various purposes, Feshanghary [8] has hybridized HSA with sequential quadratic programming for engineering optimization problems. Some of the application such as blocking permutation flow shop scheduling problem [9], total flow time on a flow shop [10] and water resource engineering problems are also solved using many variants of HS algorithm. In the mean time some the research has been carried out on the mathematical foundations and characteristics of HSA. Das et. al. [11] and Gao et. al. [12] have developed some exploratory & convergence behavior of HSA based on expected variance of the harmonies. These articles provide a significant mathematical formulation of HSA.

In the current article the exploratory power of some the variants are analyzed and discussed along with their convergence. To understand the exploration power of HSA, potential exploration range of the improvised harmony is defined. This study provides the exact nature of exploration of HSA variants due to improvised harmonies. A class of competent HS algorithms has been collected from the literature and based on their designing of the improvised harmony vector; the potential exploration range is designed to analyze the possible exploration of the harmonies during a run.

2   Potential Exploration Range of Harmony Search

To understand the exploration power of HS algorithm due to improvised harmony vector a novel approach is defined. The proposed methodology provides a measure of exploitation ability of a harmony search algorithm. Let $X_i = (x_i^1, x_i^2, \dots x_i^{HMS})$ is the initially generated harmony where i varies from $1: Dim$. The very first harmony search proposed by Geem et. al. [1] follows the following strategy for generation of new harmony $X'(x_1, x_2, \dots x_{Dim})$.

If $X_i = (x_i^1, x_i^2, \dots x_i^{HMS})$ is the ith harmony vector from all the initialized harmony vectors then the improvised harmony is generated based on the following rule:

$$x_i' = \begin{cases} x_i \in \{x_i^1, x_i^2, \dots x_i^{HMS}\} \ if \ probability \ HMCR \\ new \ x_i \qquad\qquad if \ probability \ (1 - HMCR) \end{cases} \tag{1}$$

This follows the pitch adjustment

$$x_i' = \begin{cases} x_i' \pm rand(0,1). bw \ if \ probability \ PAR \\ x_i' \qquad\qquad\quad if \ probability \ (1 - PAR) \end{cases} \tag{2}$$

Eq. (1) and (2) provides the implementation of new improvised harmony and all the parameters are the standard parameters of the HS algorithm. This new improvised harmony is the key vector for providing a good exploration to the entire search space. In this continuation to measure the total volume searched by the harmonies due to improvised harmony, an exploration range equation is proposed for original harmony search. The following algorithm is used to design the potential exploration range equation for harmony search

Initialize the harmony X_i of size HMS

Generate improvised harmony X' with concern strategy of the particular harmony search variant

$for\ j = 1: Max\_itr$
$\quad\quad for\ i = 1: HMS$
$\quad\quad r(i,:) = Improvised\ harmony\ (X') - Harmony\ (x_i)$
$\quad\quad end\ of\ i$

$$R^j = |\prod_{k=1}^{HMS} (r(k,:))|;$$

$end\ of\ j$

In the next section a brief idea of few state-of-the-art harmony search algorithms is presented based on their strategy of improvised harmony generation.

2.1 Harmony Search Algorithm [1]

Based on the need and new ideas many variants and hybrid versions of HS algorithm has been proposed in last decade throughout the globe. State-of-the-art of HS algorithms are collected to understand their exploration and convergence power.

It was the very first inception of harmony search by Geem et. al. [1]. The idea of generation of improvised harmony for this version of harmony search is explained in previous section. Based on the algorithm defined in section 2, let the following equation provides the potential exploration range for Harmony Search.

$$R_1 = [r_1^1, r_2^1, \dots r_{Max\_iter}^1] \quad\quad\quad (3)$$

where r_i^1 is the maximum volume covered by improvised harmony at i^{th} iteration.

2.2 Global Best Harmony Search [5]

Mahamed and Madhavi [5] proposed a global best harmony search algorithm which was designed on the line of Particle Swarm Optimization. It follows the following approach for the generation of improvised harmony

$for\ each\ i = 1: N$

do
        if $U(0,1) \leq HMCR$ *then* /* *memory consideration* */
  begin
        $x'_i = x'_j$, *where* $j{\sim}U(1,2,...,HMS)$.
  if $U(0,1) <= PAR(t)$ *then* /* *pitch adjustment* */
  begin
  $x'_i = x_k^{best}$ *where best is the index of the best harmony in the HM and*
  $k{\sim}U(1,N)$
  endif
  else /* *random selection* */
  $x'_i = LB_i + rand(0,1)(UB_i - LB_i)$
  endif
done

Again, based on the algorithm defined in section 2, the potential exploration range for global best harmony search will be defined as

$$R_2 = [r_1^2, r_2^2, ..., r_{Max\_iter}^2] \tag{4}$$

where r_i^2 is the maximum volume covered by improvised harmony at i^{th} iteration.

2.3 *Improved Harmony Search [2]*

Mahadavi et. al. [2] proposed an improved harmony search algorithm. The major finding of the improved harmony search is to use dynamic values of PAR and bw in place of their fixed values as coined in original harmony search. The dynamic choices of PAR and bw are made based on the following two equations

$$PAR(gn) = PAR_{min} + \frac{(PAR_{max} - PAR_{min})}{NI} gn \tag{5}$$

$$bw(gn) = bw_{max}\exp{(c.gn)} \tag{6}$$

Where $c = \frac{\ln\left(\frac{bw_{min}}{bw_{max}}\right)}{NI}$

PAR : pitch adjusting rate for each generation
PAR_{min} : minimum pitch adjusting rate
PAR_{max} :maximum pitch adjusting rate
NI: number of solution vector generations
gn: generation number
With this change of choice of PAR and bw values, the proposed potential search range and potential search volume of improved harmony search will be

$$R_3 = [r_1^3, r_2^3, ..., r_{Max\_iter}^3] \tag{7}$$

where r_i^3 is the maximum volume covered by improvised harmony at i^{th} iteration.

2.4 Self Adaptive Global Best HS Algorithm [3]

Kuan-ke et. al. [3] proposed a self adaptive global best harmony search algorithm. The major modification was to provide a self adaptive nature to the parameters of global best harmony search algorithm [5]. The parameter values $HMCR$ and PAR are advised to choose statistically as well as based on harmony memory. The choice of bw is provided by using the following formulation

$$bw = \begin{cases} bw_{min} + \dfrac{(bw_{max} - bw_{min})}{NI} 2t \; if \; t < NI/2 \\ bw_{min} \quad if \; t \geq NI/2 \end{cases} \tag{8}$$

Where bw_{min} and bw_{max} are the bounds for bw, t is the current iteration and NI is the total number of iterations. All the parameter values are exactly inherited as they are in the original article. Based on this approach the defined potential search range and potential search volume for self adaptive Harmony search algorithm will be

$$R_4 = [r_1^4, r_2^4, ..., r_{Max\_iter}^4] \tag{9}$$

where r_i^4 is the maximum volume covered by improvised harmony at i^{th} iteration. In the next section the potential search range and volume is calculated and discussed over various benchmark problems.

3   Experimental Results of Potential Exploration Range

In order to understand the exploration power of HS algorithms on experimental problems, the following problems are solved with the HS algorithms and detailed study of the potential exploration range is discussed based on the results. The benchmark problems and experimental setup is presented in Table 1.

Table 1 Benchmark problems and experimental setup

Sr. No.	Problems	Dim	Algorithm	HMS	HMCR	PAR	bw
1.	Sphere		HS [1]	100			
2.	Rastrigin	20	GBHS [5]	100	As per the original setup		
3.	Schewfel		IHS [2]	100			
4.	Greiwank		SGHS [3]	100			

The maximum iteration (Max_iter) is set 10000. The values of $HMCR, PAR, bw, bw_{min}$ and bw_{max} used in corresponding algorithms is taken as suggested by the authors of their original articles. The size of HMS is fixed for all the algorithms which is 100. The search range for all the functions is kept $[-100, 100]$.

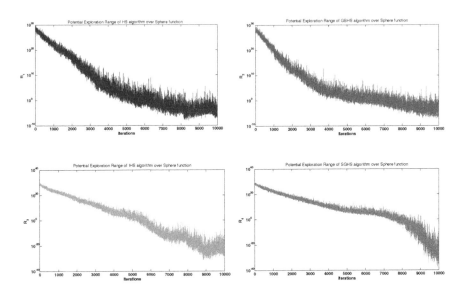

Fig. 1 Potential exploration behavior of HS, GBHS, IHS and SGHS over Sphere function with 20D against iterations

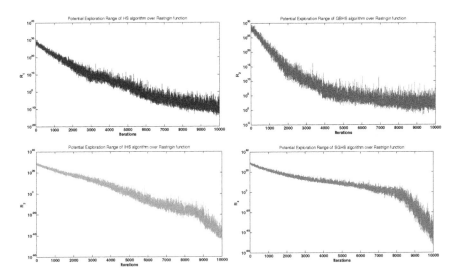

Fig. 2 Potential exploration behavior of HS, GBHS, IHS and SGHS over Rastrigin function with 20D against iterations

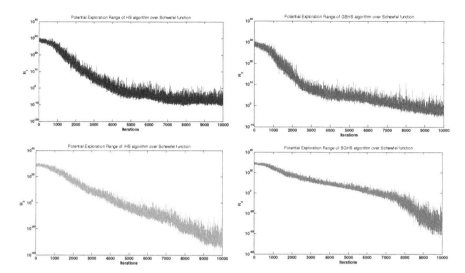

Fig. 3 Potential exploration behavior of HS, GBHS, IHS and SGHS over Schewfel function with 20D against iterations

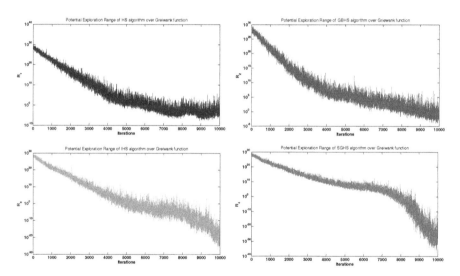

Fig. 4 Potential exploration behavior of HS, GBHS, IHS and SGHS over Griewank function with 20D against iterations

4   Expected Population Variance

Das et. al. [11] provides a formula for the calculation of expected population variance. This formula is designed based on the theoretical findings and statistical measures of HS algorithm. Eq. 10 presents this formula

$$Exp\big(Var(Y)\big) = \frac{n}{n-1}.\,[HMCR.\,Var(X) + HMCR.\,(1 - HMCR).\,\bar{X}^2$$
$$+ \frac{1}{3}HMCR.\,PAR.\,bw^2 + \frac{a^2}{3}.\,(1 - HMCR)] \tag{10}$$

Where Y is the intermediate harmony and X is Harmony population.

An empirical study based on this equation is performed. The expected variance of improved harmony search [2] is evaluated and a plot of expected variance of the New Harmony is presented in Fig. 5

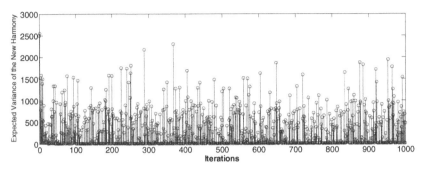

Fig. 5 Expected variance of the New Harmony against iterations

The expected variance of the new improvised harmony depicted in Fig 5, justifies the exploration ability of the improved harmony search as it was analyzed with the help of potential exploration range of the same.

5   Discussions

The exploration range of the each algorithm is plotted against iteration for all four functions with fixed number of Harmony Size. Fig. 1 provides the plot of the maximum volume covered by HS, IHS, GBHS and SGHS algorithms against the progressive iterations. It is observed that the volume covered by HS and GBHS algorithms is little more than other two algorithms but the IHS and SGHS are having better convergence approach towards a point. This analysis provides an idea that over unimodal function the behavior if HS algorithms is alike. In Fig. 2 it has been observed that the behavior of SGHS is slightly better than other three algorithms since it is constantly having a better exploration rate in comparison to others. The important point is before execution point the chaotic nature of the SGHS gives a good edge over others because this kind of behavior is helpful for the algorithms to clear the local minima. The Fig. 3 repeats the same kind of behavior for each algorithm, but on careful observation it is observed that HS and GBHS are having good exploration rate but their converging ability is poor in comparison to other two algorithms. The strong converging nature of SGHS and IHS provides a better convergence which is of the order 10^{-40}. In Fig. 4, again it has been observed that the exploration of SGHS is very excellent, the continuous better value of the exploration shows its strong ability over multimodal optimization problems. The other three algorithms are also having the satisfactory performance. On the other hand the convergence of SGHS and IHS is again much better than other two.

Over all it is observed that the exploration and convergence power of original HS algorithm is good while its convergence is not up to the mark when we deal with multimodal problems. The supremacy of the SGHS is well recorded on each kind of the problems and shows an edge over other HS algorithms based on this study.

6   Conclusions

In this article a novel approach is proposed to measure the exploration ability of HS algorithms based on the new improvised harmony. The potential exploration range is defined for original HS algorithm as well as three other improved variants of the harmony search. The exploration ability and convergence is measured by calculating the volume explored by the improvised harmony in each algorithm. The idea of potential exploration range is tested numerically over four benchmark problems and the results are plotted against each iteration. The plot of expected variance of the population is also plotted against iterations which justify the availability of the designed potential exploration range. Based on this study the exploration ability of SGHS is found better than other select algorithms. In future this theory may be applied to check the exploration ability of other HS variants and the idea may be extended by incorporating theoretical finding and some more function can be tested.

Acknowledgement This work was supported by SERB, Department of Science and Technology, Govt. of India, National Institute of Technology Uttarakhand and National Research Foundation (NRF) of Korea under a grant funded by the Korean government (MSIP) (NRF-2013R1A2A1A01013886).

References

1. Geem, Z.W., Kim, J.H., Loganathan, G.V.: A new heuristic optimization algorithm: harmony search. Simulation **76**(2), 60–68 (2001)
2. Mahdavi, M., Fesanghary, M., Damangir, E.: An improved harmony search algorithm for solving optimization problems. Applied Mathematics and Computation **188**(2), 1567–1579 (2007)
3. Pan, Q.K., Suganthan, P.N., Tasgetiren, M.F., Liang, J.J.: A self-adaptive global best harmony search algorithm for continuous optimization problems. Applied Mathematics and Computation **216**(3), 830–848 (2010)

4. Wang, C.M., Huang, Y.F.: Self-adaptive harmony search algorithm for optimization. Expert Systems with Applications **37**(4), 2826–2837 (2010)
5. Mahamed, G.H.O., Mahdavi, M.: Global-best harmony search. Applied Mathematics and Computation **198**(2), 643–656 (2008)
6. Lee, K.S., Geem, Z.W.: A new meta-heuristic algorithm for continuous engineering optimization: harmony search theory and practice. Computer Methods in Applied Mechanics and Engineering **194**(36), 3902–3933 (2005)
7. Das, S., Mukhopadhyay, A., Roy, A., Abraham, A., Panigrahi, B.K.: Exploratory power of the harmony search algorithm: analysis and improvements for global numerical optimization. IEEE Transactions on Systems, Man, and Cybernetics, Part B: Cybernetics **41**(1), 89–106 (2011)
8. Fesanghary, M., Mahdavi, M., Jolandan, M.M., Alizadeh, Y.: Hybridizing harmony search algorithm with sequential quadratic programming for engineering optimization problems. Computer Methods in Applied Mechanics and Engineering **197**(33), 3080–3091 (2008)
9. Wang, L., Pan, Q.K., Tasgetiren, M.F.: A hybrid harmony search algorithm for the blocking permutation flow shop scheduling problem. Computers & Industrial Engineering **61**(1), 76–83 (2011)
10. Wang, L.: An improved adaptive binary harmony search algorithm. Information Sciences **232**, 58–87 (2013)
11. Wang, L., Pan, Q.K., Tasgetiren, M.F.: Minimizing the total flow time in a flow shop with blocking by using hybrid harmony search algorithms. Expert Systems with Applications **37**(12), 7929–7936 (2010)
12. Zou, D., Gao, L., Wu, J., Li, S.: Novel global harmony search algorithm for unconstrained problems. Neurocomputing **73**(16), 3308–3318 (2010)

Design of the Motorized Spindle Temperature Control System with PID Algorithm

Lixiu Zhang, Teng Liu and Yuhou Wu

Abstract The thermal error of the motorized spindle has great influence on the accuracy of the NC machine tools. In order to reduce the thermal error, the increase type PID control algorithm is adopted for the control system which can make the temperature controlled in reasonable range. In accordance with the control object features, the Application of the control algorithm is realized in MCU System. Finally, the experiments are carried out which verified the validity and effectiveness of the design of temperature control system by analysis of the experimental data.

Keywords PID control · Temperature measurement · Control algorithm · Motorized spindle

1 Introduction

PID control is one of the earliest control strategies. Today, most of the industrial control loop still use PID control or improved PID control strategy [1]. In the PID control model, three factors was considered include the system errors , the error change and error accumulation and it was widely used in industrial process control for the simple algorithm, good robustness and high reliability. And especially suitable for the accurate mathematical model and deterministic control system [2].

In the actual running environment of the motorized spindle, the heat exchange between the inner system and the external is difficult to control. The interference of other heat sources is also unable to accurate calculation [3]. It is important to control the temperature of the motorized spindle to reduce the thermal error. To achieve control accuracy a temperature controlled cooling control system based on PID has been developed.

L. Zhang · T. Liu(✉) · Y. Wu
Shenyang Jianzhu University, Shenyang 110168, China
e-mail: 1240059679@qq.com

© Springer-Verlag Berlin Heidelberg 2016
J.H. Kim and Z.W. Geem (eds.), *Harmony Search Algorithm,*
Advances in Intelligent Systems and Computing 382,
DOI: 10.1007/978-3-662-47926-1_7

63

2   PID Temperature Control System Theory and Principle Analysis

2.1   PID Temperature Control System Theory

PID controller is a linear controller [4], and it constitutes control deviation e(t) according to the given value r(t) and the actual output value y(t).

$$e(t) = r(t) - y(t) \tag{1}$$

Through the linear combination, Proportional, Integral and Derivative of the deviation e(t) constitute control quantity, and to control the controlled object. So called PID control. Its control law is:

$$U(t) = K_P[e(t) + \frac{1}{T_I} \int_0^T e(t)dt + T_D \frac{de(t)}{dt}] \tag{2}$$

In the equation,

K_P : proportional coefficient

T_I : integral time constant

T_D : Derivative time constant

2.2   Principle Analysis of PID Temperature Control System

2.2.1   General Structure of Temperature Control System

System consists of 1 set of PID controller, cooling water valve, temperature sensor, and cooling water flow meter, etc. The overall structure of the system is shown in Fig. 1

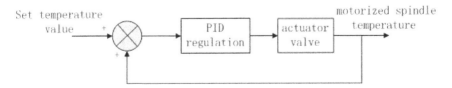

Fig. 1 The general structure of based on PID temperature control system

Fig. 1 is based on the PID temperature control system block diagram. In the temperature control system, it can take advantage of continuous PID control more complex operations, to achieve the requirements of the motorized spindle temperature control. Motorized spindle cooling control system are designed on the basis of motorized spindle temperature as main parameters, cooling water flow rate as adjustable parameters, with temperature regulator output as adjusted value of the cooling water flow regulator with broken accidentally alarm function [5]. Site controller is based on AT89C52 microcontroller as the core temperature controller.

2.2.2   PID the Temperature Control System Control Theory

The main controlling unit: When measured by thermocouple acquisition temperature deviation to the given value, the PID control based on the measurement signal control signals as the next section, adjusting unit of a given value [6].Deputy controlling unit: The unit is cooling water regulation, given the value of the signal from the upper level PID adjustment output. By comparing the values of cooling water flux with the self-defining values, PID, which is based on deviation values, for proportion(P), integral (I) and differential (D) operation, to output a control signal, to the control actuator (cooling water flow regulating valve), so as to achieve the aim of the automatic control of cooling water flow rate. The whole adjustment system can automatically control the cooling water flow according to a certain flow rate, and finally achieve the purpose of controlling the temperature of motorized spindle automatically.

3   PID Control Algorithm

There are three simple PID control algorithms which include incremental algorithm, position algorithm and differential ahead algorithm. The incremental PID control algorithm is easy to get better control effect by weighting treatment which good for the system security system safety operation. So, the incremental algorithm is used in this system.

3.1   PID Incremental Algorithm

T is now the sampling period, At a series of sampling points KT represents a continuous time t, In order to sum data to replace the integral approximately. After the first order, take the difference to replace the integral approximately [7]. To do the following approximate transformation:

$$T=KT \tag{3}$$

$$\int_0^t e(t) = T\sum_{j=0}^k e(jT) = T\sum_{j=0}^k e(T) \tag{4}$$

$$\frac{de(t)}{dt} = \frac{e(KT)-e[(K-1)T]}{T} = \frac{e(K)-e(K-1)}{T} \tag{5}$$

Among them, T is the sampling period, e(k) for the system of the deviation of the sampling time of the K, e(k-1) for the system of the deviation of the sampling time of the K-1, K for sample serial number, k=0, 1, 2,...Take the above type (4) and (5) into equation (2), You can get the discrete PID expression:

$$U(K) = K_P\{e(k) + \frac{T}{T_I}\sum_{j=0}^{k}e(j) + \frac{T_D}{T}[e(k)-e(k-1)]\} \qquad (6)$$

Take ΔU(k)= U(k)- U(k-1), Then

$$\Delta U(K) = a_0 e(k) + a_1 e(k-1) + a_2 e(k-2) \qquad (7)$$

Among   them, $a_0 = KP(1+T/T1+TD/T)$ $a_1 = KP(1+2TD/T)$ $a_2 = KPTD/T$

Type (7) is the incremental PID control algorithm. Its program flow chart is shown in Fig. 2 below.

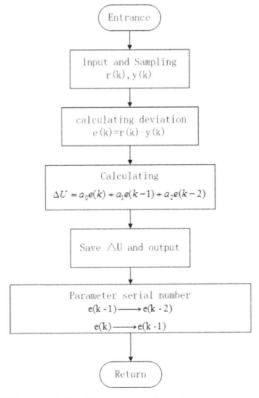

Fig. 2 Incremental PID control algorithm program flow chart

4   Temperature Control Experiment

The controlled object of this experiment is motorized spindle (no load is 10000r/min, ambient temperature is 15°C) Surface temperature is given value. On the basis of the control block diagram, intelligent PID controller controlled by programming, achieve control effect. The design of software interface include the programming of control process are used VB software programmed. The main control interface is shown as Fig. 3, the thermocouple sensor what is fixed in motorized spindle case is shown as Fig. 4.

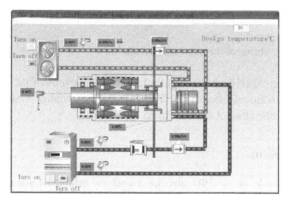

Fig. 3 The temperature test system control interface

Fig. 4 Thermocouple sensor on motorized spindle

The surface temperature of motorized spindle is set as 30°C, the temperature of the motorized spindle what is controlled by PID controller is changed as shown in Fig. 5.

Figure 5 shows that the temperature of the motorized spindle case starts at low to high and stable at about 30°C. The experimental temperature of the motorized spindle is close to the temperature of the system setting and the errors are within ±0.5°C.

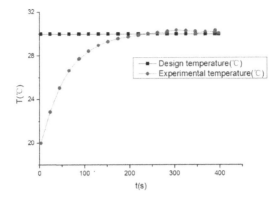

Fig. 5 Temperature changeof the motorized spindle shell

The results of the experiment are slightly less than the results of the system, but the relative error is relatively small. So, The PID cooling control system has a good control on the temperature of the motorized spindle.

5   Conclusion

The system is made up of PID algorithm and the whole adjustment system can automatically control the cooling water flow according to a certain flow rate, and finally achieving the expected goal of controlling the temperature of motorized spindle automatically. The PID cooling control system have good practical control effect.

Acknowledgements This work is supported by the natural science foundation of china under the item (grant no. 51375317), the natural science foundation project of Liaoning province under the item (grant no. 2014020069) and the major scientific and technological innovation in Liaoning province special (grant no. 201301001).

References

1. Wang, L., Song, W.: PID Control. Automation instrument **4**, 4–8 (2004)
2. Lu, G.: Research on fuzzy PID control system of temperature for electric boiler, pp. 1–10. Harbin University of Science and Technology (2007)
3. Chen, J.S., Hsu, W.Y.: Characterizations and models for the thermal growth of a motorized high speed spindle. Int. J. Mach. Tools Manuf. **43**(11), 1163–1170 (2003)
4. Wu, H.X., Shen, S.P.: Basis of theory and applications on PID control. Control engineering of China **10**(1), 37–42 (2003)
5. Cui, Q.: Research on the Temperature Control System of Continuous Annealing Furnace on RTF Section, vol. 7, pp. 2–25. Northeastern University (2011)
6. Sanchez, I., Banga, J.R., Alonso, A.A.: Temperature control in microwave combination ovens. J. Food Eng. **46**(1), 21–29 (2000)
7. Yang, L., Zhu, H., Sun, W.Y.: An improved PID incremental algorithm for saturation problem. Machine tools and hydraulic **5**, 106–108 (2002)

The Design of Kalman Filter
for Nano-Positioning Stage

Jing Dai, Peng Qu, Meng Shao, Hui Zhang and Yongming Mao

Abstract The noise signal influences the stage positioning accuracy in the process of the nano stage motion, for which designs a Kalman filter to filter out noise effectively. The model of nano-positioning stage is established. Then the motion of stage is estimated by using Kalman filtering model, and the filtering effect of Kalman filter can be observed in the Matlab. Experimental results show that Kalman filtering can effectively reduce the positioning deviation, which is less than 4nm, and positioning accuracy has been improved significantly. Kalman filter has a good effect on filtering and can meet the requirement of positioning precision for nano-positioning stage.

Keywords Nano-positioning stage · Kalman filter · Matlab

1 Introduction

With the development of the science and technologies, precision motion control technology has played an important role in the national defense industry, microelectronics engineering, aerospace, biological engineering and other fields[1,2]. The nano-positioning stage consists of driver, detector, actuator and control system[3-4]. Generally, the piezo is used for driver and the capacitive sensor is used for detector[5]. The edge of the capacitive sensor is nonlinear[6], and the signal detected by capacitive sensor is weak and signal to noise ratio is small. Hence, it is necessary to improve the signal to noise ratio by filtering.

J. Dai(✉) · P. Qu · H. Zhang · Y. Mao
Information and Control Engineering College, Shenyang Jianzhu University,
Shenyang 110168, China
e-mail: daijing6615@163.com, qupeng0314@126.com, {1693022786,42019303}@qq.com

M. Shao
Micro-Nano Detection and Motion Control Laboratory,
Shenyang Jianzhu University, Shenyang, China
e-mail: shaom81@126.com

© Springer-Verlag Berlin Heidelberg 2016   69
J.H. Kim and Z.W. Geem (eds.), *Harmony Search Algorithm,*
Advances in Intelligent Systems and Computing 382,
DOI: 10.1007/978-3-662-47926-1_8

The resonance occurs because of the inherent frequency of mechanical stage, which produces noise in the process of nano stage motion[7]. Moreover, there are also quantization noise, sampling noise and random noise etc. The traditional frequency-selective filter cannot meet the requirement of high-precision, so it is more suitable for nano stage to select the filtering algorithm that combines with the system model and has better self-adaptive. Kalman filter is relative simple in mathematical structure and is optimal linear recursive filtering method. Kalman filter is suitable for real time operation system and can be combined with the state equation model, so it is used for filtering out the noise of nano stage.

2   Establish the Model of Nano-Positioning Stage

Nano-positioning stage approximates to a linear system after nonlinear compensation. Kalman filter estimation model of nano stage is built. The motion of stage is estimated by using Kalman filtering model, which can improve the observation accuracy of stage motion. Supposing that the model state vector $X(k)$ at time point k consists of the state noise and the state transition function of the vector $X(k-1)$ at time point k-1. However, and observation vector consists of the observation noise and the observation function of the state vector $X(k)$ at time point k. Assuming system state sequence is $X(k)$ and system excitation noise is $W(k)$. The equation of state for Kalman filter model is:

$$X(k) = A(k)X(k-1) + B(k)W(k-1) \tag{1}$$

Observation sequence is defined as $Y(k)$, and observation noise is defined as $V(k)$. The observation equation is:

$$Y(k) = C(k)X(k) + V(k) \tag{2}$$

According to state equation and observation equation of the Kalman filtering, the signal model of Kalman filtering can be obtained. As it is shown in Fig. 1.

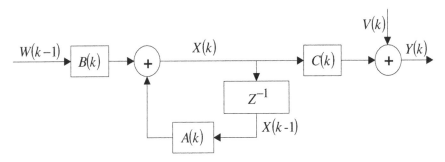

Fig. 1 The signal model of Kalman filtering

In the Fig. 1, the system excitation noise $W(k-1)$ at time point $k-1$ is input signal. The state vector $X(k-1)$ is obtained via the unit lags of the state vector $X(k)$ at time point k. The state vector $X(k)$ gets through $C(k)$, then superimposes with the observed noise $V(k)$ to obtain observation vector $Y(k)$

The statistics results determine that the state noise $W(k)$ and the observed noise $V(k)$ are Gaussian white noise of zero mean and uncorrelated. Q and R are covariance matrix corresponding to $W(k)$ and $V(k)$ respectively. $W(k)$ and $V(k)$ subject to the normal distribution: $W(k) \sim N(0,Q)$, $V(k) \sim N(0,R)$, further to get:

$$Q = \text{cov}(\omega) = E\{\omega\omega^T\} \tag{3}$$

$$R = \text{cov}(v) = E\{vv^T\} \tag{4}$$

$A(k)$, $B(k)$ and $C(k)$ are all referred as the state transformation matrix. Assume that adjustment coefficients are constant in the process of state transformation. Because of the nano-positioning stage moving on X-axis or Y-axis, it is a two-dimensional positioning stage. The stage is observed when it moves on X-axis or Y-axis. State vector $X(k)$ is expressed by the formula (5):

$$X(k) = \begin{bmatrix} L \\ dL \end{bmatrix} \tag{5}$$

In the formula (5), L is the motion displacement of the barycentre between two sampling points, dL is unit displacement in corresponding direction. $Y(k)$ is one-dimensional observation vector, namely:

$$Y(k) = [L] \tag{6}$$

The time interval T is small due to the high sampling frequency, so the stage can be considered moving at a constant speed. Among them:

$$A(k) = \begin{bmatrix} 1 & T \\ 0 & 1 \end{bmatrix} \quad B(k) = \begin{bmatrix} 0 \\ 1 \end{bmatrix} \quad C(k) = [1 \ 0] \quad Q=0.01 \quad R=4*10^{-6}$$

The state equation and observation equation of the nano-positioning stage model are obtained based on formula (1) and formula (2). Reasonable covariance matrix of state noise and observation noise are determined by the simulation, then the value of next moment can be estimated based on the current value.

3   Research on Kalman Filtering Algorithm

3.1   Description of Filtering Algorithm

The excitation noise $W(k)$ is not related with the observed noise $V(k)$. At time point k, the predictive estimation value $\hat{Y}(k \mid k-1)$ of system observed value $Y(k)$ is:

$$\hat{Y}(k \mid k-1) = C(k)\hat{X}(k \mid k-1) \tag{7}$$

$\hat{X}(k \mid k-1)$ of the formula (7) is state-step predictive value, and it can be obtained by the formula (8).

$$\hat{X}(k \mid k-1) = A(k)\hat{X}(k-1) \tag{8}$$

The observed value $Y(k)$ is obtained at time k. There is an error between $Y(k)$ and predictive estimation value $\hat{Y}(k \mid k-1)$. The predictive error is:

$$\tilde{Y}(k \mid k-1) = Y(k) - \hat{Y}(k \mid k-1) = Y(k) - C(k)\hat{X}(k \mid k-1) \tag{9}$$

The reason for this predictive error is that the state step prediction value $\hat{X}(k \mid k-1)$ and the observed values $Y(k)$ both have error. In order to obtain the state estimation value $\hat{X}(k)$ of the $X(k)$, the predictive error $\tilde{Y}(k \mid k-1)$ is used to amend the original state-step predictive value $\hat{X}(k \mid k-1)$, then there is the formula (10).

$$\hat{X}(k) = \hat{X}(k \mid k-1) + H(k)\big[Y(k) - \hat{X}(k \mid k-1)\big] \tag{10}$$

H(k) is the filtering gain matrix, which can be obtained by the formula (11).

$$H(k) = P(k \mid k-1)C^{T}(k)\big[C(k)P(k \mid k-1)C^{T}(k) + R(k)\big]^{-1} \tag{11}$$

In the formula (11), $P(k \mid k-1)$ is step budget error variance matrix, and it can be obtained by the formula (12).

$$P(k \mid k-1) = A(k)P(k-1)A^{T}(k) + B(k)Q(k-1)B^{T}(k) \tag{12}$$

The filtering error variance matrix P(k) can be expressed as:

$$P(k) = [I - H(k)C(k)]P(k \mid k-1) \tag{13}$$

Formula (8) is to predict the state estimation value at time point k based on the state estimation value at time point k-1, and formula (12) is to make the quantitative description for this prediction. From the point of view of time, the time changes from k-1 time to k time, which is achieved by the two formulas. Therefore, the equation (8) and (12) are the time update process. However, the equation (10), (11) and (13) are used to calculate the correction of the time correction value, and the observation information $Y(k)$ is used correctly and rationally by these formulas. Therefore, the equation (10), (11) and (13) are state update process. If the filtering initial values $P(0)$ and $\hat{X}(0)$ are given, the mean square error and the optimal estimation value can be calculated at any time by repeating the above steps according to the observation value at time k.

The equation (8) and (10) are called the Kalman filter equations, and the block diagram of Kalman filter can be obtained by the two formulas. It is shown in Fig. 2.

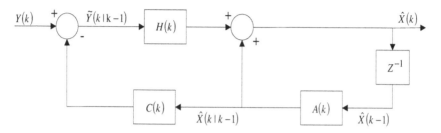

Fig. 2 The structure diagram of Kalman filter

In the Fig. 2, the observation value $Y(k)$ of the system is the input of the filter, and the state estimation value $\hat{X}(k)$ of the $X(k)$ is the output of the filter. The state estimation value $\hat{X}(k-1)$ is obtained via the unit lags of the state estimation value $\hat{X}(k)$. The estimation value $\hat{X}(k)$ gets through $C(k)$ to obtain state-step predictive value $\hat{X}(k\mid k-1)$. The predictive error $\tilde{Y}(k\mid k-1)$ is decided by the input signal $Y(k)$ and the state-step predictive value.

The filtering algorithm from formula (8) to formula (13) can be shown in Fig. 3.

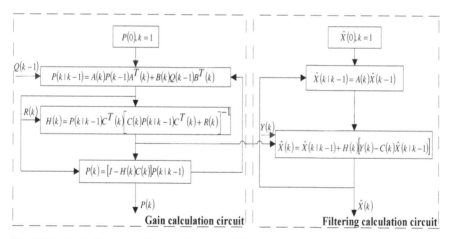

Fig. 3 The block diagram of Kalman filter algorithm

Kalman filtering algorithm consists of two computing circuits shown in Fig. 3, which are gain calculation circuit and filtering calculation circuit. Gain calculation circuit can be calculated independently, while filtering calculation circuit depends on the gain calculation circuit.

3.2   The Simulation of Kalman Filtering Algorithm in the Matlab

In the ideal case, the output displacement of piezo actuator is proportional to the applied voltage. Because of the piezo actuator driving nano stage to move, motion displacement of the nano stage is proportional to the voltage. Because of the driving voltage of nano stage ranges from -10V to +10V, however, the motion range of the stage is from -50μm to +50μm. Step signal superimposing random signal is seen as input signal. When the voltage value of the input signal is 0.4V, output waveform of the nano-positioning stage is shown in Fig. 4. Sampling frequency of the signal is 50kHz.

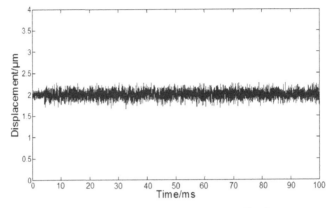

Fig. 4 Output waveform of the positioning stage without any filtering process

As can be seen from Fig. 4 in the time period of 100ms, ideal value of the output signal is 2μm. Because of the existence of random noise, the vibration of stage centers around the step signal of 2μm and the maximum deviation can be up to 400nm. The same input signal after Kalman filtering can get the output waveform of the stage, which is shown in Fig. 5.

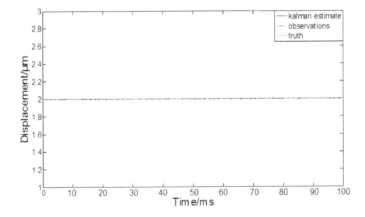

Fig. 5 Output waveform of the positioning stage after the Kalman filtering

In Fig. 5, the Kalman estimate curve and the truth curve have the maximum deviation, which is 4nm. Positioning accuracy meets the requirement of nano-positioning stage. The Kalman estimate curve and the observations curve are basic coincidence [9].

4 Conclusion

Kalman filter algorithm is applied to nano-positioning stage by establishing the system model of stage, which mostly consists of some basic formulas. The Klaman filtering results can be obtained by these basic fomulas and the filtering effect of Kalman filter can be observed in the Matlab. Experimental results show that maximum deviation of the output signal is from 400nm down to 4nm after Kalman filter, which can meet the precision requirement of stage.

Acknowledgement This work was supported by the national science and technology support project of the Ministry of Science and Technology of the People's Republic of China under project No. 2011BAK15B09.

References

1. Zhu, Y., Bazaei, A., Moheimani, S.O.R., Yuce, M.R.: Design, prototyping, modeling and control of a MEMS nanopositioning stage. In: American Control Conference, pp. 2278–2283 (2011)
2. Yu, R.H., Cai, J.X., Ma, H.R., Li, Y.Q.: Sorting Control System Design of New Materials Sorter Based on FPGA. Int. J. Control Autom. **7**, 253–264 (2014)
3. Wang, S.S., Fan, L.B., Chen, G., He, T.: Three-dimensional micro-displacement control system for a conductor defect detection system. Advanced Materials Research (2012). doi:10.4028/www.scientific.net/AMR.588-589.1431
4. Zhu, D.H., Li, Z.H., Tian, X.X.: Ant Colony Optimization Control in VAV System. Journal of Shenyang Jianzhu University (Natural Science) **28**, 1131–1135 (2012)
5. Li, Y., Sun, Y.F.: Development of Actuation Control System of Piezoelectric Micro-Displacement Device Based on Increment PID Algorithm. Meas. Control Tech. **30**, 40–44 (2011)
6. Dai, J., Shen, J., Shao, M., Bai, L.X.: Design and implementation of filter in the control system of the high accuracy micro-displacement platform. International Journal of Modelling, Identification and Control **21**, 82–92 (2014)
7. Zhou, Y., Yao, X.X., Pian, J.X., Su, Y.Q.: Band-stop filter algorithm research based on nano-displacement positioning system. Appl. Mech. Mater (2013). doi:10.4028/www.scientific.net/AMM.380-384.697
8. Fu, M.Y., Deng, Z.H., Yan, L.P.: Kalman filtering theory and its application in navigation system. Scientific Publishers (2010)
9. Li, K.Q.: The application research on 6-RSS parallel institution in the six-dimensional active vibration platform. doctoral thesis of Beijing Jiaotong University (2008)

Research on Adaptive Control Algorithm of Nano Positioning Stage

Jing Dai, Tianqi Wang, Meng Shao and Peng Qu

Abstract The mechanical structure and load changes of nano positioning stage cause a poor accuracy of the control system. For solving the problem, Adaptive PID control algorithm was applied to control the nano positioning stage. The model of nano positioning stage was established on the basis of Controlled Auto-regressive Moving Average model (CARMA). The parameters of controller were identified based on Recursive Extended Least Squares algorithm (RELS). The control system of nano positioning stage was steady through 4ms after parameters identification, which static error was less than 5nm. The experimental results demonstrated that adaptive PID control algorithm was able to identify the parameters of controlled object and calculate the parameters of controller. The accuracy of control system can be at nanometer resolution.

Keywords Nano positioning stage · Adaptive control algorithm · Parameters identification · PID control

1 Introduction

Modern science and technology rely increasingly on the development of precision instruments. As a key and generic technology in precision instruments, the application of nano positioning system is extremely broad, such as semiconductor processing and testing equipment, LCD/LED and other display manufacturing facilities, laser equipment, biomedical microscopy and operating instruments, all kinds of precision machine tools, aerospace technology and so on. As an essential stage for the development of nano positioning technology, nano positioning stage is required with

J. Dai(✉) · T. Wang · P. Qu
Faculty of Information and Control Engineering, Shenyang Jianzhu University,
Shenyang, Liaoning Province, China
e-mail: daijing6615@163.com, TianqiWang_21@126.com

M. Shao
Micro-Nano Detection and Motion Control Laboratory, Shenyang Jianzhu University,
Shenyang, Liaoning Province, China
e-mail: mshao@sjzu.edu.cn

© Springer-Verlag Berlin Heidelberg 2016   77
J.H. Kim and Z.W. Geem (eds.), *Harmony Search Algorithm*,
Advances in Intelligent Systems and Computing 382,
DOI: 10.1007/978-3-662-47926-1_9

high accuracy and excellent dynamic performance [1]. For meeting the requirement, an appropriate intelligent control algorithm is needed. There are many ways to control the nano positioning stage, including feed-forward control, PID control, fuzzy control, neural control and so on. Basing on a determinate hysteretic model, JUNG and KIM controlled piezoelectric actuator using feed-forward algorithm, and improved the scanning accuracy by 10 times than traditional open-loop control [2]. HANZ and EDUAROO deduced the step response rising time of piezoelectric actuator to 3-4ms by PI control algorithm [3]. Fuzzy control was used by Xiaodong Hu, Xiaotang Hu et al to study the control of XYZ micro-displacement driven by piezoelectric actuator, and improved the positioning accuracy and dynamic response of micro-displacement [4]. LU, CHEN et al researched a neural network integrated control system with adaptive and self-learning performance by combining single neural network control with PID control [5]. The neural network integrated control system achieved a high positioning accuracy. Feedforward control and PID control algorithm are convenient, but difficult to get good dynamic properties and positioning accuracy when the load changes. The performances of fuzzy control and neural network integrated control are better; however, algorithms are more complex [6]. Adaptive PID control algorithm is the synthesis of adaptive control and PID control algorithm [7]. In this paper, Adaptive PID control algorithm solves the issue caused by load changes.

2   Design of Nano Positioning Stage Control System

In this paper, the moving range of nano positioning stage driven by piezoelectric actuator is 100μm x 100μm. Drive voltage is -10V to 10V, and control two-axis displacement of X and Y independently. The structure of nano positioning stage control system is shown in Fig. 1. Analog-to-digital converter (ADC) converts actual position signal detected by capacitance sensor to feedback signal y(k) [8]. Error signal e(k) is obtained by comparing y(k) with r(k)which is given by personal computer (PC). PID controller receives the control parameters and error signal, and then calculates the controlled variable u(k). Thus, PID controller can be able to take stage to the commanded position by controlling piezoelectric actuator. In the system, identification of stage model parameters and controller parameters calculation are completed by PC.

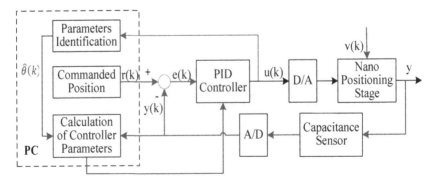

Fig. 1 The structure diagram of nano positioning stage control system

3   Adaptive PID Control Algorithms

Controlled autoregressive moving average model (CARMA) is an auxiliary model which is established based on the measurable information [9]. By choosing its parameters, makes the output of auxiliary model approach the actual output, and achieves the consistent estimate of system parameters. We use actual position signal and control variable to build CARMA model for nano positioning stage. The CARMA model can be expressed in equation (1)

$$A(z^{-1})y(k) = B(z^{-1})u(k-d) + v(k) \tag{1}$$

where d is lag coefficient, v(k) and y(k) are disturbance and actual position at k moment, respectively. $A(z^{-1})$、 $B(z^{-1})$ and v(k) are expressed in equation (2), equation (3), and equation (4), respectively.

$$A(z^{-1}) = 1 + a_1 z^{-1} + \cdots + a_n z^{-n} \tag{2}$$

$$B(z^{-1}) = b_0 + b_1 z^{-1} + \cdots + b_m z^{-m} \tag{3}$$

$$v(k) = C(z^{-1})e_w(k) \tag{4}$$

$$C(z^{-1}) = 1 + c_1 z^{-1} + \cdots + c_i z^{-i} \tag{5}$$

3.1   Parameters Identification

Least square estimation method is used to identify the parameters of auxiliary model, including recursive least squares (RLS), recursive extended least squares (RELS), generalized least squares(GLS) and so on. When disturbance v(k) is colored noises, the identification of RLS method is biased estimation, either RELS or GLS can obtain the unbiased estimation of the parameters. Due to the complexity of GLS method, RELS method is used in this paper. Based on feedback y(k) and control variable u(k), parameters estimator module online estimates the parameters of auxiliary model $\hat{\theta}(k)$ which is the coefficient of $A(z^{-1})$ and $B(z^{-1})$ using RELS method [10].

Recursive extended least squares:

$$\begin{cases} K = \dfrac{P(k-1)\varphi(k)}{\beta + \varphi^T(k)P(k-1)\varphi(k)} \\ P(k) = \dfrac{P(k-1) - K(k)\varphi^T(k)P(k-1)}{\beta} \\ \hat{\theta}(k) = \hat{\theta}(k-1) + K(k)\left[y(k) - \varphi^T(k)\hat{\theta}(k-1)\right] \end{cases} \tag{6}$$

where β is a forgetting factor, initial value is $P_0 = a^2 I$ I (a is sufficiently large positive number), $\varphi^T(k)$ consist of feedbacks and control variables filtered by a linear system.

$$\varphi^T(k) = [y'(k-d), y'(k-d-1), \cdots y'(k-d-n_F) \qquad (7)$$
$$u'(k-d), \cdots, u'(k-d-n_G)]$$

Parameters of auxiliary model is

$$\hat{\theta}(k) = [a_1, a_2, \cdots a_n, b_0, b_1, \cdots b_m] \qquad (8)$$

So, the parameters $\hat{A}(z^{-1})$ and $\hat{B}(z^{-1})$ are found.

The auxiliary model of nano positioning stage will vary under load changes. For meeting the requirement of real-time identification, RELS method is used. Fig. 2 shows the identification parameters curve of nano positioning stage. The sampling cycle of experiment is 0.2ms, and the parameters are identified exactly after 5ms.

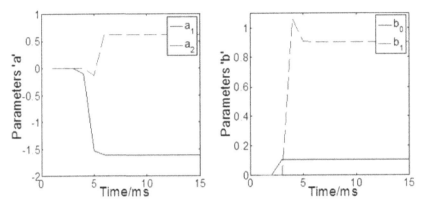

Fig. 2 The identification parameters curve of nano positioning stage

3.2 Calculation of Controller Parameters

According to Diophantine equations, there are $G(z^{-1})$ and $F(z^{-1})$ meet formula (9).

$$A(z^{-1})G(z^{-1}) + z^{-d}B(z^{-1})F(z^{-1}) = C(z^{-1})A_m(z^{-1}) \qquad (9)$$

where $C(z^{-1})$ is polynomial filtering, $C(z^{-1})A_m(z^{-1})$ denotes the desired closed loop poles polynomials. The module of controller parameters calculation solves the Diophantine equations using $\hat{A}(z^{-1})$ and $\hat{B}(z^{-1})$ to get $\hat{F}(z^{-1})$ and $\hat{G}(z^{-1})$. The updated parameters and control variable of PID controller is shown in equation (10).

$$u(k) = \frac{H(z^{-1})r(k) - \hat{G}(z^{-1})y(k)}{\hat{F}(z^{-1})} \qquad (10)$$

where r(k) is commanded position, $H(z^{-1})$ is usually preferable to 1.

4 The Experiments of Control Simulation

Fig. 3 and Fig. 4 are the square-wave response of nano positioning stage in X axis before and after the changes of stage model in adaptive PID control system. In Fig. 5, the control accuracy of adaptive PID control system for square-wave signal can be achieved to 3.7nm. The steady-state characteristics are very well. When stage model varies in Fig. 6, the control system identifies the new parameters and calculates a new control variable automatically. After the process of identification, the system comes back into the stable, and static error is 5.3nm. The control accuracy has a difference of 1.6nm with before changes.

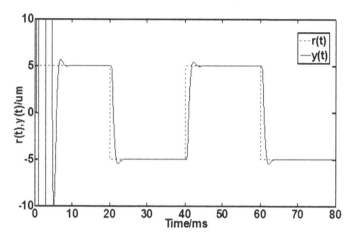

Fig. 3 The square-wave response of adaptive PID control system

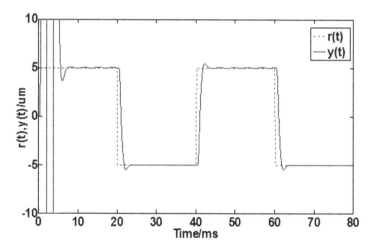

Fig. 4 The square-wave response of adaptive PID control system after the changes of stage model

5   Conclusions

The loads of nano positioning stage are variable. If the load changes, the control system should update stage model parameters and calculate the new controller parameters. PID controllers receive the new controller parameters and control the system. In simulations, the system comes back into stable after changes of stage model and makes the static error less than 5.3nm. In conclusion, adaptive PID control algorithm realizes the real-time control.

Acknowledgement This work was supported by the national science and technology support project of the Ministry of Science and Technology of the People's Republic of China under project No. 2011BAK15B09.

References

1. Xie, X.H., Du, R.X., Qiang, S.: Design and simulation of a nano-scale micro positioning stage. International Journal of Modelling, Identification and Control (2015). doi:10.1504/IJMIC.2009.027019
2. Jung, S.B., Kim, S.W.: Improvement of scanning accuracy of PZT piezoelectric actuators by feed-forward model-reference control. Precis Eng., 49–55 (1994)
3. Hanz, R., Eduaroo, A.M., Lucca, A.: Characterization of nonlinearities in a piezoelectric positioning device. In: IEEE International Conference on Control Applications, pp. 717–720 (1997)
4. Hu, X.D., Hu, X.T., Yong, Y.: Study on the performance of 3D monolithic ultra-micropositioning system. Journal of China Mechanical Engineering **11**, 1201–1203 (1999)
5. Lu, Z.Q., Chen, D.J., Kong, H.F. et al.: Neuron net control and it's application in precision feed mechanism of machine tool driven by PZT. In: IEEE International Conference on Industrial Technology, pp. 80–82 (1994)
6. Ma, B., Meng, Q.B., Feng, Y. et al.: The controller design for dc motor speed system based on improved fuzzy PID. In: International Conference on Intelligent System and Applied Material, pp. 466–467, 1246–1250 (2012)
7. Wang, Y.G., Tian, Z.P.: Adaptive PID controllers with robustness specifications. International Journal of Modelling, Identification and Control **20**, 148–155 (2013)
8. Dai, J., Jing, S., Meng, S., et al.: Design and implementation of filter in the control system of the high accuracy micro-displacement platform. International Journal of Modelling, Identification and Control **21**, 82–92 (2014)
9. Feng, D.: System identification Part D: Auxiliary model identification idea and methods. Journal of Nanjing University of Information Science and Technology: Natural Science Edition **4**, 289–318 (2011)
10. Feng, Z., Lu, Y.H., Feng, Q., et al.: Self-tuning fuzzy adaptive PID pitch control of wind power systems. International Conference on Mechatronics, Applied Mechanics and Energy Engineering **394**, 404–409 (2013)

Two Frameworks for Cross-Domain Heuristic and Parameter Selection Using Harmony Search

Paul Dempster and John H. Drake

Abstract Harmony Search is a metaheuristic technique for optimizing problems involving sets of continuous or discrete variables, inspired by musicians searching for harmony between instruments in a performance. Here we investigate two frameworks, using Harmony Search to select a mixture of continuous and discrete variables forming the components of a Memetic Algorithm for cross-domain heuristic search. The first is a single-point based framework which maintains a single solution, updating the harmony memory based on performance from a fixed starting position. The second is a population-based method which co-evolves a set of solutions to a problem alongside a set of harmony vectors. This work examines the behaviour of each framework over thirty problem instances taken from six different, real-world problem domains. The results suggest that population co-evolution performs better in a time-constrained scenario, however both approaches are ultimately constrained by the underlying metaphors.

Keywords Harmony search · Hyper-heuristics · Combinatorial optimisation · Metaheuristics · Memetic algorithms

1 Introduction

Harmony Search (HS) is a population-based metaheuristic technique introduced by Geem et al [1], inspired by the improvisation process of musicians. HS is an Evolutionary Algorithm (EA) which seeks to optimise a given set of parameters,

P. Dempster(✉) · J.H. Drake
School of Computer Science, University of Nottingham Ningbo, Ningbo 315100, China
e-mail: {paul.dempster,john.drake}@nottingham.edu.cn

J.H. Drake
ASAP Research Group, School of Computer Science, University of Nottingham,
Jubilee Campus, Wollaton Road, Nottingham NG8 1BB, UK
e-mail: drakejohnh@gmail.com

© Springer-Verlag Berlin Heidelberg 2016   83
J.H. Kim and Z.W. Geem (eds.), *Harmony Search Algorithm*,
Advances in Intelligent Systems and Computing 382,
DOI: 10.1007/978-3-662-47926-1_10

analogous to musicians searching for a 'harmonious' combination of musical notes. Using a short-term memory, a population of vectors representing a set of decision variables is maintained and updated over time. At each step, a new vector is generated and compared to the existing vectors within the memory based on a given quality measure, and if the new vector is deemed to be of better quality than one of the existing vectors it replaces that vector within the memory.

Over recent years, many variants of HS have been proposed and applied to combinatorial optimisation problems including educational timetabling [2], flow shop scheduling [3] and the travelling salesman problem [1]. For the interested reader, a survey of the applications HS has been applied to is provided by Manjarres et al. [4].

Cowling et al. [5] introduced the term 'hyper-heuristic' to the field of combinatorial optimisation, defining hyper-heuristics as '*heuristics to choose heuristics*'. Unlike traditional search and optimisation techniques, hyper-heuristics operate over a space of low-level heuristics or heuristic components, rather than directly over a search space of solutions. More recently, changes in research trends have led to a variety of hyper-heuristic approaches being developed for which this original definition does not provide the scope to cover. Burke et al. [6,7] offered a more general definition covering the two main classes of hyper-heuristics, selection hyper-heuristics (e.g. [8,9]) and generation hyper-heuristics (e.g. [10,11]):

'A hyper-heuristic is a search method or learning mechanism for selecting or generating heuristics to solve computational search problems.'

Hyper-heuristics have been applied successfully to a variety of problems such as bin packing [12], dynamic environments [13], examination timetabling [14,15], multidimensional knapsack problem [9,16], nurse scheduling [14], production scheduling [17], sports scheduling [18] and vehicle routing [19].

In this paper we investigate two frameworks for using Harmony Search to choose the components and parameter settings of a selection hyper-heuristic applied to multiple problem domains.

2   Selection Hyper-Heuristics and Cross-Domain Heuristic Search

Many optimisation problems create a search space which is too large to enumerate exhaustively to check every possible solution. A variety of heuristic and metaheuristic methods have been used previously to solve such problems. A disadvantage of these methods is the lack of flexibility to solve different types of problem instances or problem class. Typically this will result in the need to tune the proposed method each time a new type of problem is encountered. The goal of cross-domain heuristic search is to develop methods which are able to find high quality solutions consistently over multiple problem domains, operating over a search space of low-level heuristics. Hyper-heuristics, as introduced previously, are high-level search methodologies that operate at a higher level of abstraction than traditional search and optimisation techniques [20], searching a space of heuristics rather than solutions. Single-point based selection hyper-heuristics typically rely on two components, a heuristic selection method and a move acceptance criteria [21]. Hyper-heuristics using this framework iteratively select and apply low-level heuristics to a single solution with a decision made at each step as to whether to accept the move. This process continues until some termination criteria is met.

The experiments in this paper will use the HyFlex framework [22] as a benchmark for comparison. Originally developed to support the first cross-domain heuristic search challenge (CHeSC2011) [23], HyFlex provides a common software interface with which high-level search methodologies can be implemented and compared. In addition, due to the nature of the framework, a direct comparison can be made to a large number of existing methods from the literature. The core HyFlex framework provides support for six widely studied real-world problems: MAX-SAT, one-dimensional bin packing, personnel scheduling, permutation flow shop, the travelling salesman problem and the vehicle routing problem. For each of these problem domains, a set of low-level heuristics is implemented, defining the search space within which a selection hyper-heuristic can operate. Each of these low-level heuristics belongs to one of four categories: mutation, hill-climbing, ruin-recreate or crossover. All low-level heuristics take a single solution as input and give a single solution as output, with the exception of crossover where two solutions are required for input. The number of low-level heuristics of each type varies, depending on the problem domain being considered. For the sake of our experimentation in this paper, we consider ruin-recreate heuristics within the set of mutation operators as both take a solution as input, perform some perturbation and return a new solution with no guarantee of quality. In addition to this, two continuous variables $\in [0, 1]$ are used to modify the behaviour of certain low-level heuristics. The *intensity of mutation* parameter defines the extent to which a mutation or ruin-recreate low-level heuristic modifies a given solution, with a value closer to 1 indicating a greater amount of change. The *depth of search* parameter relates to computational effort afforded to the hill-climbing heuristic, with a value closer to 1 indicating that the search will continue to a particular depth limit.

2.1   Memetic Algorithms within the HyFlex Framework

In addition to the single-point framework described above, it is also possible to implement population-based methods within the HyFlex framework. A Genetic Algorithm (GA) evolves a population of solutions to a problem, using crossover and mutation operators to recombine and modify solutions, inspired by the natural process of evolution and the concepts of selection and inheritance. A Memetic Algorithm (MA) [24] is an extension of a general GA which introduces *memes* [25] into the evolutionary process, where a meme is a 'unit of cultural transmission'. A basic MA simply introduces a hill-climbing phase into a GA after crossover and mutation have been applied. MAs are not the only approach to include an explicit hill-climbing phase into a search method. Özcan et al. [21] tested a number of frameworks for selection hyper-heuristics. Their work found that improved performance could be achieved on a set of benchmark functions by applying a hill-climber following the application of a perturbative low-level heuristic. Ochoa et al. [26] implemented an MA variant within HyFlex, an Adaptive Memetic Algorithm, showing better performance than all of the CHeSC2011 entrants on instances of the vehicle routing problem.

3   Harmony Search-Based Frameworks for Heuristic and Parameter Selection

In this paper we test two frameworks to select the components of an MA. In this case, HS is operating as a hyper-heuristic, selecting which low-level heuristics and parameter values to use and apply to solutions to different problem domains. The first framework is based on a traditional selection hyper-heuristic, operating on a single solution, whereas the second is a population-based approach much closer to the traditional idea of an MA. The core of both of the frameworks tested relies on HS evolving vectors representing a choice of operator for each component and their associated parameter values. Each harmony corresponds to a set of three low-level heuristics, which are applied sequentially to a particular solution in the manner of an MA (i.e. crossover, followed by mutation then hill-climbing), and two associated parameter settings. The Harmony Memory (HM), of size HMS, within each framework is as follows:

$$
HM = \begin{bmatrix} C^1 & M^1 & H^1 & I^1 & D^1 \\ C^2 & M^2 & H^2 & I^2 & D^2 \\ \vdots & \vdots & \vdots & \vdots & \vdots \\ C^{HMS} & M^{HMS} & H^{HMS} & I^{HMS} & D^{HMS} \end{bmatrix}
$$

where C, M and H correspond to discrete choices of crossover, mutational and hill-climbing heuristics for the current problem domain. I and D represent continuous variables indicating the values of the *intensity of mutation* and *depth of search* parameters for these heuristics.

At each step a vector is generated ('improvised') and applied to a particular solution. The new vector is generated depending on the harmony memory consideration rate (*HMCR*) parameter, denoting the probability using existing information from memory as opposed to constructing a new vector from scratch. With probability *HMCR*, each of the individual values of decision variables in the new vector are taken randomly from one of the existing solutions in HM. In order to maintain some diversity within the search, the pitch adjustment rate (*PAR*) parameter defines the probability of modifying each of the two continuous parameters in the newly generated vector by a particular fret width (*FW*). In our experiments, the *PAR* does not affect the discrete decision variables. With probability $(1 - HMCR)$ a completely new vector is generated with random valued assigned to each variable. The components of the new vector are then applied to a solution, in a manner defined by one of the frameworks described below, with the new vector replacing the poorest quality vector in the HM if it is of better quality.

3.1 Single-Point Search Framework

This framework evolves a population of vectors, with each applied to a single solution and updated based on the performance observed. At each step a new harmony is improvised, then each harmony in HM and the newly generated harmony are all separately applied to the current solution. If the solution generated by the new harmony is better than any of the solutions generated by existing harmonies, it replaces that solution in the HM. In the case that any solution generated is better than the previous best solution, the best solution is updated to become that solution. Where a second input solution is required for crossover low-level heuristics, a solution is chosen at random from the set of solutions generated in the previous set of intermediate solutions. In the initial iteration, the second solution is generated randomly using the methods defined in the HyFlex framework. This method is similar to the *Greedy* selection hyper-heuristic of Cowling et al. [5] from their early work in selection hyper-heuristics. An overview of this framework is shown in Figure 1.

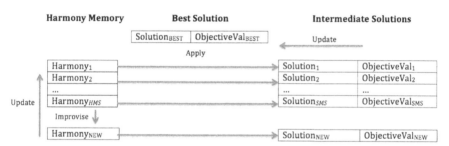

Fig. 1 Single-point search framework

3.2 Population-Based Co-evolutionary Framework

The second framework is very similar to the one proposed and applied to examination timetabling by Anwar et al. [27]. Rather than operating on a single solution, a population of solutions is maintained in a Solution Memory (SM) in addition to the population of heuristics and parameters managed using HS. In our experimentation the size of SM (Solution Memory Size (*SMS*)) and HMS are equal, with the harmony used to produce a particular solution contained at the same index in HM as the solution in SM. In the first instance, the contents of HM and SM are populated randomly, with each harmony in HM applied once to the solution in the corresponding index of SM, with random solutions taken from SM for crossover purposes. Following this initialisation period, a new harmony is improvised, either based on harmonies in memory or generated from scratch depending on whether a randomly generated variable $U(0,1)$ is less than or equal to the value of *HMCR*. The low-level heuristics in this harmony are then applied in order to a solution from the SM as described above. In the case that the new harmony is based on an existing harmony in memory, one input solution is taken from SM at the same index as the crossover low-level heuristic was taken from HM, with a second solution (as required for crossover low-level heuristics) taken randomly. Where the harmony has been improvised randomly, two random solutions from SM are used as input for the crossover low-level heuristic. Following the application of crossover, the mutation and hill-climbing (with the parameter values for *intensity of mutation* and *depth of search*) are then applied to the resulting solution as defined by the new harmony. If the quality of the final solution is better than an existing solution in SM, the new solution replaces the existing solution in SM and the new harmony replaces the harmony at the corresponding index in HM. This general framework is shown in Figure 2.

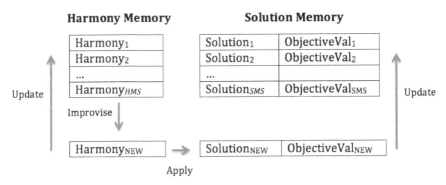

Fig. 2 Population-based co-evolutionary framework

3.3 Experimental Settings

As introduced in Section 2, the HyFlex framework will be used to compare the two proposed frameworks. We will use the 30 instances tested in the CHeSC2011 competition, with 5 instances taken from each of the 6 problem domains. All problem domains are minimization problems, with a lower objective value indicating better performance. In line with the competition, each hyper-heuristic is allowed a notional 10 minutes computation time as calculated by a benchmark provided by the organisers. Each instance is run 31 times, with the median and minimum values of all runs reported. In all of our experiments, *HMS* and *SMS* are set to 10, *HMCR* is 90%, *PAR* is 30% and *FW* is 0.2. The setting of *FW*, in particular, to 0.2 is due to the underlying nature of the HyFlex parameter values.

4 Computational Results

4.1 Direct Comparison of Single-Point and Population-Based Frameworks

Firstly we will directly compare the objective values obtained by the single-point and population-based variants described in Section 3.1 and Section 3.2. Table 1

Table 1 Median and minimum objective value obtained by Harmony Search MA variants

Instance	Single-point		Population-based	
	Median	Min	Median	Min
SAT0	**29**	20	**29**	**19**
SAT1	**57**	46	**57**	**40**
SAT2	**42**	**18**	**42**	19
SAT3	**30**	22	32	**15**
SAT4	**17**	**13**	20	14
BP0	0.076754703	0.063377656	**0.063257737**	**0.055634589**
BP1	0.011526797	0.007252416	**0.008133224**	**0.006673**
BP2	0.027774455	0.013815935	**0.014645751**	**0.012569296**
BP3	0.111261704	0.109816678	**0.10926376**	**0.108950592**
BP4	0.045656986	0.037023311	**0.029869874**	**0.023851741**
PS0	33	21	**27**	**20**
PS1	10688	10243	**9850**	**9556**
PS2	3327	3200	**3232**	**3130**
PS3	1830	1485	**1685**	**1443**
PS4	355	**320**	**350**	**320**
FS0	6287	6252	**6267**	**6237**
FS1	26847	26755	**26813**	**26754**
FS2	6367	6323	**6345**	**6303**
FS3	11436	11382	**11409**	**11336**
FS4	26658	26579	**26639**	**26534**
TSP0	48286.76001	**48194.9201**	**48194.9201**	**48194.9201**
TSP1	21308223.69	21076767.44	**20794298.01**	**20729164.65**
TSP2	6853.294645	6823.87626	**6822.783065**	**6798.088796**
TSP3	67692.9665	66570.50774	**67050.91033**	**66423.61825**
TSP4	**53699.5647**	**52272.74022**	54049.52449	52561.08631
VRP0	91485.99596	86722.503	**62722.29739**	**60850.09874**
VRP1	14395.55092	**13317.12627**	**13383.59972**	13334.93593
VRP2	201800.2381	147303.8795	**148281.8064**	**144058.3475**
VRP3	21672.79408	**20658.96394**	**21658.21568**	20658.96815
VRP4	178263.89	166947.0722	**147932.5119**	**146313.0592**

shows the median and minimum values observed over 31 runs for each of the 30 problem instances tested. The best values for each instance are marked in **bold**. From this table we can quickly see that the population-based version is performing better in the majority of cases in terms of both median and minimum objective value observed. Overall the population method achieves a better median value for 24 of the 30 instances tested and a better minimum value in 23 of the 30 instances. Whilst both methods show similar performance in the SAT instances, the objective values obtained are poor when compared to the literature standard. This will be discussed further in the following section.

4.2 Comparison to CHeSC2011 Entrants

Following the competition, the results were provided for the competition entries over a subset of the problems of all six problem domains. Methods are ranked using a scoring system inspired by the one used in Formula One motor racing (2003-2009).

Table 2 Rankings obtained by Harmony Search MA variants compared to CHeSC2011 entrants

Hyper-heuristic	Total	Hyper-heuristic	Total
AdapHH	181	AdapHH	173.5
VNS-TW	133	VNS-TW	130.5
ML	131.5	ML	129.5
PHUNTER	93.25	PHUNTER	87.75
EPH	89.75	EPH	83.75
NAHH	75	HAHA	71.083
HAHA	73.75	NAHH	71
ISEA	69	**POP-MA**	**65.5**
KSATS-HH	66.5	KSATS-HH	64.5
HAEA	53.5	ISEA	63
ACO-HH	39	HAEA	47.833
GenHive	36.5	ACO-HH	36.33
DynILS	26	GenHive	32.5
SA-ILS	24.25	SA-ILS	23.25
XCJ	22.5	XCJ	21.5
AVEG-Nep	21	AVEG-Nep	21
GISS	16.75	DynILS	21
SelfSearch	7	GISS	16.75
SP-MA	**6**	SelfSearch	5
MCHH-S	4.75	MCHH-S	4.75
Ant-Q	0	Ant-Q	0

For each of the 30 problem instances, the best performing hyper-heuristic of those currently being compared is awarded 10 points, the second best 8 points with each further method allocated 6, 5, 4, 3, 2, 1 and 0 points respectively. In the case there are more than 8 methods being compared, all methods ranked > 8th are awarded 0 points. The two HS hyper-heuristic variants can be compared directly against the CHeSC2011 entrants using this scoring system. Table 2 shows the relative ranking of each variant against the 20 entrants to CHeSC2011, where SP-MA is the single-point framework and POP-MA is the population-based variant.

When compared to the CHeSC2011 entrants, the population-based framework scores 65.5 points finishing 8th overall, with the single point approach scoring 6 points and finishing 19th out of 21. As this is a relative ranking system, it shows that not only is POP-MA outperforming SP-MA, in at least some problem domains or instances it is outperforming some of the leading entrants to the competition, taking 7.5 points away from the winning hyper-heuristic AdapHH. A breakdown of the points per problem domain scored by each method is given in Table 3.

Table 3 Formula One points obtained by each method in each problem domain

Domain	SP-MA	POP-MA
SAT	0	0
Bin Packing	0	0
Personnel Scheduling	0	**4.5**
Flow Shop	3	**16**
TSP	3	**30**
VRP	0	**15**

As can be seen, neither framework performed well in SAT or Bin Packing. The population-based framework performed particularly well in TSP, coming 3rd in this domain and beating all other hyper-heuristics in one particular instance. Given that the best results of the single-point framework also came in TSP, it suggests that the heuristics supplied for that domain produce good solutions when combined via MA. Figure 3 plots the number of points scored by each of the twenty competition entrants to CHeSC2011 and POP-MA for the TSP. An interesting observation when using this framework is that the number of heuristics that could be applied per run within the time limit varied dramatically. This is due to both the nature of the heuristics in a particular domain, and the evolved values of the heuristic parameters (e.g., between 415000 and 450 heuristic applications were observed for the first two TSP instances). Since HS requires a new harmony to be improvised each iteration and MA requires all three of cross-over, mutation, and hill-climbing, the high-level of retention of solutions by the population-based approach (keeping the best HMS − 1 solutions rather than throwing away the HMS − 1 worst as with single-point) may be the reason for the comparatively good Personnel Scheduling score, as this domain typically had low heuristic application counts. It is also worth noting that as this ranking system only rewards the top 8 entrants, a score of 0 does not necessarily indicate that a method is one of the worst of the 21 being compared, only that it is not one of the top hyper-heuristics for that instance.

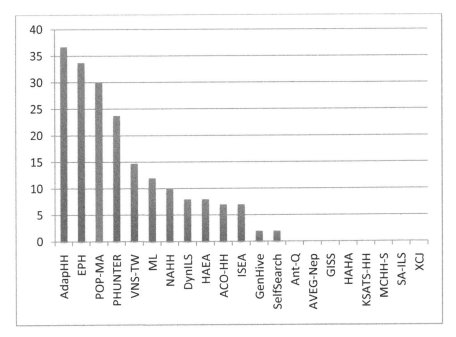

Fig. 3 Formula One points scored in the TSP problem domain by POP-MA and CHeSC2011 entrants

5   Conclusions and Future Work

In this work we have presented two frameworks, using Harmony Search (HS) to select the components and parameters for a Memetic Algorithm (MA) operating over multiple problem domains. The first framework is similar to a traditional greedy single-point selection hyper-heuristic, applying each set of low-level heuristic and parameter combinations in the Harmony Memory (HM) to a current best-of-run solution before updating HM and the best-of-run solution if an improvement in solution quality is found. The second is a population-based approach which co-evolves a set of solutions to the problem at the same time as evolving a set of harmonies representing the operator and parameter choices of an MA applied to the solutions. Our results indicate that the population-based approach is able to achieve significantly better results than the single-point approach. This method performed particularly well on instances of the Travelling Salesman Problem, outperforming all but two of the twenty entrants to CHeSC2011.

We intend to improve on this work in future by extending the framework in a number of ways. Currently the representation of each harmony is very rigid, due to the decision of conforming to the structure of an MA. More flexible representations are possible, allowing for a greater combination of low-level heuristics and more freedom with the order in which they are applied. Another constraining factor in the population-based framework is the decision to limit the Solution Memory Size and Harmony Memory Size to the same value. Future work will look at dynamic strategies for managing the length of these two components.

References

1. Geem, Z.W., Kim, J.H., Loganathan, G.V.: New heuristic optimization algorithm: Harmony search. Simulation **76**(2), 60–68 (2001)
2. Al-Betar, M.A., Khader, A.T.: A harmony search algorithm for university course timetabling. Ann. Oper. Res. **194**(1), 3–31 (2012)
3. Wang, L., Pan, Q.K., Tasgetiren, M.F.: Minimizing the total flow time in a flow shop with blocking by using hybrid harmony search algorithms. Expert Syst. Appl. **37**(12), 7929–7936 (2010)
4. Manjarres, D., Landa-Torres, I., Gil-Lopez, S., Ser, J.D., Bilbao, M., Salcedo-Sanz, S., Geem, Z.: A survey on applications of the harmony search algorithm. Eng. Appl. Artif. Intell. **26**(8), 1818–1831 (2013)
5. Cowling, P.I., Kendall, G., Soubeiga, E.: A hyperheuristic approach to scheduling a sales summit. In: Burke, E., Erben, W. (eds.) PATAT 2000. LNCS, vol. 2079, pp. 176–190. Springer, Heidelberg (2001)
6. Burke, E.K., Hyde, M., Kendall, G., Ochoa, G., Özcan, E., Woodward, J.: A classification of hyper-heuristics approaches. In: Handbook of Metaheuristics, 2nd edn., pp. 49–468 (2010)
7. Burke, E.K., Hyde, M., Kendall, G., Ochoa, G., Özcan, E., Qu, R.: Hyper-heuristics: A survey of the state of the art. J. Oper. Res. Soc. **64**(12), 1695–1724 (2013)
8. Drake, J.H., Özcan, E., Burke, E.K.: An improved choice function heuristic selection for cross domain heuristic search. In: Coello, C.A., Cutello, V., Deb, K., Forrest, S., Nicosia, G., Pavone, M. (eds.) PPSN 2012, Part II. LNCS, vol. 7492, pp. 307–316. Springer, Heidelberg (2012)
9. Drake, J.H., Özcan, E., Burke, E.K.: Modified choice function heuristic selection for the multidimensional knapsack problem. In: Sun, H., Yang, Chin-Yu., Lin, C.-W., Pan, J.-S., Snasel, V., Abraham, A. (eds.) Genetic and Evolutionary Computing. AISC, vol. 329, pp. 225–234. Springer, Heidelberg (2015)
10. Drake, J.H., Hyde, M., Ibrahim, K., Özcan, E.: A genetic programming hyper-heuristic for the multidimensional knapsack problem. Kybernetes **43**(9–10), 1500–1511 (2014)
11. Drake, J.H., Kililis, N., Özcan, E.: Generation of VNS components with grammatical evolution for vehicle routing. In: Krawiec, K., Moraglio, A., Hu, T., Etaner-Uyar, A., Hu, B. (eds.) EuroGP 2013. LNCS, vol. 7831, pp. 25–36. Springer, Heidelberg (2013)
12. López-Camacho, E., Terashima-Marín, H., Ross, P.: A hyper-heuristic for solving one and two-dimensional bin packing problems. In: Proceedings of the 13th Annual Conference Companion on Genetic and Evolutionary Computation, pp. 257–258 (2011)
13. Kiraz, B., Uyar, A.S., Özcan, E.: Selection hyper-heuristics in dynamic environments. J. Oper. Res. Soc. **64**(12), 1753–1769 (2013)

14. Burke, E.K., Kendall, G., Soubeiga, E.: A tabu-search hyper-heuristic for timetabling and rostering. J. Heuristics 9(6), 451–470 (2003)
15. Özcan, E., Misir, M., Ochoa, G., Burke, E.K.: A reinforcement learning – great deluge hyper-heuristic for examination timetabling. International Journal of Applied Metaheuristic Computing 1(1), 39–59 (2010)
16. Drake, J.H., Özcan, E., Burke, E.K.: A case study of controlling crossover in a selection hyper-heuristic framework using the multidimensional knapsack problem. Evolutionary computation (2015)
17. Fisher, H., Thompson, G.: Probabilistic learning combinations of local job-shop scheduling rules. In: Factory Scheduling Conference, Carnegie Institute of Technology (1961)
18. Gibbs, J., Kendall, G., Özcan, E.: Scheduling english football fixtures over the holiday period using hyper-heuristics. In: Schaefer, R., Cotta, C., Kołodziej, J., Rudolph, G. (eds.) PPSN XI. LNCS, vol. 6238, pp. 496–505. Springer, Heidelberg (2010)
19. Garrido, P., Castro, C.: Stable solving of cvrps using hyperheuristics. In: Proceedings of the 11th Annual Conference on Genetic and Evolutionary Computation, pp. 255–262 (2009)
20. Burke, E., Kendall, G., Newall, J., Hart, E., Ross, P., Schulenburg, S.: Hyper-heuristics: an emerging direction in modern search technology. International series in operations research and management science, 457–474 (2003)
21. Özcan, E., Bilgin, B., Korkmaz, E.E.: Hill climbers and mutational heuristics in hyperheuristics. In: Runarsson, T.P., Beyer, H.-G., Burke, E.K., Merelo-Guervós, J.J., Whitley, L., Yao, X. (eds.) PPSN 2006. LNCS, vol. 4193, pp. 202–211. Springer, Heidelberg (2006)
22. Ochoa, G., Hyde, M., Curtois, T., Vazquez-Rodriguez, J. A., Walker, J., Gendreau, M., et al.: Hyflex: a benchmark framework for cross-domain heuristic search. In: Evolutionary Computation in Combinatorial Optimization, pp. 136–147 (2012)
23. Burke, E.K., Gendreau, M., Hyde, M., Kendall, G., McCollum, B., Ochoa, G., Parkes, A.J., Petrovic, S.: The cross-domain heuristic search challenge – an international research competition. In: Coello, C.A. (ed.) LION 2011. LNCS, vol. 6683, pp. 631–634. Springer, Heidelberg (2011)
24. Moscato, P.: On evolution, search, optimization, genetic algorithms and martial arts: Towards memetic algorithms. Caltech concurrent computation program. C3P Report, 826 (1989)
25. Dawkins, R.: The Selfish Gene. Oxford University Press, Oxford (2006)
26. Ochoa, G., Walker, J., Hyde, M., Curtois, T.: Adaptive evolutionary algorithms and extensions to the hyflex hyper-heuristic framework. In: Coello, C.A., Cutello, V., Deb, K., Forrest, S., Nicosia, G., Pavone, M. (eds.) PPSN 2012, Part II. LNCS, vol. 7492, pp. 418–427. Springer, Heidelberg (2012)
27. Anwar, K., Khader, A.T., Al-Betar, M.A., Awadallah, M.: Harmony search-based hyper-heuristic for examination timetabling. In: 2013 IEEE 9th International Colloquium on Signal Processing and its Applications (CSPA), pp. 176–181 (2013)

Part II
Large Scale Applications of HSA

An Improved Harmony Search Algorithm for the Distributed Two Machine Flow-Shop Scheduling Problem

Jin Deng, Ling Wang, Jingnan Shen and Xiaolong Zheng

Abstract In this paper, an improved harmony search (IHS) algorithm is proposed to solve the distributed two machine flow-shop scheduling problem (DTMFSP) with makespan criterion. First, a two-stage decoding rule is developed for the decimal vector based representation. At the first stage, a job-to-factory assignment method is designed to transform a continuous harmony vector to a factory assignment. At the second stage, the Johnson's method is applied to provide a job sequence in each factory. Second, a new pitch adjustment rule is developed to adjust factory assignment effectively. The influence of parameter setting on the IHS is investigated based on the Taguchi method of design of experiments, and numerical experiments are carried out. The comparisons with the global-best harmony search and the original harmony search demonstrate the effectiveness of the IHS in solving the DTMFSP.

Keywords Harmony search · Distributed flow-shop scheduling · Decoding rule · Pitch adjustment

1 Introduction

The two machine flow-shop scheduling problem has been widely studied during the past few decades [1-10]. Recently, the distributed shop scheduling has attracted more and more attention [11-18], which is considered under the globalization environment to improve the production efficiency and economic benefits in the multi-plant companies. However, to the best of our knowledge, there is no

J. Deng(✉) · L. Wang · J. Shen · X. Zheng
Tsinghua National Laboratory for Information Science and Technology (TNList),
Department of Automation, Tsinghua University, Beijing 100084,
People's Republic of China
e-mail: {dengj13,chenjn12,zhengxl11}@mails.tsinghua.edu.cn, wangling@tsinghua.edu.cn

© Springer-Verlag Berlin Heidelberg 2016
J.H. Kim and Z.W. Geem (eds.), *Harmony Search Algorithm*,
Advances in Intelligent Systems and Computing 382,
DOI: 10.1007/978-3-662-47926-1_11

published work that directly addresses the distributed two machine flow-shop scheduling problem (DTMFSP). The DTMFSP is to allocate jobs to suitable factories and to determine a reasonable processing sequence in each factory so as to optimize certain objectives, such as the makespan criterion.

The classical two machine flow-shop scheduling problem with makespan criterion can be solved by Johnson's algorithm [1] in polynomial time. Nevertheless, the DTMFSP is more complex than the distributed single machine scheduling problem (DSMSP) that is a special version of the parallel machine scheduling problem (PMSP). The DTMFSP is NP-hard, as the PMSP is NP-hard [19]. Therefore, it is significant to develop effective and efficient approaches for solving the DTMFSP.

Inspired by the search procedure for better harmonies in the musical performance, the harmony search (HS) algorithm [20] is one of the population-based meta-heuristics. Different from the genetic algorithm (GA) that utilizes only two parents for generating the offspring, the HS generates a new harmony vector by considering all of the vectors in the harmony memory. Due to its simplicity and easy implementation, the HS algorithm has been applied to many optimization problems, such as engineering optimization [21], vehicle routing [22], truss structures design [23], water network design [24], electrical distribution network reconfiguration [25], and shop scheduling [26]. Numerical comparisons showed that the HS algorithm was faster than the GA [27, 28]. In this paper, an improved HS (IHS) algorithm will be proposed to solve the DTMFSP. To be specific, a harmony is represented as a real vector, and a two-stage decoding rule is developed to convert the continuous vector to a feasible schedule, and a new pitch adjustment rule is designed to adjust factory assignment effectively. The influence of parameter setting is investigated and numerical results are provided. The comparative results demonstrate the effectiveness of the IHS.

The remainder of the paper is organized as follows. The DTMFSP is described in Section 2. The IHS algorithm for the DTMFSP is introduced in Section 3. In Section 4, the influence of parameter setting is investigated, and numerical results and comparisons are provided. Finally, we end the paper with some conclusions and future work in Section 5.

2   Problem Description

Notions:

n: the total number of jobs.
f: the total number of factories.
n_k: the number of jobs in the factory k.
$O_{i,j}$: the j-th operation of job i.
$p_{i,j}$: the processing time of $O_{i,j}$.
$C_{i,j}$: the completion time of $O_{i,j}$.
π^k: the processing sequence in factory k.
$\pi = \{\pi^1, \pi^2, ..., \pi^f\}$: a certain schedule.

The DTMFSP can be described as follows. There are n jobs to be processed in f identical factories, where each factory has two machines. Job i has two operations $\{O_{i,1}, O_{i,2}\}$ to be processed one after another. Operation $O_{i,j}$ is executed on machine j with processing time $p_{i,j}$. Once a job is assigned to a factory, it cannot be transferred to other factories.

For the DTMFSP, the makespan $C_{\max}$ of a certain schedule can be calculated in the following way:

$$C_{\pi^k(1),1} = P_{\pi^k(1),1}, k = 1, 2, \dots, f \tag{1}$$

$$C_{\pi^k(i),1} = C_{\pi^k(i-1),1} + P_{\pi^k(i),1}, k = 1, 2, \dots, f; i = 2, 3, \dots, n_k \tag{2}$$

$$C_{\pi^k(1),2} = C_{\pi^k(1),1} + P_{\pi^k(1),2}, k = 1, 2, \dots, f \tag{3}$$

$$C_{\pi^k(i),2} = \max\{C_{\pi^k(i),1}, C_{\pi^k(i-1),2}\} + P_{\pi^k(i),2}, k = 1, 2, \dots, f; i = 2, 3, \dots, n_k \tag{4}$$

$$C_{\max} = \max C_{\pi^k(n_k),2}, k = 1, 2, \dots, f \tag{5}$$

3   IHS for DTMFSP

3.1   Original HS

In the original HS algorithm, each harmony denotes a solution, represented by a D-dimension real vector $X_i = \{x_i(1), x_i(2), \dots, x_i(D)\}$. The HS algorithm contains four parameters: the harmony memory size (MS), the harmony memory consideration rate (P_{CR}), the pitch adjusting rate (P_{AR}) and the distance bandwidth (dB). The harmony vectors are stored in the harmony memory (HM) as $\{X_1, X_2, \dots, X_{MS}\}$ to generate new harmony vectors. According to [22], P_{CR} and P_{AR} help the algorithm find globally and locally improved solutions, respectively. P_{AR} and dB are important in fine-tuning the solution vectors and in adjusting convergence rate.

The procedure of the HS algorithm can be simply described as follows:

Step 1. Set parameters MS, P_{CR}, P_{AR}, dB and the stopping criterion.

Step 2. Initialize the HM and calculate the objective function value $F(X_i)$ for each harmony X_i.

Step 3. Improvise a new harmony X_{new} from the HM as Fig. 1. When improvising a new harmony, three rules are applied: a memory consideration, a pitch adjustment and a random selection.

Step 4. Update the HM as $X_w = X_{new}$ if X_{new} is better, where X_w represents the worst harmony in the HM.

Step 5. If stopping criteron is not satisfied, go to Step 3.

For $j = 1 : D$
   If $(rand_1 < P_{CR})$ /* memory consideration */
     $x_{new}(j) = x_a(j), a = \{1, 2, \ldots, MS\}$
     If $(rand_2 < P_{AR})$ /* pitch adjustment */
       $x_{new}(j) = x_{new}(j) + \mathrm{sgn}(rand_3 - 0.5) \times dB$
     End if
   Else /* random selection */
     $x_{new}(j) = x_{min}(j) + (x_{max}(j) - x_{min}(j)) \times rand_4$
   End if
End for
Note:
1. $rand_1$–$rand_4$ are random numbers generated uniformly between 0 and 1;
2. sgn(.) is a sign function that returns -1, 0 or 1.

Fig. 1 Improvising a new harmony in HS

3.2 Encoding and Decoding

To solve the DTMFSP, a harmony in the IHS is represented by an n-dimension decimal vector. To determine the factory assignment for each job and the processing sequence in each factory, a two-stage decoding rule is developed.

At the first stage, a job-to-factory assignment method is designed to map a continuous harmony vector to a factory assignment. To be specific, the search range (0, 1] is partitioned into f intervals (0, $1/f$], ($1/f$, $2/f$] … ((f–1)/f, 1]. If the value of the j-th dimension belongs to ((k–1)/f, k/f], job j is assigned to factory k. For an example with 2 factories and 5 jobs, suppose the harmony vector is {0.54, 0.22, 0.78, 0.16, 0.93}. Job 1 is assigned to factory 2 because 0.54 belongs to (1/2, 1]. Similarly, job 2 and job 4 are assigned to factory 1, while job 3 and job 5 are assigned to factory 2.

Since the two machine flow-shop scheduling problem with makespan criterion can be solved by the Johnson's algorithm [1], the Johnson's algorithm is applied at the second stage to provide a job sequence for each factory.

With such a decoding rule, a real harmony vector can be converted to a feasible schedule, and then its objective value can be calculated.

3.3 Initial Harmony Memory

To generate a harmony memory with enough diversity, all MS initial harmony vectors are randomly generated.

For each harmony vector, the value of each dimension is a random number that is uniformly generated between 0 and 1.

3.4 Improvise New Harmony

To adjust the factory assignment effectively, a new pitch adjustment is developed as follows.

When adjusting the value of $x_{new}(j)$, it first allocates job j to the position that is determined by the Johnson's algorithm in every factory. Then, it chooses factory k with the earliest completion time. The value of $x_{new}(j)$ is calculated as follows:

$$x_{new}(j) = \quad (k - rand) \, / \, f \tag{6}$$

For an example with 2 factories and 5 jobs, suppose that $x_{new}(3)$ is undergoing the pitch adjustment with the precondition of $x_{new}(1) = 0.54$ and $x_{new}(2) = 0.22$. First, suppose that job 3 is allocated to factory 1 and factory 2, respectively. Assume that the processing sequences obtained by Johnson's algorithm of the two factories are $\pi^1 = \{2, 3\}$ and $\pi^2 = \{3, 1\}$. Then, job 3 will be assigned to factory 1 if the completion time of π^1 is smaller than that of π^2; otherwise, job 3 will be assigned to factory 2. Finally, the value of $x_{new}(3)$ is calculated by formula (6).

The pseudo-code of improvising a new harmony in the IHS is illustrated in Fig. 2.

```
For j = 1 : D
    If (rand₁ < P_CR) /* memory consideration */
        x_new(j) = x_a(j), a = {1, 2, ..., MS}
        If (rand₂ < P_AR) /*new pitch adjustment */
            Find the factory k that can process job j with the earliest completion time;
            x_new(j) = (k − rand₃) / f
        End if
    Else /* random selection */
        x_new(j) = x_min(j) + (x_max(j) − x_min(j)) × rand₄
    End if
End for
Note: rand₁–rand₄ are uniform random number between 0 and 1.
```

Fig. 2 Improvising a new harmony in HIS for DTMFSP

3.5   Update Harmony Memory

After a new harmony vector X_{new} is generated, it will be compared with the worst harmony vector X_w in the HM, and then a greedy selection is employed.

That is, if X_{new} is better than X_w, then X_{new} will replace X_w and become a new member of the HM.

4   Numerical Results and Comparisons

4.1   Experimental Setup

To evaluate the performance of the IHS, 100 random instances are generated. Table 1 lists the ranges of parameters for generating the instances. The IHS is coded in C language, and all the tests are run on a PC with an Intel(R) core(TM) i5-3470 CPU @ 3.2GHz / 8GB RAM under Microsoft Windows 7. The stopping criterion is set as $0.1 \times n$ seconds CPU time.

Table 1 Ranges of Parameters for Instances

Parameter	Range
f	$\{2, 3, 4, 5, 6\}$
n	$\{20, 50, 100, 200\}$
$p_{i,j}$	$U(1, 100)$

4.2   Parameter Setting

The IHS contains three parameters: MS, P_{CR} and P_{AR}. To investigate their influence on the performance of the IHS, the Taguchi method of design of experiments (DOE) is carried out with a moderate-sized instance F4_11 (i.e. $f = 4$, $n = 100$). Four levels are considered for each parameter as in Table 2.

For each parameter combination, the IHS is run 10 times independently. With the orthogonal array $L_{16}(4^3)$, Table 3 lists the resulted average makespan value as response value (RV). Then, the response value and the rank of each parameter are calculated and listed in Table 4, and the main effect plots are shown in Fig. 3.

Table 2 Factor Levels

Factor	Factor levels			
	1	2	3	4
MS	5	10	15	20
P_{CR}	0.7	0.75	0.8	0.85
P_{AR}	0.7	0.75	0.8	0.85

Table 3 Orthogonal Array and RV Values

Experiment number	Factor levels			RV
	MS	P_{CR}	P_{AR}	
1	1	1	1	1200.0
2	1	2	2	1200.0
3	1	3	3	1200.0
4	1	4	4	1200.3
5	2	1	2	1200.0
6	2	2	1	1200.1
7	2	3	4	1200.0
8	2	4	3	1199.9
9	3	1	3	1200.0
10	3	2	4	1200.2
11	3	3	1	1200.1
12	3	4	2	1200.3
13	4	1	4	1200.4
14	4	2	3	1200.3
15	4	3	2	1200.0
16	4	4	1	1200.1

Table 4 Response Value and Rank of Each Parameter

Level	MS	P_{CR}	P_{AR}
1	1200.075	1200.100	1200.075
2	1200.000	1200.150	1200.075
3	1200.150	1200.025	1200.050
4	1200.200	1200.150	1200.225
Delta	0.200	0.125	0.175
Rank	1	3	2

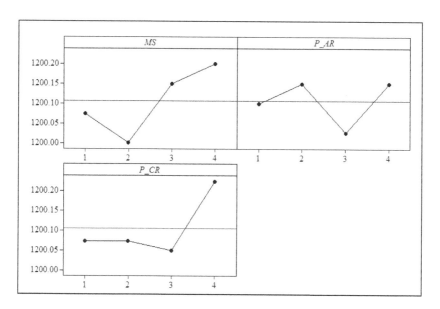

Fig. 3 Main Effect Plot

From Table 4, it can be seen that MS is the most significant to the IHS. A too small value of MS is harmful to the diversity of the harmony memory, while a too large value may slow down the convergence. The influence of P_{AR} ranks the second. Although a large value of P_{AR} is beneficial to find a proper factory for a job, it costs more computational time due to the greedy way of the pitch adjustment rule. Besides, a large value of P_{CR} is helpful to use the information of the explored solutions, but it may lead to a premature convergence. According to the above analysis, a good choice of parameter combination is recommended as $MS = 10$, $P_{CR} = 0.8$ and $P_{AR} = 0.8$, which will be used in the following tests.

4.3 Comparative Results

To the best of our knowledge, there is no published work addressing the DTMFSP. Therefore, we compare the IHS with the original HS and the global-best HS (GHS) [30]. For the HS, we set $MS = 5$, $P_{CR} = 0.9$, $P_{AR} = 0.3$ and $dB = 1/f$. For the GHS, we set $MS = 5$, $P_{CR} = 0.9$, $P_{AR}^{min} = 0.01$ and $P_{AR}^{max} = 0.99$ as suggested in [30].

For each instance, the above algorithms are run 10 times independently. The Best, average (Ave) and standard deviation (SD) values obtained by the algorithms are listed in Tables 5-9 in solving the problems with different number of factories.

From Tables 5-9, it can be seen that the IHS outperforms the HS and the GHS, especially when $f > 3$. As f increases, the factory assignment becomes more complex and it is more difficult to find proper factories for jobs. Since the IHS can obtain better Best and Ave results, it can be concluded that the proposed new pitch adjustment rule is effective to adjust the factory assignment in solving the DTMFSP.

Besides, the SD values of the IHS are also better than those of the HS and the GHS, which implies that the IHS is more robust than the HS and GHS.

Table 5 Results of the Algorithms $(f = 2)$

Instance		IHS			HS			GHS		
No.	n	Best	Ave	SD	Best	Ave	SD	Best	Ave	SD
1	20	582	582	0.00	582	582	0.00	582	582	0.00
2	20	564	564	0.00	564	564	0.00	564	564	0.00
3	20	529	529	0.00	529	529	0.00	529	529	0.00
4	20	529	529	0.00	529	529	0.00	529	529	0.00
5	20	497	497	0.00	497	497	0.00	497	497	0.00
6	50	979	979	0.00	979	979	0.00	979	979	0.00
7	50	1319	1319	0.00	1319	1319	0.00	1319	1319	0.00
8	50	1356	1356	0.00	1356	1356	0.00	1356	1356	0.00
9	50	1267	1267	0.00	1267	1267	0.00	1267	1267	0.00
10	50	1238	1238	0.00	1238	1238	0.00	1238	1238	0.00
11	100	1519	1519	0.00	1519	1519	0.00	1519	1519	0.00
12	100	2389	2389	0.00	2389	2389	0.00	2389	2389	0.00
13	100	2721	2721	0.00	2721	2721	0.00	2721	2721	0.00
14	100	2616	2616	0.00	2616	2616	0.00	2616	2616	0.00
15	100	2614	2614	0.00	2614	2614	0.00	2614	2614	0.00
16	200	2688	2688	0.00	2688	2688	0.00	2688	2688	0.00
17	200	5312	5312	0.00	5312	5312	0.00	5312	5312	0.00
18	200	5257	5257	0.00	5257	5257	0.00	5257	5257	0.00
19	200	5023	5023	0.00	5023	5023	0.00	5023	5023	0.00
20	200	5296	5296	0.00	5296	5296	0.00	5296	5296	0.00

Table 6 Results of the Algorithms ($f = 3$)

Instance		IHS			HS			GHS		
No	n	Best	Ave	SD	Best	Ave	SD	Best	Ave	SD
1	20	393	393	0.00	393	393.5	0.50	393	393.5	0.50
2	20	360	360	0.00	360	360	0.00	360	360	0.00
3	20	362	362	0.00	362	362.1	0.30	362	362.1	0.30
4	20	401	401	0.00	401	401	0.00	401	401	0.00
5	20	342	342	0.00	342	342.8	0.75	342	342.9	1.14
6	50	854	854	0.00	854	854.4	0.49	854	854.3	0.64
7	50	817	817	0.00	817	817.9	0.30	817	817.7	0.46
8	50	924	924	0.00	924	925.1	0.54	924	925.1	0.70
9	50	908	908	0.00	908	908	0.00	908	908	0.00
10	50	775	775	0.00	775	775.8	0.60	775	775.4	0.49
11	100	1700	1700	0.00	1700	1700	0.00	1700	1700	0.00
12	100	1873	1873	0.00	1873	1873.	0.49	1873	1873.8	0.60
13	100	1664	1664	0.00	1664	1664.	0.40	1664	1664.2	0.40
14	100	1678	1678	0.00	1678	1680.	1.36	1678	1679.4	1.02
15	100	1682	1682	0.00	1682	1682.	0.46	1682	1682.4	0.49
16	200	3575	3575	0.00	3575	3575	0.00	3575	3575	0.00
17	200	3614	3614	0.00	3614	3614.	0.40	3614	3614.2	0.40
18	200	3468	3468	0.00	3469	3469.	0.83	3468	3469	0.63
19	200	3559	3559	0.00	3559	3560.	0.54	3559	3559.8	0.40
20	200	3449	3449	0.00	3449	3450	0.63	3449	3450.1	0.54

Table 7 Results of the Algorithms ($f = 4$)

Instance		IHS			HS			GHS		
No	n	Best	Ave	SD	Best	Ave	SD	Best	Ave	SD
1	20	238	238.2	0.40	239	242.2	2.23	240	242.1	1.37
2	20	259	259.8	0.40	260	263.5	3.07	263	264.9	1.04
3	20	303	303	0.00	303	304.8	1.40	304	305.8	1.40
4	20	313	313.3	0.46	313	313.4	0.66	313	313.4	0.49
5	20	291	291	0.00	291	292.3	0.90	291	293.3	2.61
6	50	735	735	0.00	737	738.4	0.80	737	737.6	0.80
7	50	630	630	0.00	634	638.5	2.33	633	636.7	2.53
8	50	583	583	0.00	585	589.1	2.26	585	588.2	1.78
9	50	610	610	0.00	612	613	0.77	611	612.9	1.04
10	50	666	666	0.00	669	671.8	1.54	670	671.5	1.28
11	100	1200	1200.	0.40	1201	1202.	1.17	1200	1202.8	1.66
12	100	1293	1293	0.00	1295	1296.	1.30	1295	1296.9	1.14
13	100	1304	1304.	0.50	1305	1307.	1.42	1305	1307.5	1.28
14	100	1305	1305	0.00	1309	1310.	1.00	1306	1309.4	2.15
15	100	1277	1277	0.00	1281	1281.	0.75	1279	1281	0.89
16	200	2439	2439	0.00	2443	2449.	3.45	2443	2449.4	4.20
17	200	2456	2456.	0.49	2460	2465.	3.64	2458	2465.3	4.71
18	200	2529	2529.	0.54	2532	2536.	2.79	2533	2535.9	2.12
19	200	2448	2448	0.00	2449	2451.	1.19	2450	2452.3	1.49
20	200	2542	2542.	0.40	2544	2547.	2.06	2544	2547.5	1.50

Table 8 Results of the Algorithms ($f = 5$)

Instance		IHS			HS			GHS		
No	n	Best	Ave	SD	Best	Ave	SD	Best	Ave	SD
1	20	227	227.3	0.46	229	232	2.14	231	233.4	2.24
2	20	234	235	0.45	236	239.3	1.55	235	240.7	4.12
3	20	244	244	0.00	244	246.6	2.62	244	247.2	2.71
4	20	272	22.2	0.40	275	277.9	2.47	275	277.8	2.32
5	20	234	234.9	0.30	236	240.4	2.65	238	239.9	1.22
6	50	545	545.8	0.60	547	552.4	2.33	546	551.4	2.06
7	50	537	537.8	0.40	547	553.8	3.19	544	548.6	3.67
8	50	519	519.9	0.54	525	525.5	0.67	522	524.2	1.94
9	50	534	534.6	0.49	542	548.7	3.85	542	548.5	3.64
10	50	465	465	0.00	473	478.9	3.11	469	477	4.27
11	100	1007	1007	0.00	1024	1030.	4.40	1029	1031.4	2.11
12	100	1044	1044	0.00	1057	1064.	4.17	1060	1063.3	2.37
13	100	982	982	0.00	988	989.9	1.30	985	990.1	3.18
14	100	988	988.8	0.75	1004	1009.	2.97	1005	1011	3.85
15	100	1092	1092.	0.66	1096	1100.	2.47	1097	1100.3	2.37
16	200	2079	2079	0.00	2086	2090.	2.99	2089	2093.4	3.47
17	200	2131	2132	0.45	2140	2151.	5.63	2143	2153.1	5.91
18	200	2087	2087	0.00	2094	2098.	3.99	2095	2098.4	2.69
19	200	2050	2050	0.00	2073	2083.	5.16	2069	2082.7	6.94
20	200	2017	2017	0.00	2029	2037	4.15	2029	2039.2	5.53

Table 9 Results of the Algorithms ($f = 6$)

Instance		IHS			HS			GHS		
No	n	Best	Ave	SD	Best	Ave	SD	Best	Ave	SD
1	20	173	173	0.00	173	177.5	3.29	174	177.8	3.22
2	20	235	236.4	0.66	235	238.2	3.49	235	239.4	2.73
3	20	194	194	0.00	194	197.5	3.91	194	197.6	2.65
4	20	218	220.5	1.20	219	224.5	3.04	219	223.4	2.58
5	20	232	232.1	0.30	234	237.3	2.15	236	237.9	1.76
6	50	438	439	0.77	449	453.1	2.62	443	449.6	2.84
7	50	426	427.6	1.02	433	439.3	3.00	428	435.5	4.84
8	50	391	391.7	0.64	406	414	3.49	405	410.1	4.01
9	50	442	442	0.00	462	465	1.61	459	461.8	1.94
10	50	459	461.5	1.28	474	477.9	3.27	465	472.3	4.05
11	100	904	904.6	0.49	925	936.1	4.89	926	934.9	4.93
12	100	787	787	0.00	798	806.9	5.59	796	807.7	5.59
13	100	867	867	0.00	880	884.9	3.42	882	886.9	2.51
14	100	874	874.1	0.30	901	911	5.78	900	910.1	4.68
15	100	878	879.1	0.70	915	916.7	1.19	904	916.9	5.80
16	200	1786	1786.	0.49	1801	1805.	5.17	1799	1805.8	4.19
17	200	1724	1724	0.00	1740	1744.	3.11	1739	1745.3	4.05
18	200	1662	1662	0.00	1698	1703.	3.46	1689	1700.2	6.65
19	200	1707	1709.	1.02	1724	1737.	8.75	1731	1739.7	5.51
20	200	1677	1677.	0.50	1717	1728	5.23	1715	1728	6.84

5 Conclusions

This paper proposed an improved harmony search algorithm for solving the distributed two machine flow-shop scheduling problem. According to the characteristics of the problem, a two-stage decoding rule for the decimal vector based representation and a new pitch adjustment rule for the HS were designed. The influence of parameter setting was investigated, and the effectiveness of the IHS was demonstrated by numerical comparisons. Future work could focus on generalizing the harmony search algorithm for other types of the distributed shop scheduling problems and developing multi-objective HS algorithms for the problems with multiple scheduling criteria.

Acknowledgment This research is partially supported by the National Key Basic Research and Development Program of China (No. 2013CB329503), the National Science Foundation of China (No. 61174189), and the Doctoral Program Foundation of Institutions of Higher Education of China (20130002110057).

References

1. Johnson, S.M.: Optimal two- and three-stage production schedules with setup times included. Nav. Res. Logist. Quart. **1**, 61–68 (1954)
2. Ku, P.S., Niu, S.C.: On Johnson's two-machine flow shop with random processing times. Oper. Res. **34**, 130–136 (1986)
3. The two-machine total completion time flow shop problem: Della Croce, F., Narayan, V., Tadei, R. Eur. J. Oper. Res. **90**, 227–237 (1996)
4. Sayin, S., Karabati, S.: Theory and Methodology: a bicriteria approach to the two-machine flow shop scheduling problem. Eur. J. Oper. Res. **113**, 435–449 (1999)
5. Croce, Della: F., Ghirardi, M., Tadei, R.: An improved branch-and-bound algorithm for the two machine total completion time flow shop problem. Eur. J. Oper. Res. **139**, 293–301 (2002)
6. Pan, J.C.H., Chen, J.S., Chao, C.M.: Minimizing tardiness in a two-machine flow-shop. Comput. Oper. Res. **29**, 869–885 (2002)
7. Toktaş, B., Azizoğlu, M., Köksalan, S.K.: Two-machine flow shop scheduling with two criteria: maximum earliness and makespan. Eur. J. Oper. Res. **157**, 286–295 (2004)
8. Blazewicz, J., Pesch, E., Sterna, M., Werner, F.: Metaheuristic approaches for the two-machine flow-shop problem with weighted late work criterion and common due date. Comput. Oper. Res. **35**, 574–599 (2008)
9. Ng, C.T., Wang, J.B., Cheng, T.C.E., Liu, L.L.: A branch-and-bound algorithm for solving a two-machine flow shop problem with deteriorating jobs. Comput. Oper. Res. **37**, 83–90 (2010)
10. Kasperski, A., Kurpisz, A., Zieliński, P.: Approximating a two-machine flow shop scheduling under discrete scenario uncertainty. Eur. J. Oper. Res. **217**, 36–43 (2012)
11. Jia, H.Z., Nee, A.Y.C., Fuh, J.Y.H., Zhang, Y.F.: A modified genetic algorithm for distributed scheduling problems. J. Intell. Manuf. **14**, 351–362 (2003)

12. Moon, C., Seo, Y.: Evolutionary algorithm for advanced process planning and scheduling in a multi-plant. Comput. Ind. Eng. **48**, 311–325 (2005)
13. Chan, F.T.S., Chung, S.H., Chan, P.L.Y.: Application of genetic algorithms with dominant genes in a distributed scheduling problem in flexible manufacturing systems. Int. J. Prod. Res. **44**, 523–543 (2006)
14. Naderi, B., Ruiz, R.: The distributed permutation flowshop scheduling problem. Comput. Oper. Res. **37**, 754–768 (2010)
15. Behnamian, J.: Fatemi Ghomi, S.M.T.: The heterogeneous multi-factory production network scheduling with adaptive communication policy and parallel machine. Inf. Sci. **219**, 181–196 (2013)
16. Hatami, S., Ruiz, R., Andrés-Romano, C.: The distributed assembly permutation flowshop scheduling problem. Int. J. Prod. Res. **51**, 5292–5308 (2013)
17. Wang, S.Y., Wang, L., Liu, M., Xu, Y.: An effective estimation of distribution algorithm for solving the distributed permutation flow-shop scheduling problem. Int. J. Prod. Econ. **145**, 387–396 (2013)
18. Naderi, B., Azab, A.: Modeling and heuristics for scheduling of distributed job shops. Expert Syst. Appl. **41**, 7754–7763 (2014)
19. Garey, M.R., Johnson, D.S.: "Strong" NP-completeness results: motivation, examples, and implications. J. ACM **25**, 499–508 (1978)
20. Geem, Z.W., Kim, J.H., Loganathan, G.V.: A new heuristic optimization algorithm: harmony search. Simul. **76**, 60–68 (2001)
21. Lee, K.S., Geem, Z.W.: A new meta-heuristic algorithm for continuous engineering optimization: harmony search theory and practice. Comput. Methods Appl. Mech. Eng. **194**, 3902–3933 (2005)
22. Geem, Z.W., Lee, K.S., Park, Y.: Application of harmony search to vehicle routing. Am. J. Appl. Sci. **2**, 1552 (2005)
23. Kaveh, A., Talatahari, S.: Particle swarm optimizer, ant colony strategy and harmony search scheme hybridized for optimization of truss structures. Comput. Struct. **87**, 267–283 (2009)
24. Geem, Z.W.: Particle-swarm harmony search for water network design. Eng. Optim. **41**, 297–311 (2009)
25. Rao, Srinivasa: R., Narasimham, S.V.L., Ramalinga Raju, M., Srinivasa Rao, A.: Optimal network reconfiguration of large-scale distribution system using harmony search algorithm. IEEE Trans. Power Syst. **26**, 1080–1088 (2011)
26. Wang, L., Pan, Q.K., Tasgetiren, M.F.: A hybrid harmony search algorithm for the blocking permutation flow shop scheduling problem. Comput. Ind. Eng. **61**, 76–83 (2011)
27. Lee, K.S., Geem, Z.W., Lee, S.H., Bae, K.W.: The harmony search heuristic algorithm for discrete structural optimization. Eng. Optim. **37**, 663–684 (2005)
28. Mahdavi, M., Fesanghary, M., Damangir, E.: An improved harmony search algorithm for solving optimization problems. Appl. Math. Comput. **188**, 1567–1579 (2007)
29. Baker, K.R.: Introduction to Sequencing and Scheduling. Wiley, New York (1974)
30. Omran, M.G.H., Mahdavi, M.: Global-best harmony search. Appl. Math. Comput. **198**, 643–656 (2008)

Hybrid Harmony Search Algorithm for Nurse Rostering Problem

Yabing Nie, Bing Wang and Xianxia Zhang

Abstract This paper addresses the nurse rostering problem (NRP), whose objective is to minimize a total penalty caused by the roster. A large number of constraints required to be considered could cause a great difficulty of handling the NRP. A hybrid harmony search algorithm (HHSA) with a greedy local search is proposed to solve the NRP. A personal schedule is divided into several blocks, in which a subset of constraints is considered in advance. Based on these blocks, the pitch adjustment and randomization are carried out. Every time a roster is improvised, a coverage repairing procedure is applied to make the shift constraints satisfied, and the greedy local search is used to improve the roster's quality. The proposed HHAS was tested on many well known real-world problem instances and competitive solutions were obtained.

Keywords Harmony search algorithm · Greedy local search · Nurse rostering problem

1 Introduction

The nurse rostering problem (NRP) is a complex combinatorial problem whose objective is to produce rosters which satisfy all hard constraints while taking into account soft constraints. Hard constraints are those constraints that must be satisfied at all costs for the NRP. If some requirements are desirable but could be violated, these requirements are usually referred to as soft constraints for the NRP. The solution quality of NRP could be evaluated by estimating soft constraints.

A wide variety of approaches have been developed to solve the NRP, including exact algorithms, meta-heuristic algorithms and others [1]. Due to the computational

Y. Nie · B. Wang(✉) · X. Zhang
School of Mechatronic Engineering and Automation, Shanghai University,
Shanghai 200072, China
e-mail: susanbwang@shu.edu.cn

© Springer-Verlag Berlin Heidelberg 2016
J.H. Kim and Z.W. Geem (eds.), *Harmony Search Algorithm,*
Advances in Intelligent Systems and Computing 382,
DOI: 10.1007/978-3-662-47926-1_12

complexity of large scale NRPs, meta-heuristic algorithms were more popular than exact algorithms. Various meta-heuristic algorithms were developed to solve the NRP, including Genetic Algorithm [2], Scatter Search [3], Tabu Search [4], Simulated Annealing [5] and many others. The hybridizations of these algorithms have also attracted the interest of the researches. Post and Veltman [6] proposed a hybrid genetic algorithm, in which a local search was carried out after each generation of the genetic algorithm to make improvement. Burke et al. [7] developed a hybrid variable neighborhood search (VNS) which created an initial schedule by an adaptive ordering technique in advance and then run variable VNS based on this initial schedule. Burke et al. [8] developed an IP-based VNS which used an IP to first solve a small problem including the full set of hard constraints and a subset of the soft constraints. All the remaining constraints were then satisfied by a basic VNS.

The NRP is commonly described and solved by three views: a nurse-day view, a nurse-task view and a nurse-shift pattern view [9]. The nurse-day view and nurse-task view are similar. Accordingly, the NRP is solved by generating assignments of nurses to shift for each day of the planning period, such as [3, 6, 7, 8]. However, in the nurse-shift pattern view, rosters are built by constructing personal schedules for each individual nurse (by allocating one of his/her feasible shift patterns to him/her), such as [2, 4, 5].

HSA is a new population-based algorithm mimicking the improvisation process of searching for a perfect state of harmony measured by aesthetic standards. Since developed by Geem et al. [10], it was successfully used in various optimization problems such as university timetabling [11], vehicle routing [12] and so on. HSA was also used to solve the NRP, which was usually solved by the nurse-task view, such as [13, 14]. However, Hadwan et al. [15] adapted a harmony search algorithm to solve the NRP by the nurse-shift pattern view.

In this work, rosters were also produced by the nurse-shift pattern view, but with different pitch adjustment operator and different way of handling the coverage constraint from [15]. Moreover, a greedy local search was carried out after each generation of the HSA to make improvement, which was similar to [6].

The remainder of this paper is organized as follows. In section 2, the nurse rostering problem is described. In Section 3, we first briefly present the classic HSA and then detail our HHSA. Experiment was conducted to investigate the algorithm developed in Section 4. In section 5, we draw conclusions on the success of this HHSA.

2   Problem Description

Here, we use an ORTEC NRP as a case to study. The problem was fully described in [7], except for several specific individual rostering constraints which were not explicitly listed in the academic papers but were embedded in the solutions published on the NRP benchmark web site [16]. The length of planning period is one month (31 days), from Wednesday 1st January to Friday 31st January 2003. Four shift types (i.e. early, day, late and night) are assigned to 16 nurses. The nurses have three different contracts: standard 36, standard 32 and standard 20. For completeness, all constraints that need to be satisfied are presented briefly in the following.

The problem has the following hard constraints:

(HC1) The number of people on each shift of each day must exactly be the specified level. Both overstaffing and understaffing are not allowed.

(HC2) A nurse can work at most one shift per day.

(HC3) A nurse can work at most 3 night shifts per period of 5 consecutive weeks.

(HC4) A nurse can work at most 3 weekends per 5 week period.

(HC5) After a series of consecutive night shifts at least two days off are required.

(HC6) The number of consecutive night shifts is at most 3.

(HC7) No isolated night shift is allowed in a personal schedule.

(HC8) Each nurse has his/her maximum number of working days.

(HC9) The number of consecutive shifts is at most 6.

(HC10) No late shifts for one particular nurse.

(HC11) There is minimum time limit between shifts.

In addition, the problem has the following soft constraints:

(SC1) Either two days on duty or two days off at weekends for each nurse

(SC2) There is no night shift before a free weekend.

(SC3) For any nurse avoid stand-alone shifts.

(SC4) The number of consecutive night shifts has minimum and maximum limits.

(SC5) After a series of working days at least two free days are required.

(SC6) The number of working days per week has minimum and maximum limits.

(SC7) The number of consecutive working days has minimum and maximum limits.

(SC8) The number of consecutive early shifts has minimum and maximum limits.

(SC9) The number of consecutive late shifts has minimum and maximum limits.

(SC10) Certain shift type successions (e.g. day shift followed by early shift) should be avoided.

Hard constraints are constraints that must be satisfied at all costs so as to obtain a feasible solution. Soft constraints can be violated if necessary, but at the cost of incurring penalty. The objective of the NRP is to minimize the total penalty caused by the roster. The solution of this problem is a roster which is made up of the personal schedules of all nurses. Each personal schedule is represented as a string. The jth position in the string of nurse i is represented as follows.

$$p_{ij} = \begin{cases} "O" & \textit{nurse } i \textit{ works no shift on day } j \\ "E" & \textit{nurse } i \textit{ works on early shift on day } j \\ "D" & \textit{nurse } i \textit{ works on day shift on day } j \\ "L" & \textit{nurse } i \textit{ works on late shift on day } j \\ "N" & \textit{nurse } i \textit{ works on night shift on day } j \end{cases} \tag{1}$$

The difficulty of the NRP is due to the large number and a variety of constraints that need to be satisfied. When the nurse-shift pattern view is used to solve the NRP, constraints are usually further divided into two categories to be handled: shift constraints and nurse constraints [17]. Shift constraints specify the number of nurses required for each shift during the entire planning period, such as (HC1). Nurse constraints refer to all the restrictions on personal schedules including personal requests, personal preferences, and constraints on balancing the workload among personnel, such as (HC2) ~ (HC11) and (SC1) ~ (SC10). Because all soft constraints in our case belong to nurse constraints, thus the penalty of each personal schedule can be calculated in isolation from other personal schedules. Adding the penalties of all personal schedules up the penalty of an overall roster is obtained.

3   HHSA for the Nurse Rostering Problem

3.1   Harmony Search Algorithm

HSA is a population-based meta-heuristic algorithm, developed in an analogy with musical improvisation. In music performance, each music player improvises one note at a time. All these musical notes are combined together to form a harmony, evaluated by aesthetic standards and improved through practice after practice. In optimization, each variable is assigned a value at a time. All these values are combined together to form a solution vector, evaluated by the objective function and improved iteration by iteration. The main steps in the structure of HSA are as follows:

Step 1: Initialize the algorithm parameters.
Step 2: Initialize the harmony memory (HM).
Step 3: Improvise a new solution from the HM.
Step 4: Update the HM.
Step 5: Repeat step 3 and step 4 until the stopping criterion is satisfied.

3.1.1   Initialization of Algorithm Parameters

The parameters of HSA are specified in this step, including harmony memory size (HMS), harmony memory consideration rate (HMCR), pitch adjustment rate (PAR) and the maximum number of improvisations (NI). Both HMCR and PAR are used to improvise a new solution in step3.

3.1.2   Initialization of HM

In this step, the HM is initially stuffed with as many randomly generated solutions as the HMS. Those solutions are sorted by the values of objective function. The structure of the HM is shown as following:

$$HM = \begin{bmatrix} x^1 \\ x^2 \\ \vdots \\ x^{HMS-1} \\ x^{HMS} \end{bmatrix} = \begin{bmatrix} x_1^1 & x_2^1 & \cdots & x_{N-1}^1 & x_N^1 \\ x_1^2 & x_2^2 & \cdots & x_{N-1}^2 & x_N^2 \\ \vdots & \vdots & \vdots & \vdots & \vdots \\ x_1^{HMS-1} & x_2^{HMS-1} & \cdots & x_{N-1}^{HMS-1} & x_N^{HMS-1} \\ x_1^{HMS} & x_2^{HMS} & \cdots & x_{N-1}^{HMS} & x_N^{HMS} \end{bmatrix} \quad (2)$$

3.1.3   Improvising a New Solution From the HM

In this step, a new solution is improvised based on the following three rules: memory consideration, pitch adjustment and randomization. In the improvising procedure, as shown in Fig. 1, each variable of the new solution is assigned a value in turn.

```
For each  i ∈ [1, N]  do
    If  U(0,1) ≤ HMCR  then
        Assign a value to  x_i  using memory consideration
        If  U(0,1) ≤ PAR  then
        Pitch adjust the value obtained by memory consideration
        End if
    Else
        Assign a value to  x_i  using randomization
    End if
End for
```

Fig. 1 The Pseudo-code of Improvising a New Solution

3.1.4   Updating the HM

If the newly made solution is better than the worst one in the HM, measured in terms of the objective function value, include the new solution in the HM and exclude the existing worst solution in the HM. Otherwise, discard the new solution.

3.1.5   Checking the Stopping Criterion

Repeat step3 and step4 until the stopping criterion is satisfied. Normally the stopping criterion defines the maximum number of improvisations (NI).

3.2   The HHSA for the Nurse Rostering Problem

A key factor in the application of HSA is how the algorithm handles the constraints relating to the NRP. In the improvising procedure of proposed HHSA, each nurse is allocated a personal schedule at a time while considering nurse constraints and avoiding overstaffing. All these personal schedules are combined together to form a roster. Every time a roster is built an additional repair process is employed to assign enough people to understaffed shifts and then a greedy local search is applied to improve the roster's quality. In the following, we mainly elaborate the process of improvising new rosters in Section 3.2.1 and Section 3.2.2. The repair process and the greedy local search are presented in Section 3.2.3 and Section 3.2.4 respectively.

3.2.1   The Construction of Weekly Shift Patterns

In the improvising procedure of traditional HSA, we randomly move the variable value obtained by the memory consideration to a neighboring value via pitch adjustment with probability PAR or randomly select one variable value from possible range of values via randomization with probability (1-HMCR). However, each personal schedule, i.e. variable value, is highly constrained by nurse constraints in the NRP. Thus the personal schedules generated by general pitch adjustment or randomization are likely to violate many nurse constraints, which can cause to poor performance of the algorithm and considerable increase in iterations need to find an optimal solution. In order to improve the efficiency of HSA, we divide a personal schedule into several blocks, in which a subset of constraints is considered. Based on these blocks the pitch adjustment and randomization are carried out.

One block represents one week, in which a weekly shift pattern will be put. Fig. 2 gives an example of the relationship between blocks and a complete personal schedule. It is worth mentioning that a working week runs from Monday to Sunday in the problem. Thus lengths of the first week and last week in the January are 5 due to the structure of the month itself.

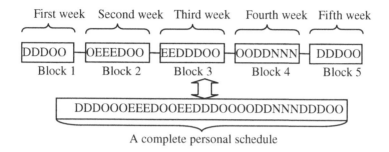

Fig. 2 The Relationship between Blocks and a Complete Personal Schedule

For each nurse, his/her valid weekly shift patterns for each week are generated as follows:

1) Generating all valid 2-day and 3-day shift sequences by systematically generating all the possible permutations of all shift types (E, D, L, N and O) over 2 days and 3 days.
2) For each individual nurse, combining the 2-day and 3-day shift sequences to form all his/her valid weekly shift patterns corresponding to each week by avoiding violating hard constraints and highly weighted soft constraints.

The reasons why we divide a complete personal schedule into several blocks by the week are following: 1) The length of a week is neither too long nor too short and thus is easy to tackle; 2) A number of constraints in our case are to restrict the shift content of a week, such as (SC1), (SC2), (SC6) and others. The violations of these constraints can be avoided during the construction of weekly shift patterns as far as possible or at least be determined to evaluate the qualities of generated weekly shift patterns. By only combining the weekly shift patterns with low penalties (≤ 50), the complete personal schedule formed is likely to have better quality and thus the solution space can be reduced effectively.

It is worth mentioning that the qualities of weekly shift patterns are just optimistic estimates, by only concerning the determined constraint violations. To elaborate, assuming that there is a weekly shift pattern 'ODOEEOO' for the second week of the month, in which there are two single 'O'. We can determine that the second 'O' violates (SC5), but we can not determine whether the first 'O' violates (SC5), because we do not know what the shifts are in the first week. In the same way, we check every soft constraint and calculate the penalty occurred by each weekly shift pattern.

3.2.2   The Improvisation of New Rosters Based on Weekly Shift Patterns

Each roster in the initial HM is generated randomly. After repaired and improved, as described in section 3.2.3 and section 3.2.4, these rosters are included and sorted in the initial HM. Then new rosters are improvised from the HM.

As described in section 2, both understaffing and overstaffing are not allowed. In the algorithm, the overstaffing is avoided during the improvising procedure and the understaffing is avoided by a coverage repairing procedure after the roster is improvised.

(1) Memory consideration
For each nurse to be scheduled, one of his/her possible personal schedules in the HM is randomly selected as his/her personal schedule in the new roster via memory consideration with probability HMCR. The shifts causing overstaffing in the selected personal schedule are set to be "O".

(2) Pitch adjustment
The personal schedule obtained by memory consideration will be pitch adjusted with probability of PAR. As described in section 3.2.1, a complete personal schedule is divided into several blocks by the week. In the pitch adjustment operator, one of these blocks will be assigned another weekly shift pattern to improve the quality of the selected personal schedule. Which block will be changed is controlled by a specific PAR range as follows:

$$the\ pitch\ adjustment \begin{cases} change\ block\ 1 & 0 \leq U(0,1) < (PAR/5) \\ change\ block\ 2 & (PAR/5) \leq U(0,1) < (2*PAR/5) \\ change\ block\ 3 & (2*PAR/5) \leq U(0,1) < (3*PAR/5) \\ change\ block\ 4 & (3*PAR/5) \leq U(0,1) < (4*PAR/5) \\ change\ block\ 5 & (4*PAR/5) \leq U(0,1) < PAR \end{cases} \tag{3}$$

After deciding which block to change, each valid weekly shift pattern causing no overstaffing will be put into this block and combined with other fixed blocks to form a neighbor personal schedule. The neighbor personal schedule with lowest penalty will be selected at last. However, if there is no weekly shift pattern could be used to form a neighbor personal schedule, add the personal schedule obtained by memory consideration to the new roster without changing.

It is worth mentioning that in order to evaluate the qualities of personal schedules if they are infeasible, the hard constraints (HC3) ~ (HC11) are also attached with very large weights in the HHSA, from 1000 to 10000. However, we emphasize that there is no violation of hard constraints allowed in the final roster.

(3) Randomization
If a personal schedule is constructed by the randomization operator, each block of the personal schedule will be randomly assigned a valid weekly shift pattern. The shifts causing overstaffing in the personal schedule are set to be "O".

Based on the three rules, personal schedules for each nurse are constructed in turn. All these personal schedules are combined together to form a new roster.

3.2.3   The Coverage Repairing Procedure

Though overstaffing has been avoided in the improvising procedure, there may still be some understaffed shifts in the new roster improvised. Thus a repair process is triggered to eliminate all the understaffed shifts by a greedy heuristic, in which each understaffed shift is added to the nurse's personal schedule whose penalty decreases the most (or increases the least if all worsen) on receiving this shift until there is no understaffing [18].

3.2.4   The Greedy Local Search

After this repair step, an efficient greedy local search is carried out on the roster to improve its quality. This greedy local search has been embedded in many methods [5, 18]. It simply swaps any pair of consecutive shifts between two nurses in the roster as long as the swaps decrease the penalty of the roster. To avoid violating shift constraints again, swaps will only be made vertically. To elaborate, Fig. 3 is used to show all the possible swaps between nurse 1 and nurse 3 within a 3-day period. The greedy search stops until no further improvement can be made. Then the improved roster is used to update the HM.

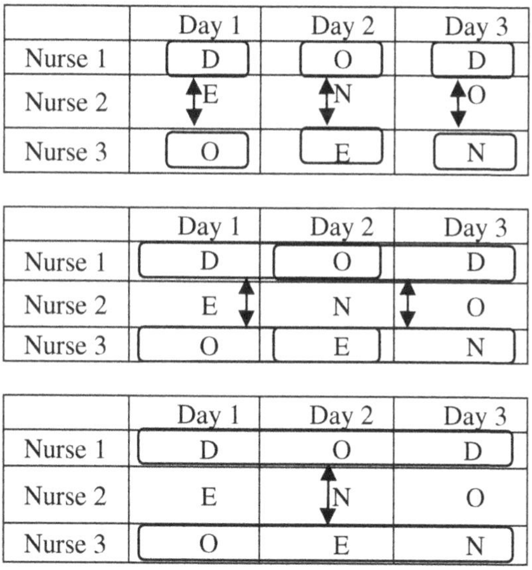

	Day 1	Day 2	Day 3
Nurse 1	D	O	D
Nurse 2	E	N	O
Nurse 3	O	E	N

	Day 1	Day 2	Day 3
Nurse 1	D	O	D
Nurse 2	E	N	O
Nurse 3	O	E	N

	Day 1	Day 2	Day 3
Nurse 1	D	O	D
Nurse 2	E	N	O
Nurse 3	O	E	N

Fig. 3 All the Possible Swaps between Nurse 1 and Nurse 3 within a 3-day Period

4   Computational Results

The proposed HHSA were tested on a real-world problem with twelve data instances, which was first described by [5]. Each instance represented a month. It is worth mentioning that these instances were designed only to produce roster for an isolated month on the assumption that previous roster was empty. The HHSA parameter values were set as following: $HMS = 5$, $HMCR = 0.95$, $PAR = 0.2$, $NI = 1000$.The experiments were performed on a PC with a Intel(R) Core(TM) i7-4790 CPU 3.60GHz processor and Windows 7 operating system. We run each of the instances 10 times. The results of the HHSA are showed in Table 1. Each result was obtained within 0.5 hours.

Table 1 Solutions obtained by the HHSA

Data	Best	Average	Worst
Jan	431	475	530
Feb	1510	1543	1575
Mar	3555	3641	3760
Apr	261	343	445
May	1900	2237	2935
Jun	10000	10187	10195
Jul	255	331	460
Aug	4381	4471	4575
Sept	271	330	465
Oct	396	479	540
Nov	1571	1620	1675
Dec	230	268	300

Table 2 Comparison of the HHSA with existing algorithms

Data	Hybrid GA [6]	Hybrid VNS [7]	Hybrid IP [8]	HHSA
Jan	775	735	460	**431**
Feb	1791	1866	1526	**1510**
Mar	2030	2010	1713	3555
Apr	612	457	391	**261**
May	2296	2161	2090	**1900**
Jun	9466	9291	8826	10000
Jul	781	481	425	**255**
Aug	4850	4880	3488	4381
Sept	615	647	330	**271**
Oct	736	665	445	**396**
Nov	2126	2030	1613	**1571**
Dec	625	520	405	**230**

Results in Table 2 demonstrate that the HHSA is able to obtain competitive results for 9 out of 12 instances compared with existing hybrid meta-heuristic algorithms. It is interesting to observe that although the constraints for each month are quite similar, the penalties for each month are quite different. These differences are caused by the structure of the month itself. For example, in the instance June, 1st June is a Sunday. Given that previous roster is empty, the nurses working on 1st June will automatically gain an unavoidable penalty by not working a complete weekend.

5 Conclusions

In this paper, a hybrid algorithm HHSA is proposed to solve the nurse rostering problem. The function of HSA is to globally and locally improvise new rosters and a greedy local search is used to improve the qualities of these rosters. In the improvising procedure of HHSA, both the pitch adjustment and randomization are carried out on high quality weekly shift patterns that are constructed in advance. By doing this, the personal schedules formed are likely to close to optimal schedule, and thus the roster consist of them can provide a good backbone, based on which the greedy local search can start from promising areas in the search space. The proposed hybrid algorithm were tested on a real-world problem with twelve data instances and obtained competitive results in a reasonable time.

Acknowledgement This work is partly supported by National Natural Science Foundation of China #61273182.

References

1. Burke, E.K., Causmaecker, P.De, Berghe, G.V., Landeghem, H.V.: The state of the art of nurse rostering. J. Sched. **7**, 441–499 (2004)
2. Aickelin, U., Dowsland, K.A.: An indirect genetic algorithm for a nurse scheduling problem. Comput. Oper. Res. **31**, 761–778 (2004)
3. Burke, E.K., Curtois, T., Qu, R., Berghe, G.V.: A scatter search methodology for the nurse rostering problem. J. Oper. Res. Soc. **61**, 1667–1679 (2010)
4. Dowsland, K.A.: Nurse scheduling with tabu search and strategic oscillation. Eur. J. Oper. Res. **106**, 393–407 (1998)
5. Burke, E.K., Li, J., Qu, R.: A pareto-based search methodology for multi-objective nurse scheduling. Ann. Oper. Res. **196**, 91–109 (2012)
6. Post, G., Veltman, B.: Harmonious personnel scheduling. In: Proceedings of 5th International Conference on Practice and Automated Timetabling, pp. 557–559 (2004)
7. Burke, E.K., Curtois, T., Post, G., Qu, R.: A hybrid heuristic ordering and variable neighbourhood search for the nurse rostering problem. Eur. J. Oper. Res. **188**, 330–341 (2008)
8. Burke, E.K., Li, J., Qu, R.: A hybrid model of integer programming and variable neighbourhood search for highly-constrained nurse rostering problems. Eur. J. Oper. Res. **203**, 484–493 (2010)

9. Cheang, B., Li, H., Lim, A., Rodrigues, B.: Nurse rostering problems—a bibliographic survey. Eur. J. Oper. Res. **151**, 447–460 (2003)
10. Geem, Z.W., Kim, J.H., Loganathan, G.V.: A new heuristic optimization algorithm: harmony search. Simulation **76**, 60–68 (2001)
11. Al-Betar, M.A., Khader, A.T.: A harmony search algorithm for university course time-tabling. Ann. Oper. Res. **194**, 3–31 (2012)
12. Geem, Z.W., Lee, K.S., Park, Y.: Application of harmony search to vehicle routing. Am. J. Appl. Sci. **2**, 1552–1557 (2005)
13. Awadallah, M.A., Khader, A.T., Al-Betar, M.A., Bolaji, A.L.: Nurse rostering using modified harmony search algorithm. Swarm, Evolutionary, and Memetic Computing **7077**, 27–37 (2011)
14. Awadallah, M.A., Khader, A.T., Al-Betar, M.A., Bolaji, A.L.: Nurse scheduling using harmony search. In: 6th International Conference on Bio-Inspired Computing: Theories and Applications, pp. 58–63 (2011)
15. Hadwan, M., Ayob, M., Sabar, N.R., Qu, R.: A harmony search algorithm for nurse rostering problems. Inform. Sciences. **233**, 126–140 (2013)
16. Curtois, T.: Personnel scheduling data sets and benchmarks. University of Nottingham. http://www.cs.nott.ac.uk/~tec/NRP/
17. Ikegami, A., Niwa, A.: A subproblem-centric model and approach to the nurse scheduling problem. Math. Program. **97**, 517–541 (2003)
18. Brucker, P., Burke, E.K., Curtois, T., Qu, R., Berghe, G.V.: A shift sequence based approach for nurse scheduling and a new benchmark dataset. J. Heuristics **16**, 559–573 (2010)

A Harmony Search Approach for the Selective Pick-Up and Delivery Problem with Delayed Drop-Off

Javier Del Ser, Miren Nekane Bilbao, Cristina Perfecto
and Sancho Salcedo-Sanz

Abstract In the last years freight transportation has undergone a sharp increase in the scales of its underlying processes and protocols mainly due to the ever-growing community of users and the increasing number of on-line shopping stores. Furthermore, when dealing with the last stage of the shipping chain an additional component of complexity enters the picture as a result of the fixed availability of the destination of the good to be delivered. As such, business opening hours and daily work schedules often clash with the delivery times programmed by couriers along their routes. In case of conflict, the courier must come to an arrangement with the destination of the package to be delivered or, alternatively, drop it off at a local depot to let the destination pick it up at his/her time convenience. In this context this paper will formulate a variant of the so-called courier problem under economic profitability criteria including the cost penalty derived from the delayed drop-off. In this context, if the courier delivers the package to its intended destination before its associated deadline, he is paid a reward. However, if he misses to deliver in time, the courier may still deliver it at the destination depending on its availability or, alternatively, drop it off at the local depot assuming a certain cost. The manuscript will formulate the mathematical optimization problem that models this logistics process and solve it efficiently by means of the Harmony Search algorithm. A simulation benchmark will be discussed to validate the solutions provided by this meta-heuristic solver and to compare its performance to other algorithmic counterparts.

J. Del Ser(✉)
TECNALIA. OPTIMA Unit, E-48160 Derio, Spain
e-mail: javier.delser@tecnalia.com

M.N. Bilbao · C. Perfecto
University of the Basque Country UPV/EHU, 48013 Bilbao, Spain
e-mail: {nekane.bilbao,cristina.perfecto}@ehu.eus

S. Salcedo-Sanz
Universidad de Alcalá, 28871 Alcalá de Henares, Spain
e-mail: sancho.salcedo@uah.es

© Springer-Verlag Berlin Heidelberg 2016
J.H. Kim and Z.W. Geem (eds.), *Harmony Search Algorithm*,
Advances in Intelligent Systems and Computing 382,
DOI: 10.1007/978-3-662-47926-1_13

Keywords Courier problem · Delayed drop-off · Hill climbing · Genetic algorithm ·
Harmony search

1 Introduction

Last-mile logistics and delivery services are nowadays growing at the pace dictated by
the similarly increasing familiarity of end users when purchasing items and goods
via Internet applications and marketplaces. This statement is buttressed by recent
surveys where e-commerce and the rapid growth of developing countries such as
China and India are highlighted as catalyzing drivers for the derivation and evolution
of distribution networks, particularly in what relates to the so-called local urban
logistics [1]. Likewise, according to the eMarketer research firm global business-to-
consumer electronic commerce sales in 2014 have increased by 19.2% from 2013 [2].
This reported upsurge undoubtedly calls for new technical approaches focused on
increasing transport efficiency at all levels of the distribution network, with emphasis
on that of the finest granularity (last-mile logistics) which, in turn, is more sensitive
to traffic eventualities.

From the mathematical perspective, transport efficiency is generally conceived as
an optimization problem with delivery route as the most widely considered decision
variable. When concentrating on last-mile logistics with regional or local depots,
problems taking this assumption are usually grouped in what are referred to as Vehicle
Routing Problems (VRP). In essence this model consists of the discovery of optimal
routes for vehicles starting and finishing at a given depot so as to deliver goods to a set
of scattered nodes under different criteria (e.g. distance, time or cost minimization).
This seminal definition of the vehicle routing problem has evolved to a wide spectrum
of alternative formulations and extensions such as the Capacitated VRP (CVRP), the
Multiple Depot VRP (MDVRP), the Periodic VRP (PVRP) and the Split Delivery
VRP (SDVRP), as well as hybridizations of these extensions with soft and hard time
constraints. The SDVRP, first defined in [3] as an instance of the VRP where the
demand can be satisfied by more than one-time delivery or by two or more vehicles,
has been tackled via Tabu Search (TS, [4, 5]), which is a meta-heuristic solver that
hinges on the construction of memory structures that describe the visited solutions
during the search process. A new logistics model encompassing soft time windows,
multi-period routing, and split delivery strategies has been recently proposed in [6]
and solved via genetically inspired heuristics. This model builds upon previous work
in [7], where the delivery of supplies in disaster areas was approached via a similar
model, but subject to hard timing constraints. As for the rest of VRP variants there are
myriads of contributions dealing with the application of evolution strategies [8, 9],
Ant Colony Optimization [10, 11], Greedy Randomized Adaptive Search Procedure
(GRASP, [12, 13, 14]), Simulated Annealing [15, 16, 17] and Variable Neighbor
Search (VNS, [18, 19]), among others. For the sake of space, the reader is referred to
the comprehensive survey in [20] for a thorough state of the art around meta-heuristics
for the VRP.

This paper deals with a particular instance of the VRP that finds its motivation in last-mile courier services, i.e. a person who delivers messages, packages,

and/or certified mail. Couriers face a given delivery commit with pickup and delivery time/location information per item. When any given item is delivered before its associated time, the courier achieves an economical reward, which is usually higher the shorter the time from the pickup to the delivery is. If arriving at the destination after the deadline of the packet at hand, the courier may still deliver it at a reduced reward due to additional costs derived from its deposit at a local depot or instead wait for the customer to be available. The problem hinges then on selecting the set of locations to be visited by the courier and the set of delivery choices that lead to a maximum reward, subject to different constraints such as the pickup times for each entry of the commit and the traveling dynamics between pair of locations. Hereafter denoted as the modified selective pickup and delivery problem (mSPDP), this paper proposes the application of Harmony Search (HS) heuristics to this optimization paradigm. The manuscript covers from the underlying mathematical formulation of the optimization problem to the description of the HS solver (encoding and operators) designed for its efficient resolution. Its performance will be discussed in comparison with other evolutionary and greedy schedulers. To the authors' knowledge, this is the first contribution in the literature dealing with HS applied to this routing scenario.

The manuscript is structured as follows: Section 2 will first elaborate on the formulation of the mSPDP problem, whereas Section 3 will delve into the proposed HS scheduler, including its encoding strategy and improvisation operators. Performance results over synthetically generated scenarios will be presented in Section 4 and, finally, Section 5 ends the paper with conclusions and future research lines.

2 Problem Formulation

The formulation of the mSPDP problem starts by conceiving the delivery commit as a set of packets $\mathcal{P} \doteq \{\mathbf{P}_p\}_{p=1}^{P}$, each characterized by: 1) pickup time and location $T_p^{\uparrow}$ and $X_p^{\uparrow}$; 2) delivery time and location $T_p^{\downarrow}$ and $X_p^{\downarrow}$; 3) destination daily availability times $T_p^{\diamond}$ (start) and $T_p^{\spadesuit}$ (end); 4) a delivery reward $R_p \geq 0$; and 5) a deposit cost $C_p \geq 0$ to be paid if the courier decides to deposit the packet at the local depot. Any given route followed by the courier can be expressed as a sequence of visited locations $\mathbf{X} = \{(X_1^{\triangleright}, T_1^{\triangleright}), (X_2^{\triangleright}, T_2^{\triangleright}), \ldots, (X_N^{\triangleright}, T_N^{\triangleright})\}$, with $X_n^{\triangleright} \in \mathcal{X} \ \forall n \in \{1, \ldots, N\}$. Upon its completion, this route gives rise to a subset of delivered packets or items $\mathcal{P}(\mathbf{X}) \subseteq \mathcal{P}$, which renders a set $\mathcal{J} \subseteq \{1, \ldots, P\}$ of delivered packet indexes with effective pickup and delivery times $T_j^{\uparrow,e}$ and $T_j^{\downarrow,e}$ ($j \in \mathcal{J}$) for each of its compounding items. The reward $R(\mathcal{P}(\mathbf{X}))$ associated to $\mathbf{X}$ will be given by

$$R(\mathcal{P}(\mathbf{X})) \doteq \sum_{j \in \mathcal{J}} R_j \cdot \mathbb{I}(T_j^{\uparrow,e} \geq T_j^{\uparrow}) \cdot \mathbb{I}(T_j^{\downarrow,e} \leq T_j^{\downarrow}), \tag{1}$$

where $\mathbb{I}(\cdot)$ is an indicator binary function taking value 1 if the condition in its argument is true and 0 otherwise. A more realistic assumption in Expression (1) includes the traveling cost per kilometer C_{km} as a result of e.g. the fuel consumption required to get from one location to another. The consideration of this additional cost can be reflected if, by a small notational abuse, the distance from location X_m to X_n is denoted as $|X_m - X_n|$, yielding a net reward $R'(\mathcal{P}(\mathbf{X}))$ given by

$$R'(\mathcal{P}(\mathbf{X})) = \sum_{j \in \mathcal{J}} \left(R_j \cdot \mathbb{I}(T_j^{\uparrow,e} \geq T_j^{\uparrow}) \cdot \mathbb{I}(T_j^{\downarrow,e} \leq T_j^{\downarrow}) \right) - \sum_{n=2}^{N} C_{km} \cdot |X_n - X_{n-1}|,$$

which, without loss of generality, accommodates any other cost source and/or accounting (e.g. transported weight dependence).

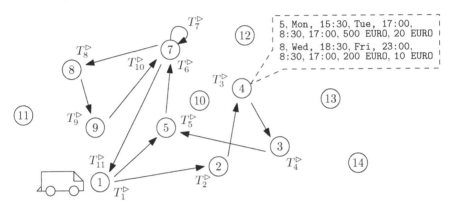

Fig. 1 Diagram representing the scenario modeled in the mSPDP problem with arrows corresponding to the courier route $\mathbf{X} = \{1, 2, 4, 3, 5, 7, 7, 8, 9, 7, 1\}$. The time of arrival at the last location of the route, denoted as $T_{11}^{\triangleright}$, should be less than T_{max}.

However, in those picked packets whose delivery deadline $T_j^{\downarrow}$ is not met the courier should decide what to do with the packet at hand depending on the availability of the destination (given by $T_j^{\Diamond}$ and $T_j^{\blacklozenge}$) and the remaining delivery commit. Two actions are assumed: 1) the courier waits for the destination to be available and delivers it, obtaining no reward for this delayed service; or 2) deposits the packet at a local depot at a cost penalty C_j. By denoting as $A_j \in \{0, 1\}$ the action selected by the courier for packet $j \in \mathcal{J}$ (0: wait for delivery after its deadline; 1: deposited at the local depot), the net reward can be rewritten as

$$R''(\mathcal{P}(\mathbf{X}, \mathbf{A})) = \sum_{j \in \mathcal{J}} \left(R_j \cdot \mathbb{I}(T_j^{\uparrow,e} \geq T_j^{\uparrow}) \cdot \mathbb{I}(T_j^{\downarrow,e} \leq T_j^{\downarrow}) \right) - \sum_{n=2}^{N} C_{km} \cdot |X_n - X_{n-1}|,$$
$$- \sum_{j \in \mathcal{J}} C_j \cdot \mathbb{I}(T_j^{\downarrow,e} > T_j^{\downarrow}) \cdot \mathbb{I}(A_j = 1), \qquad (2)$$

where $\mathbf{A} \doteq \{A_j\}_{j \in \mathcal{J}}$. In this reformulation of the net reward any packet is assumed to be automatically delivered whenever both $T_j^{\uparrow,e} \geq T_j^{\uparrow}$ and $T_j^{\downarrow,e} \leq T_j^{\downarrow}$ hold (i.e. whenever the courier arrives in time). The mSPDP problem considered in this paper searches for the optimal route of the courier, expressed as $\mathbf{X}^*$, and their associated set of actions $\mathbf{A}^*$ such that its subset of delivered packets $\mathcal{P}(\mathbf{X}^*, \mathbf{A}^*) \subseteq \mathcal{P}$ over a maximum time window T_{max} provides a maximum reward. In mathematical notation,

$$\mathbf{X}^*, \mathbf{A}^* = \arg\max_{\mathbf{X},\mathbf{A}} R''(\mathcal{P}(\mathbf{X}, \mathbf{A})), \tag{3}$$

$$\text{subject to: } T_N^{\triangleright} \leq T_{max}, \tag{4}$$

where it is implicit in the expression that effective pickup and delivery times in the overall reward associated to the route depend roughly on the sequence of visited cities, the traveling times among them and the waiting times for those packets with $A_j = 0$ (given by $T_j^{\diamond}$ and $T_j^{\blacklozenge}$). In other words, any given optimization algorithm facing the above problem should consider not only the pickup and delivery times of the packets within the commit, but also the traveling distances between the cities and their associated rewards. Likewise, it is important to note that due to the assumed selectiveness fo the courier service, not all items in the delivery commit may be delivered to their destination: some of them may be discarded due to e.g. low expected profitability of its processing when considered jointly with the required distance to reach either its pickup and/or delivery locations.

It can be proven that the mSPDP formulation yields a computationally hard optimization problem. This complexity calls for self-learning approximate solvers for its efficient solving, such as the proposed HS scheduler next described.

3   Proposed Scheme

The search for the optimal set of visited locations $\mathbf{X}^*$ that maximizes the net reward of the courier relies on the Harmony Search (HS) meta-heuristic algorithm as its algorithmic core. First coined in [21], HS is a relatively new population-based meta-heuristic that has been proven to outperform other approximative approaches in a wide portfolio of continuous and combinatorial problems and applications. A survey on different application fields to which this meta-heuristic algorithm has been applied so far can be found in [22].

Operationally HS mimics the behavior of a music orchestra when aiming at composing the most harmonious melody as measured by aesthetic standards. HS imitates a musician's improvisation process when intending to produce a piece of music with aesthetically perfect harmony. When comparing the improvisation process of musicians with optimization methods, one can realize that each musician corresponds to a decision variable; the musical instrument's pitch range refers to the alphabet of the decision variable; the musical harmony improvised at a certain time corresponds to a solution vector at a given iteration; and audience's aesthetic impression links to the fitness function of the optimization problem at hand. In a similar way to how musicians improve the melody – through variation and check – time after time, the HS algorithm progressively enhances the fitness of the solution vector in an iterative fashion. Each musician corresponds to an attribute of a given candidate solution picked from the problem solution domain, whereas each instrument's pitch and range corresponds to the bounds and constraints imposed on the decision variable.

HS is a population-based solver; it hence maintains a pool of Ψ solutions or Harmony Memory (HM), similar to the population of chromosomes in genetic algorithms. An estimation of the optimum solution is achieved at every iteration by applying a set of optimization operators to the HM, which springs a new harmony vector every time. Notwithstanding their evident similarity, the advantages of HS with respect to genetic algorithms spring from the fact that HS hybridizes concepts from population-based meta-heuristics with subtle yet important modifications (e.g. stochastically driven, uniform polygamy), and evolutionary principles (i.e. new solutions evolve from previous versions of themselves, based on the application of global-search and local-search operators). This synergy has been empirically shown to balance better between the explorative and exploitative behavior of the search procedure with respect to other standard crossover and mutation procedures.

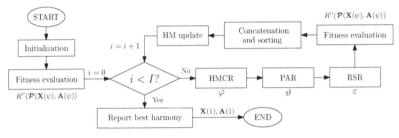

Fig. 2 Flow diagram of the HS scheduler for the mSPDP.

In the problem at hand each harmony will be composed by two different albeit related parts: 1) an N_{max}-sized integer representation of the sequence of locations $\mathbf{X}(\psi)$ to be visited by the courier during the considered time span T_{max} (with $\psi \in \{1, \ldots, \Psi\}$ indicating the index of the harmony within the HM); and 2) a binary vector $\mathbf{A}(\psi)$, each of whose entries indicate the action adopted by the courier for each of the picked packets during its journey through $\mathbf{X}(\psi)$. It is important to note that when visiting a certain location the courier may carry more than one packet, each with different destinations. The encoding approach taken in this work is to represent the set of actions for all packets by setting the length of $\mathbf{A}(\psi)$ to $P \doteq |\mathcal{P}| \; \forall \psi$. Since packets are assumed to be automatically picked when available in the location where the courier is located (i.e. no capacity constraints are assumed for the courier vehicle), entries in $\mathbf{A}(\psi)$ corresponding to non-picked packets are masked out and not considered for the HS operators. As for $\mathbf{X}(\psi)$, even though the harmony length is fixed to N_{max} a discrete event simulator determines whether the journey duration for the sequence of visited locations at hand fits in the time span T_{max} imposed as a constraint in Expression (4). Those locations in $\mathbf{X}(\psi)$ visited by the courier at times beyond T_{max} are left out by the simulator from the evaluation of the fitness $R''(\mathcal{P}(\mathbf{X}(\psi), \mathbf{A}(\psi))$.

Figure 2 illustrates the flow diagram of the HS algorithm: (i) initialization of the HM; (ii) improvisation of a new harmony; (iii) update of the HM; and (iv) returning to step (ii) until a termination criteria is satisfied. The improvisation procedure is controlled by three operators sequentially applied to each note so as to produce a new set of improvised harmonies or candidate solutions:

1) The Harmony Memory Considering Rate (HMCR), which is driven by the probabilistic parameter $\varphi \in [0, 1]$, imposes that the new value for a certain note is drawn uniformly from the values of the same note in all the remaining melodies. Following the notation introduced in this manuscript, if $\mathbf{X}(\psi) \doteq \{X(\psi, 1), X(\psi, 2), \ldots, X(\psi, N_{max})\}$ denotes the ψ-th harmony of the HM, then

$$\varphi \triangleq Pr\{X(\psi, n) \rightsquigarrow \omega_{\text{HMCR}}\}, \tag{5}$$

where ω_{HMCR} is a discrete variable distributed uniformly over the integer set $\{X(\psi, 1), \ldots, X(\psi, n-1), X(\psi, n+1), \ldots, X(\psi, N_{max})\}$. This operator is applied note by note in the HM, i.e. the above operation is repeated for every $n \in \{1, \ldots, N_{max}\}$ and $\psi \in \{1, \ldots, \Psi\}$. If $X(\psi, n) = X(\psi, n+1)$ for any given n and ψ, the discrete event simulator interprets that the courier decides to stay at the location represented by $X(\psi, n)$ until a packet is ready to be picked up at this location. A similar HMCR procedure holds with respect to the set of actions for the picked packets by considering $\mathbf{A}(\psi) \doteq \{A(\psi, 1), A(\psi, 2), \ldots, X(\psi, P)\}$ and the additional constraint that the set of possible new values must be drawn from non-masked notes.

2) The Pitch Adjusting Rate (PAR), controlled by the probability $\vartheta \in [0, 1]$, determines that the new value $\widehat{X}(\psi, n)$ for a given note value $X(\psi, n)$ is computed via a controlled random perturbation, i.e.

$$\vartheta \triangleq Pr\{X(\psi, n) \rightsquigarrow \omega_{\text{PAR}}\}, \tag{6}$$

where ω_{PAR} is a random binary variable with equal probability of taking the neighboring values around $X(\psi, n)$ in $\mathcal{X}$. In order to boost the performance of this operator, the alphabet $\mathcal{X}$ – namely, the overall set of locations – is sorted prior to its application in terms of their distance to $X(\psi, n)$. Again, it should be emphasized that this operator operates note-wise on the constituent harmonies of the HM. As for its application to $\mathbf{A}(\psi)$, given its binary alphabet the PAR operator flips its value whenever the value at hand is not masked out (e.g. the associated packet is processed by the courier).

3) The Random Selection Rate (RSR), with corresponding parameter $\varepsilon \in [0, 1]$, works in a similar fashion to the aforementioned PAR operator, the difference being that no neighborhood criterion is taken into account, i.e. the new value for $X(\psi, n)$ is drawn uniformly at random from $\mathcal{X}$. This operator is applied exclusively over $\mathbf{X}(\psi) \ \forall \psi \in \{1, \ldots, \Psi\}$.

As can be inferred from the flow diagram in Figure 2, the above three operators are sequentially applied on the pool of candidate harmonies, which produces a new, potentially refined harmony memory whose net reward $R''(\mathcal{P}(\mathbf{X}(\psi), \mathbf{A}(\psi))$ is subsequently evaluated by means of a discrete event simulator designed to cope with all timing constraints and delays incurred by the courier when following the route and set of actions imposed by each candidate solution. Once the values of their net reward have been computed, the HS-based solver concatenates the new harmony memory with that remaining from the previous iteration and keeps the best Ψ candidates, discarding the rest of harmonies. The refining process is repeated until a maximum of I iterations are completed.

4   Experiments and Results

In order to assess the performance of the proposed HS-based scheduler Monte Carlo simulations have been performed on emulated problem instances of increasing size. The set of simulated experiments aim at evincing the scalability of the produced solutions as the number of packets and cities increase over a fixed time frame of one week. Therefore, simulation scenarios will be hereafter identified as $\{|\mathcal{X}|, P\}$, with $|\mathcal{X}| \in \{4, 8, 12\}$ denoting the number of cities and $P \in \{20, 40, 60\}$ standing for the number of packets to be delivered. The focus will be then placed on 1) how first and second order statistics (mean, standard deviation and extreme values averaged over the Monte Carlo realizations) of the reward and number of delivered packets associated to the solution provided by the HS solver behave as $|\mathcal{X}|$ and P increase; and 2) the comparison of such statistical indicators to those produced by a genetic algorithm (GA) and a random-start hill-climbing procedure (HC). While the former implements a naive single-point crossover with tournament selection and uniform mutation, the latter conducts several hill-climbing explorations over the solution space based on randomly generated initial seeds, each kept running until reaching a steady state [23]. The reason for selecting both alternative meta-heuristics lies on checking whether the proposed HS scheduler performs better than other search techniques disregarding their population-based nature.

All parameters of the compared schedulers (e.g. φ, ϑ and ε for HS; mutation and crossover probability in GA) have been optimized via a grid search and the selection of the values leading to the best *average* performance over the simulated scenario. For a fair comparison the number of fitness evaluations have been kept to the same value $I = 2500$ for all the algorithms in the benchmark. In all cases pickup times for the packets compounding the delivery commit have been modeled by means of an exponential distribution with interpackage mean time equal to 3 hours, whereas delivery deadlines have been computed by

$$T_i^{\downarrow} = \lambda_i \cdot \frac{|X_i^{\uparrow} - X_i^{\downarrow}|}{V}, \tag{7}$$

where V denotes the speed of the courier vehicle and λ_i is drawn uniformly at random from the set $[1.1, 5]$ to represent the *priority* of the packet at hand. When λ_i gets close to 1, there is almost no time flexibility for the courier to deliver the packet at its corresponding destination. Accordingly, the higher λ_i is, the wider the time gap from the pickup time to the delivery deadline will be, hence allowing for a higher scheduling flexibility of the courier. The reward R_i is also set linearly yet inversely proportional to λ_i, whereas the depot cost has been set to $0.25 \cdot R_i$ monetary units. Despite these parametric assumptions, it should be clear that other models can be adopted without any loss of generality.

The discussion begins by analyzing Table 1, where statistics for the reward $R(\cdot)$ and delivered packets $P(\cdot)$ are listed for the different simulated scenarios and the three compared solvers. First it is important to note that when dealing with problems with small scales, the HC approach outperforms in what refers to mean and standard deviation of the net reward and the number of delivered packets. Note, however, that in terms of their maximum value over 30 realizations all algorithms under comparison achieve the same score (reward equal to 3547.40 monetary units and 17 delivered packets). This is certainly remarkable due to the fact that in such small scenarios the computation speed of the algorithms is high enough to permit running as many Monte Carlo experiments as desired and implement in practice the best solution obtained. Based on this rationale, the compared solvers can be considered as equally performing in the {4, 20} scenario.

Table 1 Monte Carlo statistics (mean/std/max) of the reward $\mathcal{R}''$ and delivered packets $\mathcal{P}$ for the different scenarios and schedulers under comparison.

	Scenario	HC	GA	HS
	{4, 20}	3443.81/74.22/3547.40	2995.98/271.79/3501.90	3167.68/210.19/3547.40
R''	{8, 40}	4744.15/245.84/5196.80	4914.60/613.33/6033.01	5156.01/450.14/6033.01
	{12, 60}	3958.03/212.25/4501.40	3928.01/713.15/5332.80	4110.25/645.60/5699.90
	{4, 20}	14.75/1.13/17	14.05/1.20/17	14.20/0.85/17
P	{8, 40}	20.80/1.80/25	21.20/2.11/25	22.05/3.44/26
	{12, 60}	17.65/2.43/24	15.60/3.13/22	17.75/3.75/26

By contrast, when the scales of the simulated scenario increase the computation time of the algorithms become high enough to require focusing on the *average* quality of the produced solutions. As evinced by the obtained results, HS outperforms GA and HC in terms of the average net reward and number of delivered packets. Unfortunately the relatively high standard deviation of the HS scores impacts on the statistical relevance of the differences between the algorithms under comparison. This is unveiled by a Wilcoxon Rank-Sum test applied to each pair of score sets to assess whether such samples are drawn from distributions with equal medians. While at a significance level of $\alpha = 0.15$ all Monte Carlo result sets are certified to come from distributions with different medians, this statement does not hold when $\alpha = 0.05$.

5 Conclusions and Future Research Lines

This paper has elaborated on a HS-based scheduler for a variant of the so-called Selective Pickup and Delivery Problem that further considers the possibility to drop packets at local depots for the final customer at a certain cost penalty. The problem resides not only on the route discovery for the courier to pick up and deliver packets depending on timing constraints (due to traveling, availability of the packet and delivery deadlines), but also on whether the courier decides to wait for the customer to be available or instead, leaves the packet off at the depot and continues the service. The manuscript has mathematically formulated the problem, for whose efficient solving a scheduler based on the Harmony Search algorithm has been proposed and described in detail. The harmony encoding strategy considers both the set of visited locations $\mathbf{X}$ and the actions $\mathbf{A}$ taken for each processed packet. Simulation results have shed light on the superior performance of the proposed solver with respect to other heuristics, more notably when the scales of the simulated scenario increase. The statistical relevance of the obtained results has been certified for acceptable confidence levels.

Future research will be conducted towards extending the functionalities of the discrete event simulator utilized to simulate the courier service. In particular the focus will be placed on dynamically changing traveling conditions (e.g. variable vehicle speed) that could impact on the courier schedule. Weight constraints will be also studied and incorporated to the problem formulation.

Acknowledgments This work has been partially supported by the Basque Government under the ETORTEK Program (grant reference IE14-382).

References

1. Jones Lang Lasalle IP, Inc.: E-commerce boom triggers transformation in retail logistics. White paper (2013)
2. eMarketer: Worldwide B2C Ecommerce: Q3 2014 Complete Forecast. Research report (2014)
3. Dror, M., Trudeau, G.: Savings by Split Delivery Routing. Transportation Science **23**, 141–145 (1989)
4. Archetti, C., Speranza, M.G., Hertz, A.: A Tabu Search Algorithm for the Split Delivery Vehicle Routing Problem. Transportation Science **40**(1), 64–73 (2006)
5. Ho, S.C., Haugland, D.: A Tabu Search Heuristic for the Vehicle Routing Problem with Time Windows and Split Deliveries. Computers and Operations Research **31**(12), 1947–1964 (2004)
6. Lin, Y.H., Batta, R., Rogerson, P.A., Blatt, A., Flanigan, M.: A Logistics Model for Delivery of Prioritized Items: Application to a Disaster Relief Effort. Technical report. New York: University of Buffalo (2009)
7. Balcik, B., Beamon, B.M., Smilowitz, K.: Last Mile Distribution in Humanitarian Relief. Journal of Intelligent Transportation Systems: Technology, Planning, and Operations **12**(2), 51–63 (2008)

8. Mester, D., Bräysy, O., Dullaert, W.: A Multi-Parametric Evolution Strategies Algorithm for Vehicle Routing Problems. Expert Systems with Applications **32**, 508–517 (2007)
9. Alba, E., Dorronsoro, B.: Computing Nine New Best-so-far Solutions for Capacitated VRP with a Cellular Genetic Algorithm. Information Processing Letters **98**, 225–230 (2006)
10. Doerner, K.F., Hartl, R.F., Lucka, M.: A Parallel Version of the D-ant Algorithm for the Vehicle Routing Problem. Parallel Numerics, 109–118 (2005)
11. Li, X., Tian, P.: An ant colony system for the open vehicle routing problem. In: Dorigo, M., Gambardella, L.M., Birattari, M., Martinoli, A., Poli, R., Stützle, T. (eds.) ANTS 2006. LNCS, vol. 4150, pp. 356–363. Springer, Heidelberg (2006)
12. Layeb, A., Ammi, M., Chikhi, S.: A GRASP Algorithm Based on New Randomized Heuristic for Vehicle Routing Problem. Journal of Computing and Information Technology **1**, 35–46 (2013)
13. Suárez, J.G., Anticona, M.T.: Solving the capacitated vehicle routing problem and the split delivery using GRASP metaheuristic. In: Bramer, M. (ed.) IFIP AI 2010. IFIP AICT, vol. 331, pp. 243–249. Springer, Heidelberg (2010)
14. Chaovalitwongse, W., Kim, D., Pardalos, P.M.: GRASP with a New Local Search Scheme for Vehicle Routing Problems with Time Windows. Journal of Combinatorial Optimization **7**, 179–207 (2003)
15. Wang, C., Zhao, F., Mu, D., Sutherland, J.W.: Simulated annealing for a vehicle routing problem with simultaneous pickup-delivery and time windows. In: Prabhu, V., Taisch, M., Kiritsis, D. (eds.) APMS 2013, Part II. IFIP AICT, vol. 415, pp. 170–177. Springer, Heidelberg (2013)
16. Chen, S., Golden, B., Wasil, E.: The Split Delivery Vehicle Routing Problem: Applications, Algorithms, Test Problems, and Computational Results. Networks **49**, 318–329 (2007)
17. Czech, Z.J., Czarnas, P.: Parallel Simulated Annealing for the Vehicle Routing Problem with Time Windows. In: Euromicro Workshop on Parallel, Distributed and Network-based Processing, pp. 376–383 (2002)
18. Pirkwieser, S., Raidl, G.R.: Multilevel variable neighborhood search for periodic routing problems. In: Cowling, P., Merz, P. (eds.) EvoCOP 2010. LNCS, vol. 6022, pp. 226–238. Springer, Heidelberg (2010)
19. Kytöjoki, J., Nuortio, T., Bräysy, O., Gendreau, M.: An Efficient Variable Neighborhood Search Heuristic for Very Large Scale Vehicle Routing Problems. Computers & Operations Research **34**, 2743–2757 (2007)
20. Gendreau, M., Potvin, J.-Y., Bräumlaysy, O., Hasle, G., Løkketangen, A.: Metaheuristics for the Vehicle Routing Problem and Its Extensions: A Categorized Bibliography. Operations Research/Computer Science Interfaces **43**, 143–169 (2008)
21. Geem, Z.W., Kim, J.H., Loganathan, G.V.: A New Heuristic Optimization Algorithm: Harmony Search. Simulation **76**(2), 60–68 (2001)
22. Manjarres, D., Landa-Torres, I., Gil-Lopez, S., Del Ser, J., Bilbao, M.N., Salcedo-Sanz, S., Geem, Z.W.: A Survey on Applications of the Harmony Search Algorithm. Engineering Applications of Artificial Intelligence **26**(8), 1818–1831 (2013)
23. Russell, S.J., Norvig, P.: Artificial Intelligence: A Modern Approach. Upper Saddle River. Prentice Hall (2004)

Dandelion-Encoded Harmony Search Heuristics for Opportunistic Traffic Offloading in Synthetically Modeled Mobile Networks

Cristina Perfecto, Miren Nekane Bilbao, Javier Del Ser, Armando Ferro and Sancho Salcedo-Sanz

Abstract The high data volumes being managed by and transferred through mobile networks in the last few years are the main rationale for the upsurge of research aimed at finding efficient technical means to offload exceeding traffic to alternative communication infrastructures with higher transmission bandwidths. This idea is solidly buttressed by the proliferation of short-range wireless communication technologies (e.g. mobile devices with multiple radio interfaces), which can be conceived as available opportunistic hotspots to which the operator can reroute exceeding network traffic depending on the contractual clauses of the owner at hand. Furthermore, by offloading to such hotspots a higher effective coverage can be attained by those operators providing both mobile and fixed telecommunication services. In this context, the operator must decide if data generated by its users will be sent over conventional 4G+/4G/3G communication links, or if they will instead be offloaded to nearby opportunistic networks assuming a contractual cost penalty. Mathematically speaking, this problem can be formulated as a spanning tree optimization subject to cost-performance criteria and coverage constraints. This paper will elaborate on the efficient solving of this optimization paradigm by means of the Harmony Search meta-heuristic algorithm and the so-called Dandelion solution encoding, the latter allowing for the use of conventional meta-heuristic operators maximally preserving the locality of tree representations. The manuscript will discuss the obtained

C. Perfecto(✉) · M.N. Bilbao · A. Ferro
University of the Basque Country (UPV/EHU), E-48013 Bilbao, Spain
e-mail: {cristina.perfecto,nekane.bilbao,armando.ferro}@ehu.eus

J. Del Ser
TECNALIA. OPTIMA Unit, E-48160 Derio, Spain
e-mail: javier.delser@tecnalia.com

S. Salcedo-Sanz
Universidad de Alcalá, E-28871 Alcalá de Henares, Spain
e-mail: sancho.salcedo@uah.es

© Springer-Verlag Berlin Heidelberg 2016
J.H. Kim and Z.W. Geem (eds.), *Harmony Search Algorithm*,
Advances in Intelligent Systems and Computing 382,
DOI: 10.1007/978-3-662-47926-1_14

133

simulation results over different synthetically modeled setups of the underlying communication scenario and contractual clauses of the users.

Keywords Traffic offloading · Dandelion code · Harmony search

1 Introduction

According to the Cisco Visual Networking index [1] the latterly increase in mobile data traffic is not but a foretaste of what awaits for the coming next 5 years, with a predicted tenfold traffic volume boost at the global scale. With broadband 4G still under deployment, many voices have called into question the capacity of existing networks to cope with such traffic volumes. As a consequence, the search for alternative mechanisms to divert traffic to wired networking infrastructure has become a major concern for network operators with limited radio spectrum. In this context mobile data offloading – i.e. the use of alternative network technologies for delivering data originally targeted for e.g. cellular networks when it becomes saturated – will play a decisive role to ensure optimal usage of available radio resources and load balancing among radio interfaces [2].

Existing offloading solutions include femtocells, Wi-fi hotspots or opportunistic communications. Opportunistic mobile data offload was first proposed by [3] to deliver data between mobile terminals through peer to peer communications such as Bluetooth, Wi-fi Direct and, more recently, LTE-advanced device-to-device (D2D) communications. Still, application, device, subscriber and operator awareness are required for making real-time decisions regarding selective offloading and thus effectively managing the overall process [4, 5]. In this work we address network selection as an optimization problem where the net benefit of the network operator is to be maximized subject to a number of QoS and radio coverage constraints. To the best of our knowledge this is the first work that applies advanced meta-heuristic solvers so as to maximize operators revenue considering the use of incentives. Previous related contributions hinge on utility theory, and do not consider any incentive mechanism, such as [6] or the cost function based approach in [7].

The paper is structured as follows: Section 2 delves into the mathematical formulation of the optimization problem. Section 3 details the proposed approach with an insight into Dandelion codes and the Harmony Search algorithm. Section 4 discusses simulation results and finally, Section 5 concludes the paper.

2 Problem Formulation

The mathematical formulation of the constrained mobile traffic offloading problem stems from the analogy of a opportunistic communication network with a directed graph represented by a tree rooted on the node representing the Base Station (BS) of the network under consideration. This tree can be encoded by means of the Dandelion encoding scheme described in Section 3.1. Let N denote the total number of nodes deployed in the network, where those nodes labeled as $0, 1, \ldots, N-2$ represent mobile devices located at coordinates $\{\mathbf{p}_j\}_{j=0}^{N-2}$ over an area of dimension $X_{MAX} \times Y_{MAX}$, and node $N-1$ stands for the BS located at position $\mathbf{p}_{N-1} \in \mathbb{R}^2$. Such nodes are assumed to be equipped with a dual wireless interface, one for transmitting directly to the BS (via e.g. 4G or any other cellular service alike) and the other for shorter-range communications that will be used for tethering (correspondingly, Bluetooth or any protocol within the IEEE 802.11 family of standards). For the sake of simplicity, radio coverage areas for both interfaces will be assumed constant and circularly shaped with radii $R_{\divideontimes}$ and $R_{\odot} \geq R_{\divideontimes}$, respectively. The finite coverage area of these radio interfaces gives rise to a $N \times N$ symmetric binary coverage matrix $\mathbf{E}$, each of whose compounding entries e_{ij} for $j \neq N - 1$ is set to 1 if node i is inside the tethering coverage radio $R_{\divideontimes}$ centered in node j and $e_{ij} = 0$ otherwise. If $j = N - 1$, e_{ij} will equal 1 if node i is inside the coverage radio $R_{\odot}$ centered on the BS node, and $e_{ij} = 0$ otherwise.

Any solution for the problem here tackled can be conceived by means of a rooted tree which accommodates different notational representations, such as the Dandelion code. However, for the sake of ease in the formulation of the problem the solution tree will be denoted as a binary $N \times N$ connectivity matrix $\mathbf{X}$, where each constituent $x_{i,j}$ (with $i, j \in \{0, 1, \ldots, N - 1\}$) takes on value 1 if a direct link from node i to node j exists in the represented network layout, and 0 if node i is not connected to j anyhow. For notational convenience it is forced that $x_{i,i} = 0\ \forall i \in \{0, \ldots, N-1\}$. The network operator obtains a net reward computed as the difference between the income associated to the contract of user i and the fractional costs due to 1) expenditures required for the operation of the network infrastructure if user i connects to the network, i.e. $x_{i,N-1} = 1$; or 2) the incentive paid to the tethering user j to which the traffic generated by user i is offloaded, i.e. $x_{i,j} = 1$. If costs associated to each option are denoted as $C_{\odot}$ and $C_{\divideontimes}$ and the contract income for all users is assumed constant and equal to $I_{\odot}$, the overall net benefit of the network operator can be expressed as

$$B(\mathbf{X}) \doteq (N - 1) \cdot I_{\odot} - \sum_{i=0}^{N-2} C_{\odot} \cdot x_{i,N-1} - \sum_{j=0}^{N-2} C_{\divideontimes} \cdot \mathbb{I}\left(\sum_{i=0}^{N-2} x_{i,j} > 0\right), \quad (1)$$

where $\mathbb{I}(\cdot)$ is an auxiliary indicator function taking value 1 if its argument is true and 0 otherwise. A more realistic model of the incentives paid to a tethering user considers its value as a function of the number of tethered users, yielding

$$B(\mathbf{X}) = (N - 1) \cdot I_{\odot} - \sum_{i=0}^{N-2} C_{\odot} \cdot x_{i,N-1} - \sum_{j=0}^{N-2} C_{\divideontimes} \cdot \sum_{i=0}^{N-2} x_{i,j}, \quad (2)$$

with linearity of the cost model with the number of users being assumed for simplicity. It is important to remark that since operator-governed offloading is considered, tethering is exploited by the operator of the network as a means to increase its net benefit, not by the end users as a method to reduce their expenditure. Based on these definitions, the problem undertaken in this work can be cast as finding the optimal network layout X^* such that

$$X^* = \arg\max_{\mathbf{X}} B(\mathbf{X}) \tag{3}$$

subject to the following constraints:

$$\sum_{j=0}^{N-1} x_{ij} = 1, \ \forall i \in \{0, 1, \ldots, N-2\}, \tag{4}$$

$$d_{i,j} \doteq \|\mathbf{p}_i - \mathbf{p}_j\|_2 \leq \begin{cases} R_\odot & \text{if } x_{i,N-1} = 1, \\ R_{\maltese} & \text{if } x_{i,j} = 1, \end{cases} \tag{5}$$

$$\exists \langle i, N-1 \rangle, \not\exists \langle i, i \rangle \ \forall i \in \{0, \ldots, N-2\}, \tag{6}$$

$$|\langle i, N-1 \rangle| \leq \alpha, \ \forall i \in \{0, \ldots, N-2\}, \tag{7}$$

$$\sum_{i=0}^{N-2} x_{i,j} \leq \beta, \ \forall j \in \{0, \ldots, N-2\}, \tag{8}$$

where constraint (4) implies that each node in the solution tree should be connected only to one other node; inequality (5) imposes that the distance $d_{i,j}$ between nodes i and j should be less than the associated coverage radius of the utilized radio interface; $\langle i, j \rangle$ in expression (6) denotes the path from node i to node j traced through $\mathbf{X}$; Expression (7) sets the maximum length (hops) α of the path from every node to the BS; and finally the inequality in (8) establishes a maximum number of nodes β that can connect through tethering.

3   Proposed Scheme

The search for the network configuration $\mathbf{X}^*$ that maximizes the net benefit $B(\mathbf{X})$ of its operator by providing direct or opportunistically tethered coverage to all nodes builds upon the use of the Harmony Search (HS) meta-heuristic algorithm. This section describes the application of the HS in our synthetically modeled mobile network deployment, which requires an specific encoding of possible solutions –the dandelion encoding– and the inclusion of tree repairing methods in order to fulfill problem's constraints.

3.1   Dandelion Codes for Tree Encoding

A Cayley code is a bijective mapping between the set of all labeled unrooted trees of n nodes and tuples of $n-2$ node labels. In other words, each tree can be represented by a unique Cayley encoded string and vice-versa. By virtue of their bijective nature, in general the broad family of Cayley codes features several properties that make them specially suitable for efficiently evolving tree structures via crossover and mutation operators of Evolutionary Algorithms (EA): full coverage, zero-bias and perfect-feasibility [8]. However, there are only certain Cayley codes that maintain the unique correspondence between strings and trees needed for an efficient representation of the tree solution space. Among them, a small subset have a high locality. High-locality or transformation Cayley codes include the Blob Code, the Dandelion Code and the Happy Code first described in [9]. Together with five other Dandelion-like codes, such variants complete the set of eight codes that fall into this category. All in all, Dandelion codes have been proven [10] to satisfy the five properties enunciated by [8], hence validating their efficiency to represent trees for evolutionary solvers. For the sake of completeness a brief explanation of the Dandelion decoding and encoding processes will be next given.

The so-called "fast algorithm" contributed by Picciotto is the most intuitive and computationally efficient Dandelion decoding procedure for a tree composed by n nodes and represented by the Dandelion code $C = \{C_2, C_3, \ldots, C_{n-1}\}$. Both linear-time encoding and decoding algorithms have been widely utilized in the literature for all Dandelion-like codes [11]. The decoding procedure produces an output tree $T \in \Gamma_n$, with Γ_n denoting the set of possible trees interconnecting n nodes. To this end the following steps are followed:

- Step 1: a $2 \times n - 2$ matrix $\mathbf{A}_c$ is built by inserting the integer set $\{2, 3, 4, \ldots, n-1\}$ in the first row and the elements of C in the second row. For the exemplifying code $C = \{6, 2, 1, 7, 3, 7, 3, 8\}$, $\mathbf{A}_c$ results in

$$\mathbf{A}_c = \begin{bmatrix} 2\ 3\ 4\ 5\ 6\ 7\ 8\ 9 \\ 6\ 2\ 1\ 7\ 3\ 7\ 3\ 8 \end{bmatrix} \tag{9}$$

- Step 2: define $f_C : [2, n-1] \rightarrow [1, n]$ such that $f_C(i) = C_i$ for each $i \in [2, n-1]$. Note that $f_C(i)$ corresponds to the i-th position C_i of the code.
- Step 3: cycles associated to f_C are computed as $\{Z_1, Z_2, \ldots, Z_L\}$. In the example 2 cycles, namely (2 6 3) and (7), are obtained. Provided that b_l denotes the maximum element in Z_l (with $l \in \{1, \ldots, L\}$), cycles are then reordered such that b_l is set as the rightmost element of Z_l, and that $b_l > b_{l'}$ if $l < l'$. In words, cycles are circularly shifted so that the largest element is the rightmost and sorted so that cycle maxima decreases from left to right. In the example this step yields $\{Z_2, Z_1\} = \{(7), (3\ 2\ 6)\}$.
- Step 4: a list π of the elements in $\{Z_1, Z_2, \ldots, Z_L\}$ is composed in the order they occur in the cycle list, from the first element of Z_1 to the last entry of Z_L, i.e. $\pi = \{(1)(7)(3\ 2\ 6)(10)\}$.
- Step 5: the tree $T \in \Gamma_n$ corresponding to C is constructed by arranging a set of n isolated nodes labeled with the integers from 1 to n. A path from node 1 to node n will be constructed by traversing the list π from left to right. An edge will be included between nodes i and C_i for every $i \in \{2 \ldots, n-1\}$ not occurring in π. The tree corresponding to the Dandelion code $C = (6, 2, 1, 7, 3, 7, 3, 8)$ is the tree given in Figure 1.

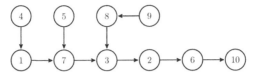

Fig. 1 Tree structure represented by $C = (6, 2, 1, 7, 3, 7, 3, 8)$

Analogously, given a tree $T \in \Gamma_n$ represented by an adjacency list or connectivity matrix to reverse above procedure and find corresponding Dandelion code:

- Step 1: intermediate nodes on the path from 1 to n in T are listed. Following the example of the encoding procedure (Figure 1), this results in 7, 3, 2, 6.
- Step 2: the list is split into cycles by searching for cycle limit elements, i.e. elements larger than all elements to their right. Limit elements for π in the example list are 7 and 6 and thus cycles are (7) and (3 2 6).
- Step 3: the matrix A_C corresponding to tree T is built by generating its first row with elements $2, 3, 4, \ldots, n - 1$ and the second row with the cycle information from the previous step.
- Step 4: Set $C_{i+1} = \text{succ}(i)$ for every $i \in [2, n-1] \notin$ cycles. Then, the Dandelion code C corresponding to tree T is given by the contents of bottom row of A_C. Dandelion code for tree T is $C = (6, 2, 1, 7, 3, 7, 3, 8)$.

3.2 Harmony Search Algorithm

The evolutionary solver utilized to evolve the tree corresponding to **X** will be the Harmony Search algorithm, first introduced in [12] and ever since applied to a plethora of domains and application scenarios modeled by combinatorial and real-valued hard optimization problems [13]. In a similar fashion to other population-based meta-heuristics, HS maintains a set of solution vectors or harmonies, which is referred to as Harmony Memory or HM in the related literature. Such harmonies are iteratively processed by means of operators that emulate the collaborative music composition process of jazz musicians in their attempt at improvising harmonies under a measure of musical quality or aesthetics. Following this jargon, notes played by the musicians represent the values of the optimization variables, whereas their aesthetic quality plays the role of the objective function to be optimized. The improvisation procedure, composed by several combination and randomization operators, is repeated until a stop criterion is met, e.g. a maximum number of iterations is attained.

In the problem at hand each harmony represents a potential tree solution that comprises all deployed mobile nodes and the BS node. At this point it is important to note that the encoding approach taken in this work only guarantees the tree nature of every produced solution code; such trees, once decoded, do not necessarily meet the constraints imposed in Expressions (4) to (8). For this reason, an additional tree repairing stage is added to the original HS flow diagram depicted in Figure 2 prior to every metric/fitness evaluation step. Therefore, the overall flow diagram is composed by the initialization of the HM, followed by the iterative improvisation of a new harmony and the update of the HM depending on the fitness of the newly produced solution until the maximum number of iterations is reached. The improvisation procedure is controlled by 1) the Harmony Memory Considering Rate HMCR ∈ [0, 1], which sets the probability that the new improvised value for a note is selected from the values of the same note in the remaining harmonies within the HM; and 2) the Pitch Adjustment Rate PAR ∈ [0, 1], which correspondingly establishes the probability that the value for a given note is replaced with that of any of its neighboring values. The notion of neighborhood in the problem at hand adopts a strict proximity criterion based on computed distance $d_{i,j}$ between nodes: therefore, any note subject to PAR operation will be replaced with its closest or second closest neighboring nodes with equal probability.

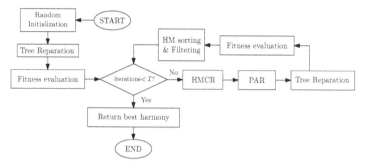

Fig. 2 Flow diagram of the HS application for Dandelion-encoded Traffic Offloading

A third improvisation operator, the Random Selection Rate RSR, dictates the likelihood to pick the new value for a certain note to all the alphabet instead of limiting its scope to the neighborhood of the value to be replaced. This operator, however, is not implemented in our case as an independent stage. The rationale is that our proposed tree repairing procedure allows by itself for an equivalent randomizing behavior whenever a produced tree is partly composed by links between nodes that are not within coverage distance of each other. The tree repairing process will be treated in detail in Subsection 3.3.

The HMCR procedure ensures that good harmonies are considered as elements of new candidate solutions, while the PAR and our repairing procedure allow both neighboring or completely new notes/links to eventually become part of the new harmonies/trees. These three operators work together to provide HS with a good balance between global exploration and local exploitation.

3.3 Tree Repair Procedure

Due to the constrained nature of the problem, all candidate harmonies subject to HS operators must undergo a repairing procedure to ensure the tree structure and feasibility of such solutions prior to their fitness evaluation. While the use of Dandelion

codes to encode candidate solutions in HM guarantees that all codes, both original
or newly generated ones as a result of HMCR and PAR operation, can be successfully reverted to its corresponding tree, there is no certainty whether such trees will
respect the imposed coverage constraints. The tree repair procedure replaces links
between nodes that fall out of coverage with links to nearby nodes, and encompasses
the following steps for each harmony in the HM:

1) Decode harmony to its corresponding connectivity matrix $\mathbf{X}$.
2) Check, $\forall i, j \in \{0, 1, \ldots, N - 1\}$ such that $x_{i,j} = 1$ in matrix $\mathbf{X}$, if its corresponding e_{ij} in coverage matrix $\mathbf{E}$ is 1. If this does not hold, for each occurrence:

   2.1) Delete the link by setting $x_{i,j} = 0$.
   2.2) Find all j' from the coverage matrix $\mathbf{E}$ such that $e_{i,j'} = 1$ (i.e. within
        coverage of i).
   2.3) Randomly pick one within all j' found in 2.2), and incorporate the resulting
        link to the connectivity matrix by setting $x_{i,j'} = 1$. The randomness behind
        the selection of the new link can be conceived as a structural-preserving
        pseudo-RSR operation.

Once the above procedure has finished, the connectivity matrix of the repaired
tree fulfills the coverage restrictions, but may not be suitable for Dandelion encoding.
The reason being that isolated subtrees and/or loops may appear if e.g. no node is
connected to BS. A final structural check needs to be done which, in case of failure,
will result in the fitness of the harmony being strongly penalized so as to force its
elimination in subsequent iterations of the algorithm. Otherwise the repaired harmony
will replace its original in the prevailing HM.

3.4 Fitness Evaluation

According to the problem formulation elaborated in Section 2, the variable to maximize is the net benefit of the network operator. In summary, this net profit is calculated
as the difference between the cumulative revenues that originate from contractual
services to clients and expenses derived from the operational costs associated to 1)
the provision of such services; and 2) the incentives paid to certain users for tethering
other nodes. This incentive should be considered as a fair rewarding mechanism in
regards to the accounting balance of the operator: expenses should be assumed in
the form of contractual discounts to those nodes providing tethering service to other
nodes, in exchange for an increasing satisfaction of their users, less operational costs
due to the more expensive licenses in medium-range (e.g. cellular) communications
and eventually, new client share thanks to the extension of its effective coverage.

The relation between the operational costs of having nodes directly connected to
the BS versus those of having nodes tethered to others and, in the later case, how
are those cost calculated lie at the core of the fitness evaluation for the algorithm.
As such, the fitness of the newly improvised solutions is evaluated in a two-fold
fashion: fixed incentive, by which all nodes that provide tethering will be equally
rewarded independently of the number of nodes being served; and variable incentive,

under which each node providing tethering will be rewarded proportionally to the number of nodes it serves. Both cost models have been modeled and mathematically formulated in Expressions (1) and (2), respectively.

These cost metrics are further modified to reflect the limits in the maximum number of hops to reach BS and the maximum number of nodes that a tethering node can accept (Expressions (7) and (8)). The latter is deemed worse as not only affects the Quality of Service (QoS) of the user, but also increases the energy consumption and ultimately, compromises the autonomy of the device. Both restrictions are implicitly reflected in the way the cost associated to the tethering connection is calculated. First, a multiplicative factor is calculated for each harmony proportional to the number of times either limits α (hops) or β (number of tethered users) are infringed. This multiplicative factor amplifies the tethering cost of the tree under consideration by penalizing its overall fitness. Finally, whenever equal benefit $B(\mathbf{X})$ is drawn for more than one solution, the one with a higher number of direct connections between nodes and the BS will be prioritized.

4 Experimental Setup

Synthetic networks with different node layouts and coverage situations have been generated and simulated to assess the performance of the designed algorithm. To be concise, densely connected networks composed by $N - 1 = 49$ nodes randomly deployed over a 50×50 area are considered, along with a single BS located in the middle of the area. To guarantee high connectivity density, the coverage radii $R_{\odot}$ and $R_{\divideontimes}$ have been dynamically adjusted through the generation of the networks so that they increase until a minimum percentage of reachable nodes is met at all node locations. Figure 3 illustrates the selected networks for experimental purposes. Subfigures 3(a) and 3(b) correspond to a minimum 10% connectivity threshold, while subfigures 3(c) and 3(d) are characterizied by a minimum connectivity per node of 7%. The BS node is in green, blue nodes are within the coverage radio $R_{\odot}$ of the BS, and red nodes have no BS coverage. As for the links, blue and red lines represent possible links to the BS and to nearby tethering nodes, respectively. In essence Figure 3 evinces *all* possible solutions for a given network layout, out of which the proposed algorithm should infer the tree leading to a higher net profit for the network operator.

Experiments aim at evaluating the suitability to maximize the expected revenues of the operator under different cost calculation schemes. To this end, simulation scenarios will be hereafter identified as (θ, μ), with $\theta = C_{\odot}/C_{\divideontimes} \in \{1, 0.5, 2\}$ standing for the ratio between operational cost and tethering incentives, e.g. $\theta = 2$ involves $C_{\odot} = 2 \cdot C_{\divideontimes}$. On the other hand, $\mu \in \{1, 2\}$ denotes whether the net benefit is computed under a fixed ($\mu = 1$) or variable ($\mu = 2$) incentive scheme. To evaluate the statistical performance of the proposed Dandelion-encoded HS algorithm 30 Monte Carlo simulations of 500 iterations each have been completed for the 4 emulated networks. The study is focused on verifying how first and second order statistics (mean, standard deviation and extreme values averaged over the Monte Carlo realizations) of the net benefit evolve along iterations and the influence of $\{\theta, \mu\}$ on the resulting

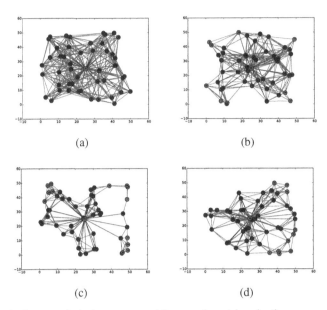

(a)   (b)

(c)   (d)

Fig. 3 Synthetic network deployments used for experimental evaluation

network topology. An upper limit of $\alpha = 3$ hops to reach the BS is set, and $\beta = 5$ users can be served at most by any tethering node.

The discussion begins by analyzing Table 1, where statistics for the maximum, mean and standard deviation of net benefit are listed for the six different simulated scenarios and four network layouts from Figure 3. As can be seen in this table, the proposed algorithm shows a good stability as the standard deviation relative to the mean is always under 1.1%. As expected, despite its fairness for the end user the variable bonus mechanism ($\mu = 2$) penalizes significantly the net revenues of operators. Regarding the effect of θ on the tree topology, the decision to rank equally-profiting solutions according to their number of direct connections to the BS has led to an equivalent behavior in scenarios $(1, 1)$ and $(0.5, 1)$, as depicted in Figure 4(a). Under this policy links to peer nodes prevail disregarding $C_\odot$ because the more the tethered nodes are, the higher the operator's benefit becomes. Fixed cost schemes ($\mu = 1$) result in flat topologies with tendency to group nodes horizontally around as few provisioning nodes as possible. In such situations connections to BS are kept to a minimum to still fulfill imposed constraints on maximum width and depth of the resulting trees, as the plot in Figure 4(c) clearly shows for the $(2, 1)$ case. An opposite behavior is featured by scenarios $(1, 2)$ and $(0.5, 2)$, where the variable bonus leads to direct connection to BS being more profitable. For this reason only those nodes outside the BS coverage resort to opportunistically tethered links (Figure 4(b)). In what relates to the $(2, 2)$ instance in Figure 4(d) similar conclusions hold as for the $(2, 1)$ case. However, since variable incentive is considered trees can grow either horizontally or vertically at the same overall cost.

Table 1 Monte Carlo statistics maximum (iteration)/mean/std of $B(\mathbf{X})$

$\{\theta, \mu\}$	Fig.3(a)	Fig.3(b)	Fig.3(c)	Fig.3(d)
$\{1, 1\}$	90500 (443)/ 89516.667/584.285	90500 (382)/ 89766.667/478.423	89500 (416)/ 88466.667/498.888	90500 (483)/ 89366.667/590.668
$\{1, 2\}$	73500 (22)/ 73500.0/0	73500 (22)/ 73500.0/0	73500 (16)/ 73500.0/0	73500 (18)/ 73500.0/0
$\{0.5, 1\}$	85000 (419)/ 83866.667/773.879	85000 (275)/ 84316.667/569.844	83500 (396)/ 82700.0/725.718	85000 (402)/ 83566.667/882.547
$\{0.5, 2\}$	73500(184)/ 72516.667/712.780	68500 (97)/ 68183.333/353.160	66500 (98)/ 66433.333/169.967	68500 (87)/ 68466.667/124.722
$\{2, 1\}$	88000 (372)/ 86683.333/908.142	89000 (372)/ 87433.333/853.750	85000 (359)/ 84150.0/660.177	87500 (447)/ 86133.333/805.536
$\{2, 2\}$	71500 (332)/ 70533.333/498.887	71500 (293)/ 70833.333/414.997	70000 (372)/ 69483.333/456.131	71000 (384)/ 69966.667/445.970

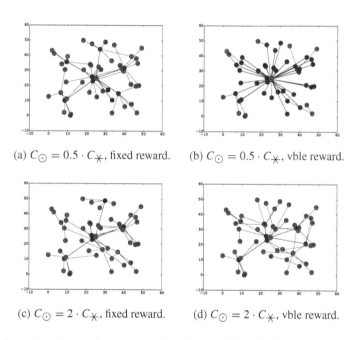

(a) $C_\odot = 0.5 \cdot C_{\not\times}$, fixed reward. (b) $C_\odot = 0.5 \cdot C_{\not\times}$, vble reward.

(c) $C_\odot = 2 \cdot C_{\not\times}$, fixed reward. (d) $C_\odot = 2 \cdot C_{\not\times}$, vble reward.

Fig. 4 Example solutions for the network in Figure 3(b) with different cost relations and incentive schemes

5 Conclusions and Future Research Lines

This paper has presented a novel algorithm inspired from Dandelion-encoded Harmony Search heuristics utilized for finding the traffic offloading network layout that leads to the maximal net benefit for the operator of the company. The algorithm optimally refines the tree structure modeling both direct connections to the base station of the wide-area network and tethered connections opportunistically served 1) to reduce the higher expenditure associated to wide-area links; and 2) to extend the effective coverage area of the operator infrastructure. The problem has been formulated by assuming different incentive mechanisms, and also addresses potential QoS issues derived from multi-hop tethering and bandwidth sharing via additionally imposed constraints. Simulation results over synthetic network deployments have been discussed, from which it is concluded that the proposed algorithm excels at maximizing the net profit of the operator while, at the same time, ensuring the fulfillment of the QoS requirements.

Future research lines will include the extension of this problem to network scenarios with more than one BS, different contractual clauses leading to user-dependent incentive mechanisms, and the consideration of policies among operators that either facilitate or block offloaded data exchanged between customers of different operators.

Acknowledgments This work has been partially funded by the Spanish Ministerio de Economia y Competitividad (MINECO) under grant TEC2013-46766-R (QoEverage).

References

1. Cisco Inc.: Visual Networking Index Global Mobile Data Traffic Forecast Update 2014–2019. White paper (2015)
2. Schumacher, A., Schlienz, J.: WLAN Traffic Offload in LTE. Rohde & Schwarz International, Application Note (2012)
3. Han, B., Hui, P., Kumar, V.A., Marathe, M.V., Pei, G., Srinivasan, A.: Cellular traffic offloading through opportunistic communications: a case study. In: ACM workshop on Challenged Networks (2010)
4. Aijaz, A., Aghvami, H.A., Amani, M.: A Survey on Mobile Data Offloading Technical and Business Perspectives. IEEE Wireless Communications **20**, 104–112 (2013)
5. Bridgewater Systems: Sharing the Load: The value of Subscriber, Service and Policy Control in Mobile Data Traffic Offload. Technical Report (2010)
6. Mota, V.F.S., Macedo, D.F., Ghamri-Doudanez, Y., Nogueira, J.M.S.: Managing the decision-making process for opportunistic mobile data offloading. In: 2014 IEEE Network Operations and Management Symposium, pp. 1–8 (2014)
7. Amani, M., Aijaz, A., Uddin, N., Aghvami, H.: On mobile data offloading policies in heterogeneous wireless networks. In: 2013 IEEE 77th Vehicular Technology Conference, pp. 1–5 (2013)
8. Palmer, C.C., Kershenbaum, A.: Representing trees in genetic algorithms. In: First IEEE Conference on Evolutionary Computation, pp. 379–384 (1994)
9. Picciotto, S.: How to Encode a Tree. Ph.D. Thesis, University of California, San Diego (1999)
10. Paulden, T., Smith, D.K.: Recent advances in the study of the dandelion code, happy code, and blob code spanning tree. In: IEEE International Conference on Evolutionary Computation, pp. 2111–2118 (2006)

11. Caminiti, S., Petreschi, R.: Parallel algorithms for dandelion-like codes. In: Allen, G., Nabrzyski, J., Seidel, E., van Albada, G.D., Dongarra, J., Sloot, P.M.A. (eds.) ICCS 2009, Part I. LNCS, vol. 5544, pp. 611–620. Springer, Heidelberg (2009)
12. Geem, Z.W., Kim, J.H., Loganathan, G.V.: A New Heuristic Optimization Algorithm: Harmony Search. Simulation **76**(2), 60–68 (2001)
13. Manjarres, D., Landa-Torres, I., Gil-Lopez, S., Del Ser, J., Bilbao, M.N., Salcedo-Sanz, S., Geem, Z.W.: A Survey on Applications of the Harmony Search Algorithm. Engineering Applications of Artificial Intelligence **26**(8), 1818–1831 (2013)

A New Parallelization Scheme for Harmony Search Algorithm

Donghwi Jung, Jiho Choi, Young Hwan Choi and Joong Hoon Kim

Abstract During the last two decades, parallel computing has drawn attention as an alternative to lessen computational burden in the engineering domain. Parallel computing has also been adopted for meta-heuristic optimization algorithms which generally require large number of functional evaluations because of their random nature of search. However, traditional parallel approaches, which distribute and perform fitness calculations concurrently on the processing units, are not intended to improve the quality of solution but to shorten CPU computation time. In this study, we propose a new parallelization scheme to improve the effectiveness and efficiency of harmony search. Four harmony searches are simultaneously run on the processors in a work station, sharing search information (e.g., a good solution) at the predefined iteration intervals. The proposed parallel HS is demonstrated through the optimization of an engineering planning problem.

Keywords Parallel computing · Processing unit · Solution quality · Harmony search · Engineering planning problem

1 Introduction

Parallel computing is to divide large computational loadings into smaller ones, distribute them under processing units in a workstation, and perform the computations

D. Jung
Research Center for Disaster Prevention Science and Technology,
Korea University, Seoul 136-713, South Korea
e-mail: donghwiku@gmail.com

J. Choi · Y.H. Choi · J.H. Kim(✉)
School of Civil, Environmental and Architectural Engineering,
Korea University, Seoul 136-713, South Korea
e-mail: y999k@daum.net, {younghwan87,jaykim}@korea.ac.kr

© Springer-Verlag Berlin Heidelberg 2016
J.H. Kim and Z.W. Geem (eds.), *Harmony Search Algorithm,*
Advances in Intelligent Systems and Computing 382,
DOI: 10.1007/978-3-662-47926-1_15

simultaneously ("in parallel"). The reduction in the computation time by parallel computing is proportional to the number of processing units in use. For example, 75% reduction in the computation time can be achieved when all processing units are utilized concurrently in a quardcore workstation and the computational loadings are equally distributed per each unit.

Parallel computing has been adopted to speed up fitness calculations in meta-heuristic algorithms [1]. However, there is no effect on the quality of solutions in the approach. Few studies used parallel implementations of an existing metaheu-ristic algorithm. Artina et al. [2]proposed new parallel structures of non-dominated sorting genetic algorithm-II (NSGA-II) [3] for multiobjective optimal design of water distribution networks. In the so-called coarse-grained parallel GA (PGA), the entire population of GA is divided into multiple subpopulations or islands whichare evolved serially and independently. Occasionally, migration phase oc-curs where some solutions are transferred to other islands. Abu-Lebdeh et al. [4] compared coarse-grained PGA and cellular PGA with respect to their perfor-mances on standard problems and a traffic control problem.

This paper proposes a new parallel structure of harmony search (HS). A HS is evolved serially and independently under each processing unit in a workstation. In migration phase, the best solutions from the HSs are compared to identify overall best solution which is broadcasted to each HS and replaced with the worst solu-tion. The proposed parallel HS (PHS) is demonstrated and compared with the coarse-grained HS without migration throughthe optimization of a multiperiod scenario planning problem of a developing area in the southwest US.

2 Parallel Harmony Search

2.1 Harmony Search

HS [5,6] was inspired by the musical ensemble. To obtain an acceptable harmony from musical instruments, the players meet and practice. At first, perfect harmony is not achieved because the rhythm and pitch of each instrument cannot be imme-diately tuned. However, continued practice to enhance the harmony enables the players to memorize the specific rhythm and pitch of each instrument, which lead to "good harmony". These sets of "good harmony" are memorized and the unac-ceptable sets are discarded as superior sets are found. The process of updating the sets of harmony continues until the best harmony is obtained. HS implements the harmony enhancement process and the sets of "good harmony" are saved to a solution space termed harmony memory (HM), which is a unique feature of HS compared to other evolutionary optimization algorithms.

HS contains a solution storage function called HM that necessitates the defini-tion of two parameters: HM considering rate (HMCR) and pitch-adjusting rate (PAR). For more details on HS, please refer to [5,6].

2.2 Parallel Harmony Search

Similar to PGA, HSs are evolved serially and independently in the proposed PHS (Fig. 1). Each HS has its own HM which is updated based the HMCR and PAR rules considered in standard HS. If new solution is better than the worst solution in HM, the latter is replaced by the former. PHS with migration has a migration phase where the best solution in each HS is delivered to a comparison island (blue box in Fig. 1) and compared to identify overall best solution. The migration phase also includes broadcasting the overall best solution to each HS except HS from which it is found and replacing the worst solution with the overall best in other HSs. If the migration phase is over, serial and independent searches are performed until the next migration. This parallel scheme of HS is intended to promptly find a feasible solution, which is generally required for the optimization of large dimensional engineering problems. For example, a near-optimal pump scheduling could be acceptable for real-time pump operation in a water distribution system (WDS) [7-11].

Various migration schemes are available in PHS. For example, unrestricted migration can be adopted where solution migration occurs between two HSs randomly matched while ring migration occurs to a neighboring HS in one direction (e.g., HS_1 to HS_2, HS_2 to HS_3). The type of migration (hierarchical or horizontal), the solution to be migrated (the best or any solution), and migration size(a single or multiple solution) affect the PHS's performance. Comparing all potential combination of the approaches and identifying the most efficient one in PHS are an interesting research topic.

In addition, the migration frequency also affects the search efficiency of PHS. The best frequency varies depending on problems. Generally speaking, frequent migration is required when fitness differences among HSs are significant.

In summary, the proposed PHS runs each HS under each processing unit in a workstation. The evolutions of the HSs are carried out independently and concurrently until the migration phase. In the migration phase, the overall best solution is

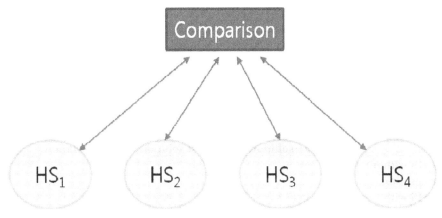

Fig. 1 Migration of PGA

identified and broadcasted to each HS to replace with the worst solution. In this study, PHS with the hierarchical migration is demonstrated and compared with PHS without migration phase in the optimization of a WDS planning problem. Therefore, the PHS without migration phase is similar to run each HS under different workstation without communication until termination.

3   Study Network

The proposed PHS is demonstrated through the planning of a developing area in the southwest US. A staged construction problem was formulated to find the most cost effective water and wastewater infrastructure design and expansion over planning periods. Total number of links is 333 and potential pump and satellite wastewater plant locations are 36 and 6, respectively. Commercial pipe sizes are from 152 mm to 1829 mm and no pipe option is allowed in the decision of pipe size. The design flow and head and the number of pump in operation under peak and average demand conditions should be determined for a pump station. Plant capacity is decided to a potential location of satellite wastewater plant. The decisions are made at three stages: 2010-2020, 2020-2030, and 2030-2050. For example, wastewater plant capacity determined at the first stage can be expanded at the later two stages (i.e., 2020 and 2030). Total number of decision variables is 1449 (=3*(333+4*36+6)).

4   Application Results

The two PHS schemes are compared with respect to three performance measures: mean, worst, and best solutions. The measures were obtained from three independent optimizations of the algorithms. In this study, the same HS parameters were used for each HS in PHS. It was confirmed that PHS with migration outperformed PHS without migration with respect to all the performance measures. The worst solution found by PHS with migration was similar to the best solution found by PHS without migration (Table 1). Search information was not transferred among HSs in the PHS without migration. PHS with migration identified and broadcasted the overall best solution every 100,000 functional evaluations (NFEs).Maximum NFEs allowed for each algorithm was 1,000,000 and thus totally 10 migrations were occurred until termination. PHS with migration found the best solution of 7.3% less cost than PHS without migration.

Table 1 Performance metrics

	Mean solution	Worst solution	Best solution
PHS without migration	1.714	1.734	1.690
PHS with migra-tion	1.649 (3.8%)	1.692 (2.5%)	1.567 (7.3%)

Fig. 2. indicates the average evolution of solution costs by the two algorithms. On average, PHS without migration reached to a feasible solution earlier than PHS with migration. However, after a feasible solution was found, the rate of improvement in the fitness (cost)was higher in the latter than the former. For example, a feasible solution was first found at the NFEs of 180,000 in PHS without migration while the iteration 288,000 was required for PHS with migration. However, total cost of the best solution was decreased as a rate of 1591 USD per NFE in the later algorithm while the rate was 1399 USD in the former algorithm.

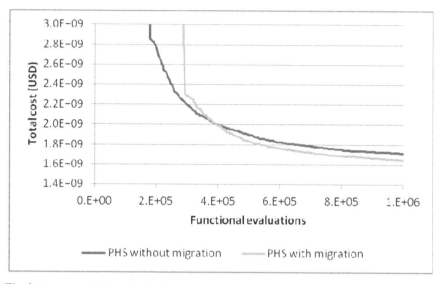

Fig. 2 Average evolution of solution costs

5   Conclusions

This study proposed a PHS to improve the solution quality of optimization. In the proposed PHS, individual HSs are evolved concurrently and independently under processors in a workstation. Migration phase occasionally occurs where the best solution in each HS is sent to comparison island to identify an overall best solution. Then, the overall solution is broadcasted to each HS and replaced with the worst solution in HS. The independent search continues until the next migration phase. The proposed PHS is demonstrated and compared with PHS without migration in the planning of water and wastewater infrastructures in a developing area in the southwest US.

Two algorithms were compared with respect to mean, worst, and best solution obtained from three independent optimizations. The results showed that PHS with migration outperformed PHS without migration with respect to all performance measures. Although the NFEs required for reaching feasible region was larger in PHS with migration than PHS without migration, the search ability of the former within the feasible region was much better than the latter, finally yielding a better final solution. For example, the best solution of the former has 7.3% less cost than that of the latter.

This study has several limitations that future research must address. First, sensitivity analysis on the migration frequency should be conducted to identify the most efficient frequency of migration. Second, various migration structures should be tested. For example, unrestricted and ring migration would improve exploration.

Acknowledgement This work was supported by the National Research Foundation (NRF) of Korea under a grant funded by the Korean government (MSIP) (NRF-2013R1A2A1A01013886) and the Korea Ministry of Environment as "The Eco-Innovation project (GT-11-G-02-001-2)"

References

1. Elfeky, E.Z., Sarker, R., Essam, D.L.: Partial decomposition and parallel GA (PD-PGA) for constrained optimization. In: IEEE International Conference on Systems, Man and Cybernetics, SMC 2008, pp. 220–227. IEEE (2008)
2. Artina, S., Bragalli, C., Erbacci, G., Marchi, A., Rivi, M.: Contribution of parallel NSGA-II in optimal design of water distribution networks. Journal of Hydroinformatics **14**(2), 310–323 (2012)
3. Deb, K., Pratap, A., Agarwal, S., Meyarivan, T.A.M.T.: A fast and elitist multiobjective genetic algorithm: NSGA-II. IEEE Transactions on Evolutionary Computation **6**(2), 182–197 (2002)
4. Abu-Lebdeh, G., Chen, H., Ghanim, M.: Improving Performance of Genetic Algorithms for Transportation Systems: Case of Parallel Genetic Algorithms. Journal of Infrastructure Systems, A4014002 (2014)
5. Geem, Z.W., Kim, J.H., Loganathan, G.V.: A new heuristic optimization algorithm: harmony search. Simulation **76**(2), 60–68 (2001)
6. Kim, J.H., Geem, Z.W., Kim, E.S.: Parameter Estimation of the Nonlinear Muskingum Model Using Harmony Search. Journal of the American Water Resources Association **37**(5), 1131–1138 (2001)
7. Jamieson, D., Shamir, U., Martinez, F., Franchini, M.: Conceptual design of a generic, real-time, near-optimal control system for water-distribution networks. Journal of Hydroinformatics **9**(1), 3–14 (2007)
8. Rao, Z., Salomons, E.: Development of a real-time, near-optimal control process for water-distribution networks. Journal of Hydroinformatics **9**(1), 25–37 (2007)
9. Pasha, M.F.K., Lansey, K.: Optimal pump scheduling by linear programming. In: Proceedings of World Environmental and Water Resources Congress American Society of Civil Engineers, Kansas City, MO, USA, pp. 395–404 (2009)
10. Pasha, M.F.K., Lansey, K.: Strategies for real time pump operation for water distribution systems. Water Distribution Systems Analysis Conference (2010)
11. Jung, D., Kang, D., Kang, M., Kim, B.: Real-time pump scheduling for water transmission systems: Case study. KSCE Journal of Civil Engineering, 1–7 (2014)

Part III
Recent Variants of HSA

Mine Blast Harmony Search and Its Applications

Ali Sadollah, Ho Min Lee, Do Guen Yoo and Joong Hoon Kim

Abstract A hybrid optimization method that combines the power of the harmony search (HS) algorithm with the mine blast algorithm (MBA) is presented in this study. The resulting mine blast harmony search (MBHS) utilizes the MBA for exploration and the HS for exploitation. The HS is inspired by the improvisation process of musicians, while the MBA is derived based on explosion of landmines. The HS used in the proposed hybrid method is an improved version, introducing a new concept for the harmony memory (HM) (i.e., dynamic HM), while the MBA is modified in terms of its mathematical formulation. Several benchmarks with many design variables are used to validate the MBHS, and the optimization results are compared with other algorithms. The obtained optimization results show that the proposed hybrid algorithm provides better exploitation ability (particularly in final iterations) and enjoys fast convergence to the optimum solution.

Keywords Harmony search · Mine blast algorithm · Hybrid metaheuristic methods · Global optimization · Large-scale problems

1 Introduction

Among optimization methods, metaheuristic algorithms have shown their potential for detecting near-optimal solutions when exact methods may fail, especially

A. Sadollah · D.G. Yoo
Research Center for Disaster Prevention Science and Technology,
Korea University, Seoul 136-713, South Korea
e-mail: ali_sadollah@yahoo.com, godqhr425@korea.ac.kr

H.M. Lee · J.H. Kim(✉)
School of Civil, Environmental, and Architectural Engineering,
Korea University, Seoul 136-713, South Korea
e-mail: {dlgh86,jaykim}@korea.ac.kr

© Springer-Verlag Berlin Heidelberg 2016   155
J.H. Kim and Z.W. Geem (eds.), *Harmony Search Algorithm*,
Advances in Intelligent Systems and Computing 382,
DOI: 10.1007/978-3-662-47926-1_16

when the global minimum is surrounded by many local minima. Hence, the need to use such approaches is understood by the optimization community.

Harmony search (HS) algorithm, developed by Geem et al. [1-3], is derived from the concepts of musical improvisations and harmony knowledge, and is a well-known metaheuristic algorithm. To date, the HS has proved its advantages over other optimization methods [4-6], and many improved versions have been developed in the literature [7-10].

In recent years, it has become clear that concentrating on a sole optimization method may be rather restrictive. A skilled combination of concepts from different optimizers can provide more efficient results and higher flexibility when dealing with large-scale problems. Thus, a number of hybrid metaheuristic algorithms have been proposed.

There are many hybrid optimization methods that employ the concept of the HS [11-14]. For instance, Kaveh and Talatahari [11] developed a hybrid optimization method for the optimum design of truss structures. Their proposed algorithm was based on a particle swarm optimization (PSO) with passive congregation (PSOPC), ant colony optimization, and the HS scheme.

Geem [12] proposed a hybrid HS incorporating the PSO concept. Known as particle swarm harmony search (PSHS), this algorithm was applied to the design of water distribution networks.

The mine blast algorithm (MBA) was developed to solve discrete and continuous optimization problems [15, 16]. The concept of the MBA was inspired by the process of exploding landmines. The results obtained by the MBA demonstrate its superiority in finding near-optimum solutions in early iterations and its fast mature convergence rate [16].

However, the exploitation (local search) ability of the MBA is not good as its exploration phase. Also, it suffers from a serious problem, that is, the MBA is almost memory-less optimizer. Though, the HS has many obvious advantages, it can be trapped in performing local search for solving optimization problems [8]. Moreover, its optimization performance is quite sensitive to its key control parameters.

Therefore, how to effectively fine-tune the key control parameters (i.e., HMS, HMCR, PAR, and bw) in the process of improvisation is a key research focus in the HS. In addition, its search precision and convergence speed are also an issue in some cases. Indeed, a reasonable balance between exploration and exploitation are beneficial to the performance of an algorithm [17].

Since, many modified and hybrid HS still cannot escape local minimum and adjust algorithm parameters effectively, so the relationship between the search mechanism of HS and the parameters is a very significant area for future research [18]. That deserves a lot more attention and this paper is thus motivated to focus on this research. Therefore, we propose the mine blast harmony search (MBHS), which embeds the HS into MBA to improve the exploitation phase in the MBA and exploration phase in the HS.

2 Mine Blast Harmony Search

The following sections provide detailed descriptions of the HS and its variants, MBA, and MBHS. The MBA and HS used in the MBHS are slightly improved.

2.1 Harmony Search Algorithm

Since the HS was first developed and reported in 2001 [1], it has been applied to various research areas and obtained considerable attention in different fields of research [6]. The HS intellectualizes the musical process of searching for a perfect state of harmony.

As musical performances search a fantastic harmony determined by aesthetic estimation, hence the optimization technique seeks a best state (global optimum) measured by an objective function value.

Further details of the HS can be found in the work of Geem et al. [1]. The main steps of the HS algorithm are summarized as below:

Step 1: Generate random vectors $(x_1, x_2, ..., x_{HMS})$ up to the harmony memory size (HMS) and store them in the harmony memory (HM) matrix:

$$HM = \begin{bmatrix} x_1^1 & \cdots & x_n^1 & f(x^1) \\ \vdots & \ddots & \vdots & \vdots \\ x_1^{HMS} & \cdots & x_n^{HMS} & f(x^{HMS}) \end{bmatrix}. \tag{1}$$

Step 2: Generate new harmony. For each component:
• With probability HMCR (harmony memory considering rate;
$0 \leq$ HMCR ≤ 1), pick a stored value from the HM: $x_i' \leftarrow x_i^{\text{int}(u(0,1) \times HMS)+1}$.
• With probability (1-HMCR), pick a random value within the allowed range.
Step 3: If the value in Step 2 came from the HM:
· With probability PAR (pitch adjusting rate; $0 \leq$ PAR ≤ 1) change x_i':

$$x_i' \leftarrow x_i' + bw \times (2rand - 1),$$

where rand is a uniformly distributed random number between zero and one and bw is the maximum change in pitch adjustment.
• With probability (1-PAR), do nothing.
Step 4: Select the best harmonies up to the HMS and consider them as the new HM matrix.
Step 5: Repeat Steps 2 to 5 until the termination criterion (e.g., maximum number of function evaluations) is satisfied.

To mention a few examples of improved versions of HS, Mahdavi et al. [8] proposed an improved HS (IHS) in which bw and PAR are not fixed values. During the optimization process, values of bw and PAR decrease and increase, respectively. This approach helps the exploitation phase of the IHS in the final iterations.

Afterwards, Geem and Sim [9] developed another improved variant, called parameter-setting-free HS (PSF-HS). In their improved method, the values of user parameters HMCR and PAR vary during the optimization process.

2.2 Mine Blast Algorithm

The MBA is inspired by the process of landmines explosion; shrapnel pieces are thrown away and collided with other landmines in the vicinity of the explosion area causing further explosions. Consider a landmine field where the goal is to clear all landmines. To clear all the landmines, the position of the most explosive mine must be determined.

This position corresponds to the optimal solution and its casualties considers as cost function [15]. Indeed, the MBA is developed to find the most explosive mine (i.e., the mine with the most casualties), and the aim is to reduce the casualties caused by the explosion of mines.

Similar to other population-based methods, the MBA requires an initial population of individuals. This population is generated by a first shot explosion. The population size is the number of shrapnel pieces (N_s) caused by an explosion. To begin, the MBA uses the lower and upper bound values (i.e., LB and UB) specified for a given problem for generating a random first shot solution (point).

At initialization step, we assume that the first shot point (X_0) is the best solution ($X_{Best} = X_0$) so far. The MBA starts with the exploration phase, which is responsible for comprehensively exploring the search space. Exploration (global search) and exploitation (local search) are the two critical steps for metaheuristic algorithms. The difference between the exploration and exploitation phases is how they affect the whole search process in finding the optimal solution.

To explore the search space from both small and large distances, an exploration factor, μ, is introduced [15]. This parameter, used in early iterations of MBA, is compared to an iteration number index (t). Explosion of a shrapnel piece triggers another landmine explosion at location $\vec{X}_t^e$. Hence, updating equations for the exploitation and exploration phases in the MBA are given in Equations (2) and (3), respectively [16]:

$$\vec{X}_t^e = \vec{X}_{Best} + \vec{d}_{t-1} \times ra\bar{n}dn^2 \times \cos\theta \qquad t \leq \mu, \qquad (2)$$

$$\vec{X}_t^e = \vec{X}_{best\_t}^e + \exp(-\sqrt{\frac{M_t}{D_t}})\vec{X}_{Best} \qquad \mu < t, \qquad (3)$$

Where $\vec{X}_{best\_t}^e$ in Equation (3) is the best exploded landmine at iteration t given as follows:

$$\vec{X}_{best\_t}^e = \vec{X}_{Best} + \vec{d}_{t-1} \times ra\bar{n}dn \times \cos(\theta) \qquad \mu < t. \qquad (4)$$

randn is normally distributed random number and d_{t-1} is distance of each shrapnel piece. The Euclidean distance (D_t) and direction (M_t) between the current and previous best landmines (X_{Best} and X_{Best-1}) in m dimensions are given by:

$$D_t = \left[\sum_{i=1}^{m} (X_i^{Best} - X_i^{Best-1})^2 \right]^{1/2}, \quad \mu < t \, , \tag{5}$$

$$M_t = \frac{F_{Best}^t - F_{Best-1}^t}{D_t}, \quad \mu < t \cdot \tag{6}$$

When the Euclidean distance in Equation (5) between the current and previous best solutions is near zero (at final iterations), the exponential term in Equation (3) is equal to zero. The shrapnel angle of incidence, denoted by θ in Equations (2) and (4), is given by:

$$\theta = k \times \Delta \quad k = 0, 1, 2, ..., N_s - 1, \tag{7}$$

where $\Delta = 360/N_s$. The value of θ ranges from 0 to 360; the resulting value of $\cos(\theta)$ ranges between -1 and 1, which generates solutions having harmonic orders. To improve MBA's global and local search abilities, the initial distance of shrapnel pieces (d_0=UB-LB) is gradually reduced at each iteration to quickly detect near location of the most explosive mine as follows:

$$\vec{d}_t = \frac{\vec{d}_{t-1}}{e^{(t/\alpha)}} \quad t = 1, 2, 3, ..., Max\_It \, , \tag{8}$$

where *Max_It* is maximum number of iteration and α is the reduction factor, the only sensitive user defined parameter of MBA, which depends on the complexity of the optimization problem. At the end of the optimization process, shrapnel distances are close to zero.

Indeed, the MBA starts with initial standard deviation named as initial distance of shrapnel pieces (d_0). By iteration continues, the MBA adaptively reduces the standard deviation in order to increase the exploitation and convergence effects.

Finally, steps of MBA are as follows:

Step 1: Choose initial parameters α, N_s (N_{pop}), and *Max_It*.
Step 2: Check the condition of the exploration factor (μ).
Step 3: If the condition of the exploration factor is satisfied, calculate the location of the exploded mine using Equation (2). Then, go to Step 8. Otherwise, continue to Step 4.
Step 4: Calculate the location of exploded landmine in the exploitation phase using Equation (4).
Step 5: Does the shrapnel piece have a lower function value than the best temporal solution? If true, archive it.
Step 6: Calculate the Euclidian distance and direction between current and previous best solutions using Equations (5) and (6).

Step 7: Calculate improved location of exploded landmine in the exploitation phase using Equation (3).

Step 8: Does the shrapnel piece have a lower function value than the best temporal solution? If true, archive it.

Step 9: Update the X_{Best} (Best=Archive).

Step 10: Reduce the distance of shrapnel pieces adaptively using Equation (8).

Step 11: Check the stopping condition. If the stopping criterion is satisfied, the MBA stops. Otherwise, return to Step 2.

2.2.1 Setting Initial Parameters of MBA

Poor choices of algorithm parameters may result in a low convergence rate and undesired solutions. The following guidelines are suggested to fine tune the user-defined parameters.

The reduction factor (α) depends on the complexity of the problem, maximum number of iteration, and problem bounds. The value of α should be chosen so that at the final iteration, the distance of shrapnel pieces is approximately zero.

It is worth mentioning that being close to zero varies from one problem to another (depends on desire accuracy and tolerance). The following formula computes a suggested value for α used in the MBA given as follows:

$$\alpha_{Suggested} = \frac{M^2 + M}{2} \times \frac{1}{\ln(\vec{d}_0 / Tol)}, \qquad (9)$$

where *Tol* is tolerance, a small value close to zero and M is maximum number of iteration. The exploration factor (μ) defines the number of iterations for the exploration phase. Increasing μ may result in getting trapped in a local minimum. For the MBA, we recommend μ be equal to the maximum number of iterations divided by five.

2.3 Mine Blast Harmony Search

Performance of HS is good at local search compared with its global search, and its convergence performance may also be an issue in some cases [18]. To overcome these drawbacks, combining the concepts and formulations of the MBA with the HS can improve the exploration and exploitation performances of both algorithms. The exploitation phase in the MBA is not as efficient as the exploration phase. Therefore, embedding the HS into the MBA can be considered to improve the exploitation phase in the MBA and exploration phase in the HS.

Since the MBA is a memory-less algorithm, almost no information is extracted dynamically during the search, whereas the HS uses memory to store information extracted during the search process (i.e., harmony memory matrix, Equation (1)).

The proposed hybrid MBHS involves two phases: (*i*) exploration phase using the strategy in the MBA and (*ii*) exploitation phase using the concepts of the HS, whereby memory consideration and pitch adjustment are employed along with the MBA operators.

For the MBHS, the updating exploitation equation in the MBA (Equation (3)) for avoiding problems with the dimension of the search space (m) is modified. Indeed, the perception of direction is replaced by moving to the best solutions in the MBHS. Hence, the new updating equations used in the MBHS are given as follows:

$$\vec{X}_t^e = \vec{X}_{Best} + \vec{d}_{t-1} \times ra\vec{n}dn \times \cos(\theta) \qquad \mu < t \,, \tag{10}$$

$$\vec{X}_t^e = \vec{X}_t^e + \exp(-\sqrt{\frac{1}{D_t}}) \times r\vec{a}nd \times \left\{ \vec{X}_{Best} - \vec{X}_{Best-1} \right\}, \qquad \mu < t \,' \tag{11}$$

In addition, the HS used in the MBHS is not the standard HS. The HS utilized in the MBHS has borrowed some features of IHS [8] and PSF-HS [9] for adaptively reducing and increasing the user parameters of HS. In this research, we also define the new concept for HM having variables size, so called dynamic harmony memory (DHM). Indeed, the HMS is not fixed parameter in the MBHS.

Increasing the value of HMS causes more exploration in the search space, and sometimes causes the optimization results to diverge. In the current hybrid MBHS, the value of HMS is changed at early and final iterations and it is fixed in between, as shown in Fig. 1.

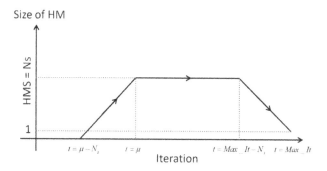

Fig. 1 Size of DHM during optimization process

For the sake of reducing the user parameters in the MBHS, value of HMS is considered to be the population size (N_s). By decreasing the value of HMS in the final iterations, further exploitation close to the current best solution can be achieved. In general, there is only one user parameter in the MBHS, the reduction factor (α) as for used in the MBA.

In addition, we assume that the bandwidth (bw) user parameter in the HS acts similarly to the distance of shrapnel pieces (d_{t-1}). Therefore, the bw has been merged with d_{t-1}, and adaptively reduces at each iteration as follows:

$$b\vec{w}_t = \frac{b\vec{w}_{t-1}}{e^{(t/\alpha)}} \qquad t = 1, 2, 3, ..., Max\_It. \qquad (12)$$

Initial values of HMCR and PAR are automatically tuned in the optimization process as given in Equations (13) and (14). The values of HMCR and PAR in HS phase are changed right after the exploration phase. The following equations describe the variation of these user parameters:

$$HMCR(t) = \begin{cases} \dfrac{t}{Max\_It} & \mu < t \\ 0 & t \le \mu \end{cases}, \qquad (13)$$

$$PAR(t) = \begin{cases} 1 - \dfrac{t}{Max\_It} & \mu < t \\ 0 & t \le \mu \end{cases}. \qquad (14)$$

In this research, we assume that values of HMCR and PAR linearly increase and decrease, respectively, at each iteration. Therefore, there is no need to tune these parameters during the optimization process. The (probability) value of HMCR goes from zero to 0.99, and from one to near zero for the PAR parameter. The reason to choose maximum value of 0.99 for the HMCR is for having one percent chance to generate random solutions at final iteration.

By progressing the optimization in the MBHS, the exploration approach decreases in importance and the exploitation phase becomes dominant ($\mu < t$). Indeed, in the final iterations, the MBHS executes only a local search near to the best current solution.

3 Numerical Optimization Results

MATLAB was used to code and implemented the algorithms. To ensure statistically significant results, 50 independent optimization runs were carried out for each test problem in this paper.

3.1 Benchmark Optimization Problems

The MBHS has been tested on eleven unconstrained benchmark functions. In order to observe the effects of proposed MBHS and having fair discussion, the MBA and HS also have been implemented for considered benchmarks. The dimensions of benchmark functions were 200 and 500. Properties of these functions are represented in Table 1.

Table 1 Properties of F1 to F11

Function	Range	Optimum ($f(x^*)$)
F_1 (Hyper Sphere)	$[-100,100]^m$	0
F_2 (Schwefel 2.21)	$[-100,100]^m$	0
F_3 (Rosenbrock)	$[-100,100]^m$	0
F_4 (Rastrigin)	$[-5,5]^m$	0
F_5 (Griewank)	$[-600,600]^m$	0
F_6 (Ackley)	$[-32,32]^m$	0
F_7 (Schwefel 2.22)	$[-10,10]^m$	0
F_8 (Schwefel 1.2)	$[-65.536,65.536]^m$	0
F_9 (Bohachevsky)	$[-15,15]^m$	0
F_{10} (Schaffer)	$[-100,100]^m$	0
F_{11} (Extended f_{10})	$[-100,100]^m$	0

Talking about maximum number of function evaluations (NFEs), considered as stopping condition in this paper, the predefined NFEs is 5000 multiple by dimension size for each function.

User parameters of MBA and MBHS were set to the recommended values for μ and α given in Section 2.2.1 and population size of 50 (*Tol.* = 1.00e-14). Accordingly, the user parameters of the HS for the considered benchmarks were: a harmony memory size of 50, and HMCR, PAR, and bw values of 0.98, 0.1, and 0.01, respectively, as suggested by [1].

The obtained statistical results (i.e., error values: $f(x) - f(x^*)$) for dimensions 200, and 500 are represented in Tables 2 and 3, respectively. The best obtained result (error) at the end of each optimization process is recorded during each run. The best, average, and worst errors and standard deviation (SD) are shown in Tables 2 and 3. In Tables 2 and 3, 0.00e+00 means 1.00e-324 (defined accuracy for zero in MATLAB).

By observing Tables 2 and 3, the MBHS considerably has reduced the error compared with its original optimizers (i.e., MBA and HS). From the obtained optimization results especially for large-scale problems, we can infer that the combination of HS with the MBA leads us to develop a hybrid optimization method having better performance and efficiency.

Furthermore, Table 4 summarizes the average error values of MBHS, MBA, and HS and compares those findings with the results using other optimizers. The PSO [19], imperialist competitive algorithm (ICA) [20], and gravitational search algorithm (GSA) [21] have been coded and implemented in this paper for comparison purposes. In this study, all error values below 1.00e-14 assume to be 0.00e+00 in Table 4. Looking at Table 4, the MBHS shows its superiority not only against the HS and MBA, also represented competitive results compared with other optimizers.

Table 2 Statistical optimization results for $m = 200$ for the MBA, HS, and MBHS

Function	Method	Best	Average	Worst	SD
	HS	1.22e+03	1.29e+03	1.44e+03	7.30e+01
F_1	MBA	7.96e-13	8.75e-13	9.66e-13	6.48e-14
	MBHS	5.68e-14	5.85e-14	5.91e-14	1.21e-14
	HS	2.38e+01	2.46e+01	2.63e+01	7.21e-01
F_2	MBA	2.25e+01	4.67e+01	8.55e+01	2.47e+01
	MBHS	6.82e-13	9.89e-13	1.71e-12	4.30e-13
	HS	4.44e+06	6.55e+06	8.02e+06	1.10e+06
F_3	MBA	2.37e+02	1.25e+03	3.87e+03	1.52e+03
	MBHS	1.98e+02	1.98e+02	1.98e+02	2.87e-02
	HS	8.30e+01	9.77e+01	1.07e+02	7.32e+00
F_4	MBA	6.40e+02	9.83e+02	1.51e+03	3.60e+02
	MBHS	0.00e+00	4.54e-14	5.68e-14	2.54e-14
	HS	1.09e+01	1.28e+01	1.50e+01	1.47e+00
F_5	MBA	9.94e-13	1.97e-03	9.86e-03	4.41e-03
	MBHS	0.00e+00	1.71e-14	2.84e-14	1.55e-14
	HS	4.34e+00	4.54e+00	4.74e+00	1.46e-01
F_6	MBA	4.18e+00	1.68e+01	2.00e+01	7.06e+00
	MBHS	1.99e-13	2.33e-13	2.56e-13	2.38e-14
	HS	2.43e+01	2.56e+01	2.69e+01	7.13e-01
F_7	MBA	4.84e+00	4.53e+02	8.00e+02	4.10e+02
	MBHS	3.36e-13	4.07e-13	4.70e-13	4.79e-14
	HS	3.45e+04	3.62e+04	3.77e+04	1.15e+03
F_8	MBA	3.30e+00	1.24e+01	2.46e+01	1.10e+01
	MBHS	1.17e-26	1.29e-26	1.40e-26	1.58e-27
	HS	1.91e+02	2.06e+02	2.25e+02	1.11e+01
F_9	MBA	7.96e+01	8.73e+01	9.55e+01	7.19e+00
	MBHS	0.00e+00	0.00e+00	0.00e+00	0.00e+00
	HS	3.87e+02	4.36e+02	4.61e+02	2.98e+01
F_{10}	MBA	1.12e+03	1.42e+03	1.71e+03	2.88e+02
	MBHS	1.09e-05	1.18e-05	1.24e-05	6.01e-07
	HS	4.09e+02	4.41e+02	4.70e+02	2.86e+01
F_{11}	MBA	1.11e+03	1.65e+03	1.88e+03	3.20e+02
	MBHS	1.06e-05	1.22e-05	1.37e-05	1.10e-06

Table 3 Statistical optimization results for m=500 using the HS, MBA, and MBHS

Function	Method	Best	Average	Worst	SD
	HS	4.89e+04	5.38e+04	5.76e+04	2.65e+03
F_1	MBA	1.99e-12	2.21e-12	2.39e-12	1.63e-13
	MBHS	5.11e-14	5.23e-14	5.68e-14	2.32e-14
	HS	5.72e+01	5.79e+01	5.85e+01	4.23e-01
F_2	MBA	1.17e+01	4.17e+01	8.59e+01	3.37e+01
	MBHS	6.82e-13	1.39e-12	2.39e-12	6.23e-13
	HS	6.67e+09	7.39e+09	7.91e+09	3.96e+08
F_3	MBA	4.91e+02	7.43e+02	1.36e+03	3.51e+02
	MBHS	4.97e+02	4.97e+02	4.97e+02	4.21e-02
	HS	8.30e+02	8.58e+02	9.13e+02	2.51e+01
F_4	MBA	2.24e-02	2.47e-02	2.95e-02	2.85e-03
	MBHS	5.22e-014	5.45e-14	5.94e-14	4.23e-14
	HS	4.42e+02	4.88e+02	5.42e+02	2.70e+01
F_5	MBA	2.90e-12	1.48e-03	7.40e-03	3.31e-03
	MBHS	2.84e-14	2.99e-14	3.12e-14	3.53e-14
	HS	1.09e+01	1.12e+01	1.14e+01	1.52e-01
F_6	MBA	1.52e+01	1.68e+01	2.00e+01	1.89e+00
	MBHS	2.84e-13	2.91e-13	2.95e-13	2.84e-13
	HS	1.92e+02	2.02e+02	2.13e+02	7.14e+00
F_7	MBA	1.20e+96	1.54e+96	1.88e+96	2.32e+95
	MBHS	4.12e-13	4.48e-13	4.79e-13	2.95e-14
	HS	4.55e+06	4.89e+06	5.11e+06	2.31e+03
F_8	MBA	6.65e+02	9.25e+02	1.11e+03	2.33e+02
	MBHS	1.62e-26	1.83e-26	2.03e-26	2.95e-27
	HS	3.83e+03	3.98e+03	4.26e+03	1.22e+02
F_9	MBA	2.08e+02	2.24e+02	2.35e+02	1.13e+01
	MBHS	0.00e+00	0.00e+00	0.00e+00	0.00e+00
	HS	1.65e+03	1.68e+03	1.71e+03	2.30e+01
F_{10}	MBA	2.46e+02	3.29e+03	5.32e+03	1.96e+03
	MBHS	1.83e-05	1.88e-05	1.94e-05	5.13e-07
	HS	1.71e+03	1.92e+03	2.10e+03	2.67e+02
F_{11}	MBA	2.96e+03	4.09e+03	4.70e+03	6.65e+02
	MBHS	1.85e-05	1.89e-05	1.98e-05	5.32e-07

Table 4 Comparison of average error values for different optimizers for F_1 to F_{11}

Function	m	PSO	ICA	GSA	HS	MBA	MBHS
F_1	200	1.10e+04	2.20e+04	1.06e-12	1.29e+03	8.75e-13	5.85e-14
	500	4.51e+09	4.18e+05	8.25e-12	5.38e+04	2.21e-12	5.23e-14
F_2	200	2.22e+01	8.97e+01	8.08e+00	2.46e+01	4.67e+01	9.89e-13
	500	2.59e+01	9.64e+01	1.1.7e+01	5.79e+01	4.17e+01	1.39e-12
F_3	200	1.07e+08	6.10e+08	1.83e+02	6.55e+06	1.25e+03	1.98e+02
	500	7.11e+08	9.96e+10	9.75e+02	7.39e+09	7.43e+02	4.97e+02
F_4	200	1.12e+03	1.62e+03	1.17e+02	9.77e+01	8.39e-03	4.54e-14
	500	3.73e+03	5.28e+03	3.62e+02	8.58e+02	2.47e-02	5.45e-14
F_5	200	9.55e+01	1.23e+02	8.00e-01	1.28e+00	1.97e-03	1.71e-14
	500	4.12e+02	3.85e+03	9.07e-01	4.88e+02	1.48e-03	2.99e-14
F_6	200	1.01e+01	1.96e+01	4.60e-09	4.54e+00	1.68e+01	2.33e-13
	500	1.06e+01	2.01e+01	9.51e-09	1.12e+01	1.68e+01	2.84e-13
F_7	200	9.04e+01	8.90e+02	2.02e-07	2.56e+01	4.53e+02	4.07e-13
	500	3.20e+02	2.35e+03	9.22e-07	2.02e+02	1.54e+96	4.48e-13
F_8	200	4.83e+05	1.06e+06	2.04e-14	3.62e+04	1.24e+01	0.00e+00
	500	4.01e+06	3.39e+07	4.64e-13	4.89e+06	9.25e+02	0.00e+00
F_9	200	9.84e+02	1.55e+03	2.44e+00	2.06e+02	8.73e+01	0.00e+00
	500	3.76e+02	2.70e+04	2.09e+01	3.98e+03	2.24e+02	0.00e+00
F_{10}	200	8.25e+02	1.82e+03	1.11e+02	4.36e+02	1.42e+03	1.18e-05
	500	2.27e+03	5.01e+03	6.57e+02	1.68e+03	3.29e+03	1.88e-05
F_{11}	200	8.18e+02	1.82e+03	1.00e+02	4.41e+02	1.65e+03	1.22e-05
	500	2.29e+03	4.97e+03	6.61e+02	1.92e+03	4.09e+03	1.89e-05

4   Conclusions

A hybrid metaheuristic optimization method has been introduced in this paper. The combination of mine blast algorithm (MBA) and harmony search (HS) algorithm produced a hybrid optimization method with excellent exploration and exploitation capabilities.

The MBA is memory-less optimization method, while the HS is memory-based algorithm. Using the advantages of MBA in global search and HS in local search and thinking about combing the HS and MBA led to develop new hybrid optimization method, so called mine blast harmony search (MBHS). Furthermore, various improvements were applied to the standard HS and MBA.

A new concept for harmony memory (HM) in HS phase, so called dynamic HM, has been proposed. Also, the perception of direction in the MBA phase has been replaced by the concept of moving toward the best solutions in the MBHS.

Eleven unconstrained benchmarks, widely used in the literature, with different design variables (i.e., from 200 to 500) have been tackled. The optimization results obtained by the proposed hybrid method show that it surpasses both the

MBA and HS in terms of solution quality and having better statistical results. Moreover, further comparisons with other optimizers indicate that the MBHS attains the optimal solution more accurately and efficiently.

Acknowledgment This work was supported by the National Research Foundation (NRF) of Korea under a grant funded by the Korean government (MSIP) (NRF-2013R1A2A1A01013886).

References

1. Geem, G.W., Kim, J.H., Loganathan, G.V.: A new heuristic optimization algorithm: harmony search. Simulation **76**(2), 60–68 (2001)
2. Kim, J.H., Geem, G.W., Kim, E.S.: Parameter estimation of the nonlinear Muskingum model using harmony search. J. of Amer. Wat. Res. Assoc. **37**(5), 1131–1138 (2001)
3. Geem, G.W., Kim, J.H., Loganathan, G.V.: Harmony search optimization: application to pipe network design. Int. J. Modelling Simul. **22**(2), 125–133 (2002)
4. Fesanghary, M., Damangir, E., Soleimani, I.: Design optimization of shell and tube heat exchangers using global sensitivity analysis and harmony search algorithm. Appl. Therm. Eng. **29**(5–6), 1026–1031 (2009)
5. Degertekin, S.: Optimum design of steel frames using harmony search algorithm. Struct. Multidiscip. Optim. **36**(4), 393–401 (2008)
6. Geem, Z.W. (ed.): Harmony Search Algo. for Structural Design Optimization. SCI, vol. 293. Springer, Berlin (2009)
7. Kim, J.H., Baek, C.W., Jo, D.J., Kim, E.S., Park, M.J.: Optimal planning model for rehabilitation of water network. Water Sci. and Technol. Water Supply **4**(3), 133–147 (2004)
8. Mahdavi, M., Fesanghary, M., Damangir, E.: An improved harmony search algorithm for solving optimization problems. Appl. Math. Comput. **188**(2), 1567–1579 (2007)
9. Geem, Z.W., Sim, K.B.: Parameter-setting-free harmony search algorithm. Appl. Math. Comp. **217**(8), 3881–3889 (2010)
10. Wang, L., Yang, R., Xu, Y., Niu, Q., Pardalos, P.M., Fei, M.: An improved adaptive binary Harmony Search algorithm. Inform. Sciences **232**, 58–87 (2013)
11. Kaveh, A., Talatahari, S.: Particle swarm optimizer, ant colony strategy and harmony search scheme hybridized for optimization of truss structures. Comput. Struct. **87**(5–6), 267–283 (2009)
12. Geem, Z.W.: Particle-swarm harmony search for water network design. Eng. Optim. **41**(4), 297–311 (2009)
13. Ayvaz, M.T., Kayhan, A.H., Ceylan, H., Gurarslan, G.: Hybridizing the harmony search algorithm with a spreadsheet 'solver' for solving continuous engineering optimization problems. Eng. Optim. **41**(12), 1119–1144 (2009)
14. Fesanghary, M., Mahdavi, M., Minary-Jolandan, M., Alizadeh, Y.: Hybridizing harmony search algorithm with sequential quadratic programming for engineering optimization problems. Comput. Methods Appl. Mech. Eng. **197**(33–40), 3080–3091 (2008)
15. Sadollah, A., Bahreininejad, A., Eskandar, H., Hamdi, M.: Mine blast algorithm for optimization of truss structures with discrete variables. Comput. Struct. **102–103**, 49–63 (2012)

16. Sadollah, A., Bahreininejad, A., Eskandar, H., Hamdi, M.: Mine blast algorithm: A new population based algorithm for solving constrained engineering optimization problems. Appl. Soft Comput. **13**(5), 2592–2612 (2013)
17. Yadav, P., Kumar, R., Panda, S.K., Chang, C.S.: An intelligent tuned harmony search algorithm for optimization. Inf. Sciences **196**, 47–72 (2012)
18. Yang, H.O., Gao, L., Li, S., Kong, X., Zou, D.: On the iterative convergence of harmony search algorithm and a proposed modification. Appl. Math. Comput. **247**(15), 1064–1095 (2014)
19. Kennedy, J., Eberhart, R.: Particle swarm optimization. In: Proc. of IEEE Int. Conf. on Neural Networks, vol. IV, pp. 1942–1948 (1995)
20. Atashpaz-Gargari, E.: Lucas, C: Imperialist competitive algorithm: An algorithm for optimization inspired by imperialistic competition. IEEE CEC **7**, 4661–4666 (2007)
21. Rashedi, E., Nezamabadi-pour, H., Saryazdi, S.: GSA: A gravitational search algorithm. Inform. Sciences **179**, 2232–2248 (2009)

Modified Harmony Search Applied to Reliability Optimization of Complex Systems

Gutha Jaya Krishna and Vadlamani Ravi

Abstract This paper proposes an Improved Modified Harmony Search Algorithm with constraint handling with application to redundancy allocation problems in reliability engineering. The performance of Improved Modified Harmony Search is being compared with that of the original Harmony Search, Modified Great Deluge Algorithm, Ant Colony Optimization, Improved Non-Equilibrium Simulated Annealing and Simulated Annealing. It is observed that Improved Modified Harmony Search requires less number of function evaluations compared to others.

Keywords Constrained optimization · Meta-heuristic · Modified harmony search algorithm · Reliability redundancy allocation problem

1 Introduction

Optimization is a process of determining a unique or multiple solutions for a given objective function and a set of constraints. When we search for an optimal solution in decision space, we encounter usually a number of local optima as well as a global optimum. There are a variety of search methods viz., complete exhaustive search and incomplete search- heuristic based search, point based search and meta-heuristic based search, modified meta-heuristic based search and hybrid meta heuristic search to name a few [9].

G.J. Krishna · V. Ravi(✉)
Institute for Development and Research in Banking Technology,
Castle Hills Road #1, Masab Tank, Hyderabad 500 057, AP, India
e-mail: {krishna.gutha,padmarav}@gmail.com

G.J. Krishna
School of Computer & Information Sciences, University of Hyderabad,
Hyderabad 500 046, AP, India

© Springer-Verlag Berlin Heidelberg 2016
J.H. Kim and Z.W. Geem (eds.), *Harmony Search Algorithm,*
Advances in Intelligent Systems and Computing 382,
DOI: 10.1007/978-3-662-47926-1_17

169

Heuristics are based on experience of search procedure pertaining only to a specific problem. They are not a general type of search procedures which can be applied to any problem at hand [15]. Meta-heuristics, in general provide a skeleton of a general procedure that can be applied to a different variety of problems. They do not pertain to only a specific set of problems. Meta-Heuristics family comprises both point based algorithms such as simulated annealing [14], tabu search [17], threshold accepting [2] and population based algorithms viz., Genetic Algorithm (GA) [22], Differential Evolution [28], Particle Swarm Optimization (PSO) [16], Ant Colony Optimization(ACO) [18], Harmony Search (HS) [10] etc. The population based algorithms are also called evolutionary algorithms. Owing to the no-free-lunch theorem [15], a number of new and hybrid algorithms proliferated in literature. These are one or more population based algorithms or point based or a combination of both population and point based meta-heuristics [4, 11, 19, 20, 21, 23, 24].

Redundancy allocation problem is of tremendous significance in reliability engineering. It is formulated as a combinatorial optimization problem. The reliability goal is achieved through a discrete set of choices made from the available parts. The reliability-redundancy allocation problem (RRAP) determines optimal component reliabilities along with the redundancy level of components in a given system to maximize the system reliability subject to several resources or cost constraints [13].

2   Literature Survey

There is much importance for optimization of complex reliable systems in reliability engineering. These problems are formulated as nonlinear programming problems [7].

Several meta-heuristics were previously employed in solving problems related to reliability engineering. Improved Non-equilibrium simulated annealing was developed by Ravi et al. [3] (INESA) for this problem. Ravi et al. [25] also contributed to development of fuzzy global optimization problems by applying threshold accepting [2]. Threshold accepting [2] is a deterministic variant of simulated annealing (SA). In recent times, Shelokar et al. [6] has applied Ant Colony Optimization (ACO) algorithm and obtained superior results compared to those of Ravi et al. [3]. Then Ravi et al. [4] applied Modified Great Deluge Algorithm to obtained superior results compared to Shelokar et al. [6].

Many hybrid meta-heuristics were reported in literature like DE and Tabu [19], differential Evolution Threshold Accepting (DETA) [20], Harmony search and PSO (HS+PSO) [21] , Harmony Search and Differential Evolution (HS-DE) [23], Heuristic particle swarm ant colony optimization (HPSACO) [24], Modified Harmony search and MGDA (MHS+MGDA) [4]. Later, Modified Harmony Search and Threshold Accepting (MHSTA) was proposed by Ravi et al. [11].

3 Modified Harmony Search

3.1 Harmony Search

Harmony search, a meta-heuristic algorithm was proposed by Geem et al. [10]. It has variety of applications in engineering sciences. This algorithm is based on building an idea upon an experience. The idea behind this meta-heuristic search procedure was borrowed from musicians. HS is explained through a discussion of musician's improvisation process. When a musician is improvising, he or she has the following possible choices [10]

1. Play a famous piece of music from memory
2. Play a similar music while adjusting the pitch slightly or
3. Compose a new music all together randomly.

The musical instrument represents a decision variable. The pitch range of the music instrument represents the range for decision variables. The practice made on the instrument represents solution vector by the harmony with thorough iterations. Aesthetics representing the improvement made during iterations happens to be the fitness value of the objective function. The technique is observed to be random in nature. Hence it is highly likely to escape local optima. This technique reduces the program execution time substantially performing a very little operation on every prospective solution. The main disadvantage of Harmony Search is that of its prolonged operational time periods (in terms of number of iterations). The solution here remains unchanged during the final stages. Hence several unproductive iterations are performed with not so genuine improvement in the solution [10].

3.2 Modified Harmony Search

In order to overcome the disadvantages of HS, modifications proposed by Choudhuri et al. [1] include: (i) The hmcr value is increased dynamically from 0 to 1 during the progress of the algorithm. (ii) When difference between best and worst solution in the harmony memory is observed to be zero, the HS algorithm is terminated.

3.3 Improved Modified Harmony Search

In the present paper the following improvements were proposed to the MHS. (i) The value of hmcr is kept dynamically increasing from 0.7 to 0.9 linearly during the execution of the program. (ii) The stopping criterion is maximum number of iterations or no change in the fitness value over a definite number of iterations [26].

The Improved Modified Harmony Search algorithm is explained as follows:

1. Generate a set of hm number of solutions randomly using Uniform Distribution and initialize harmony memory with this set .
2. Create a new solution vector with components of the solutions selected from harmony memory with a probability of hmcr such that the components when selected from the harmony memory are chosen randomly from different solutions within the harmony memory. We need to note that the value of hmcr increases linearly with the number of iterations from 0.7.
3. Perform the pitch adjustment operation by altering the variables' value by delta with a probability par ('discrete optimization problems use 'delta' value).
4. The new vector computed from the above is evaluated for its fitness by computing the objective function. If the fitness value of the new solution vector is better than that of the solutions corresponding to the best or worst fitness values in the harmony memory, then the worst values as well as the corresponding solution in the harmony memory are replaced by new value and its corresponding solution.
5. Repeat procedural steps 2 to 4 till maximum number of iterations are met or no change in the fitness value for a definite number of iterations.

There are two parameters in the IMHS, viz., (i) Harmony Memory Considering Rate (hmcr), the rate of selecting a value from harmony memory, which is set between 0..7 to 0.99. In HS, it is fixed at 0.9, but here it is gradually increased from 0.7 to 0.9. (ii) Pitch Adjusting Ratio (par), the rate of selecting a value within a neighborhood, is chosen between 0.1 and 0.5. In the case of HS, MHS and IMHS, par value is chosen to be 0.4.

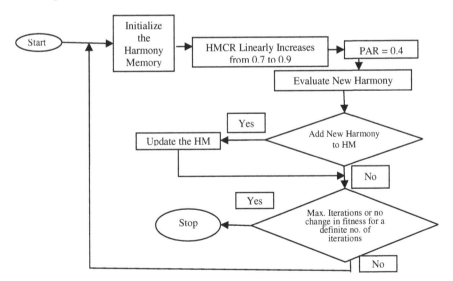

Fig. 1 Flow Chart MHS-1

4   Constraint Handling

Constraint Handling is required when optimization problem is having constraints. The region of search is divided into feasible and infeasible regions because of constraints. There are several methods for handling constraints. They are as follows: 1. Penalty functions 2. Repair operators or algorithms 3. Special representations and operators 4. Separation of constraints and objectives and 5. Hybrid Methods [27].

The common approach in evolutionary computing techniques applied to constrained problems is to use penalties. The idea of using penalties is to transform a constrained optimization problem by adding penalty for the violation of constraint. Penalty based approaches are as follows: 1. Death Penalty 2. Static Penalty 3. Dynamic Penalty 4. Adaptive Penalty [27].

4.1   Static Extinctive Penalty

The penalty factor does not depend on current generation number in any way. They remain constant during the whole run. The penalty factors are generally considered to be problem dependent. Three methods have been generally used with static penalty functions. They are (i) distance based penalties, (ii) binary penalties and (iii) extinctive penalties. Extinctive penalties are very high penalties that are used to prevent the usage of infeasible solutions [12].

Here, in this problem static extinctive penalties have been used. The idea is that a penalty value large enough is used for infeasible solutions. If the problem is of maximization of objective function, then a minimum value is used as penalty. However if the problem is of minimization, large value is used as penalty. Here a constant value, either too large or too small is used as penalty for minimization and maximization problems respectively. The consequence of this approach is that the infeasible solution will not be considered for further evaluations and will be removed [12].

5   Problem Definitions

This problem is a serial-parallel system which has subsystems and, in each there are multiple components that can be selected in parallel. System designed using these component types with known cost, reliability and weigh. So, this system design problem becomes a discrete optimization problem [7]. Figure 1 represents the N-Stage series-parallel system.

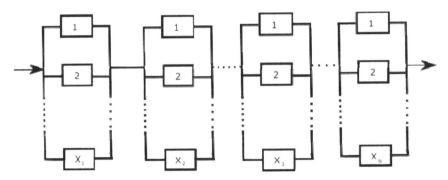

Fig. 2 Serial Parallel System

Case(i) [7]
Find the optimal x_i , i=1,2,...4

$$\text{Max } R_s = \prod_{i=1}^{N=4}[1-(1-R_i)^{x_i}]$$

Subject to

$$g_1 = \sum_{i=1}^{N=4} C_i x_i \leq 56$$

$$g_2 = \sum_{i=1}^{N=4} W_i x_i \leq 120$$

Table 1 Coefficients for Case (i)

I	1	2	3	4
R_i	0.80	0.70	0.75	0.85
C_i	1.2	2.3	3.4	4.5
W_i	5	4	8	7

Case(ii) [7]
Find the optimal x_i , i=1,2,...5

$$\text{Max } R_s = \prod_{i=1}^{N=5}[1-(1-R_i)^{x_i}]$$

Subject to

$$g_1 = \sum_{i=1}^{N=15} P_i x_j^2 \leq 110$$

$$g_2 = \sum_{i=1}^{N=5} C_i[x_i + e^{x_i/4}] \leq 175$$

$$g_3 = \sum_{i=1}^{N=5} W_i [x_i + e^{x_i/4}] \leq 200$$

Table 2 Coefficients for Case (ii)

I	1	2	3	4	5
R_i	0.80	0.85	0.9	0.65	0.75
P_i	1	2	3	4	2
C_i	7	7	5	9	4
W_i	7	8	8	6	9

Case(iii) [5]

Find the optimal x_i , i=1,2,...15

$$\text{Max } R_s = \prod_{i=1}^{N=15} [1 - (1 - R_i)^{x_i}]$$

Subject to

$$g_1 = \sum_{i=1}^{N=15} C_i x_i \leq 400$$

$$g_2 = \sum_{i=1}^{N=15} W_i x_i \leq 414$$

Table 3 Coefficients for Case (iii)

i	R_i	C_i	W_i
1	0.9	5	8
2	0.75	4	9
3	0.65	9	6
4	0.8	7	7
5	0.85	7	8
6	0.93	5	8
7	0.78	6	9
8	0.66	9	6
9	0.78	4	7
10	0.91	5	8
11	0.79	6	9
12	0.77	7	7
13	0.67	9	6
14	0.79	8	5
15	0.67	6	7

6   Results and Discussions

Here for case-i and case-ii harmony memory size (hm) is kept at 5 and for case-iii
harmony memory size (hm) is kept at 30 and maximum number of iterations is
kept at 10000 for case-i and case-ii and 250,000 for case-iii. HMCR value of HS is
kept at 0.9, linearly increased from 0 to 1 for MHS and linearly increased from 0.7
to 0.9 for IMHS. PAR value is kept constant at 0.4 for all three variants. Number
of runs is 30 each for case-i and case-ii for all three variants and 15, 15, 30 for HS,
MHS, and IMHS for case-iii. Less number of runs (15) is chosen for HS, MHS
because it takes huge computational time case-iii and also result obtained are also
not optimal for MHS.

Table 4 Results obtained for Case (i)

Algorithm	x_1	x_2	x_3	x_4	g_1	g_2	R_s	FE^a
IMHS	5	6	5	4	54.8	117.0	0.9975	401^b
MHS [11]	5	6	6	3	54.8	117.0	0.9975	1366^b
HS [10]	5	6	6	3	54.8	117.0	0.9975	1681^b
ACO [6]	5	6	5	4	54.8	117.0	0.9975	2000
INESA [3]	5	6	5	4	54.8	117.0	0.9975	NA
SA [3]	5	6	5	4	54.8	117.0	0.9975	NA
[7]	5	6	5	4	54.8	117.0	0.9975	NA

^{a} No of function evaluations (FE) , ^{b} Mean of 30 simulations

Table 5 Results obtained for Case (ii)

Algorithm	x_1	x_2	x_3	x_4	x_5	g_1	g_2	g_3	R_s	FE^a
IMHS	3	2	2	3	3	83.0	146.125	192.48	0.9045	346^b
MHS [11]	3	3	2	3	2	83.0	146.125	192.48	0.9045	1297^b
HS [10]	3	2	2	3	3	83.0	146.125	192.48	0.9045	1325^b
ACO [6]	3	2	2	3	3	83.0	146.125	192.48	0.9045	1800
INESA [3]	3	2	2	3	3	83.0	146.125	192.48	0.9045	NA
SA [3]	3	2	2	3	3	83.0	146.125	192.48	0.9045	NA
[7]	3	2	2	3	3	83.0	146.125	192.48	0.9045	NA

^{a} No of function evaluations (FE), ^{b} Mean of 30 simulations

As it is a constrained optimization problem, static extinctive penalty is used for
infeasible solutions. Here as it is a maximization problem penalty value is chosen
to be a very low value i.e. either zero. As a result, after a few iterations infeasible
solutions are removed from harmony memory and only feasible solutions remain
in memory.

IMHS yielded better results faster because the HMCR value of increased linearly from 0.7 to 0.9 than MHS as it are not increased linearly from 0 to 1. Further in the latter, a lot of function evaluations are wasted on non optimum solutions and stopping criteria is not based on fitness value unlike in MHS but rather continued till max iterations or no change in the fitness value over a definite number of iterations. IMHS also gave faster results than HS as HMCR is not fixed at 0.9 but varied linearly as result more better solutions are brought into harmony memory. IMHS combines the advantages of both MHS and HS because of which better solutions are obtained faster.

While obtaining the global optimal solutions and the global optima as reported in the literature, the average function evaluations spent by IMHS, for the three cases are 400.033, 345.2333, 28278.17, which are less than that of HS and MHS which consumed average function evaluations of 1680.9, 1324.367, 102210.7 and 1366.4, 1296.867 respectively. However, in Table 3, because MHS could not obtain global optimum, we did not present the function evaluations consumed.

Table 6 Results obtained for Case (iii)

Algo. Sol. Vector	IMHS	MHS [11]	HS [10]	MGDA [4]	ACO [6]	INESA [3]	SA [3]	Luus [5]
x_1	3	3	3	3	3	3	3	3
x_2	4	4	4	4	4	4	4	4
x_3	6	5	6	6	6	5	5	5
x_4	4	3	4	4	4	3	4	3
x_5	3	3	3	3	3	3	3	3
x_6	2	2	2	2	2	2	2	2
x_7	4	4	4	4	4	4	4	4
x_8	5	5	5	5	5	5	5	5
x_9	4	4	4	4	4	4	4	4
x_{10}	2	3	2	2	2	3	3	3
x_{11}	3	3	3	3	3	3	3	3
x_{12}	4	4	4	4	4	4	4	4
x_{13}	5	5	5	5	5	5	5	5
x_{14}	4	5	4	4	4	5	5	5
x_{15}	5	5	5	5	5	5	4	5
R_s	Opt[d]	NOpt	Opt[d]	Opt[d]	Opt[d]	NOpt[e]	NOpt[e]	NOpt[e]
FE[a]	28,377[b]	NA	102,211[c]	217,157	244,000	NA	NA	NA

[a] No of function evaluations (FE), [b] Mean of 30 simulations, [c] Mean of 15 simulations, [d] Optimum=0.945613, [e] Did not Converge to Optimum

Table 7 Function Evaluations

Method	RRAP Problem	Best NFEs[1]	Average NFEs	Worst NFEs	SD[2]	No. Runs	Max. NFEs
HS	Case - i	417	1680.9	5560	1330.231	30	10000
	Case - ii	271	1324.367	4628	1006.391	30	10000
	Case - iii	60045	102210.7	150345	29976.38	15	250000
MHS	Case - i	658	1366.4	2086	286.1513	30	10000
	Case - ii	671	1296.867	2104	297.7236	30	10000
	Case - iii	NA[3]	NA[3]	NA[3]	NA[3]	15	250000
IMHS	Case - i	138	400.0333	1837	390.0115	30	10000
	Case - ii	110	345.2333	854	207.8117	30	10000
	Case - iii	4349	28278.17	80343	19558.59	30	250000

[1] Number of function evaluations
[2] Standard deviation
[3] Didn't Converge to Optimum

7   Conclusions

We considered three discrete constrained optimization problems with four, five and fifteen variables used for RRAP's and achieved better results in terms of function evaluations using proposed IMHS. IMHS can further be hybridized with any point based algorithms like Simulated Annealing, Threshold Accepting, Great Deluge algorithms, etc., and used for both continuous and discrete optimization problems. Further variants of HS can also be tried with these three problems.

Acknowledgement The first author would like to thank Mr. Andrey Volkov, Research Intern from University of Ghent for helping him in getting acquitted with R language.

References

1. Choudhuri, R., Ravi, V., Mahesh Kumar, Y.: A Hybrid Harmony Search and Modified Great Deluge Algorithm for Unconstrained Optimization. International Journal of Comp. Intelligence Research 6(4), 755–761 (2010)
2. Dueck, G., Scheur, T.: Threshold Accepting: A General Purpose Optimization Algorithm appearing Superior to Simulated Annealing. J. Comput. Phys. 90, 161–175 (1990)
3. Ravi, V., Murthy, B.S.N., Reddy, P.J.: Non-equilibrium simulated annealing-algorithm applied to reliability optimization of complex systems. IEEE Transanctions on Reliability 46, 233–239 (1997)

4. Ravi, V.: Optimization of Complex System Reliability by a Modified Great Deluge Algorithm. Asia-Pacific Journal of Operational Research 21(4), 487–497 (2004)
5. Luus, R.: Optimization of system reliability by a new nonlinear integer programming procedure. IEEE Transactions on Reliability 24, 14–16 (1975)
6. Shelokar, P.S., Jayaraman, V.K., Kulkarni, B.D.: Ant algorithm for single and multi objective reliability optimization problems. Qual. Reliab. Eng. Int. 18, 497–514 (2002)
7. Tillman, F.A., Hwang, C.L., Kuo, W.: Optimization of System Reliability. Marcel Dekker, Inc., NewYork (1980)
8. Mohan, C., Shanker, K.: Reliability optimization of complex systems using random search techniques. Microelectronics and Reliability 28, 513–518 (1988)
9. Michel, G., Jean-Yves, P.: Handbook of Metaheuristics. Springer US (2002)
10. Geem, Z., Kim, J., Loganathan, G.: A new heuristic optimization algorithm: harmony search. Simulation 76, 60–68 (2001)
11. Maheshkumar, Y., Ravi, V.: A modified harmony search threshold accepting hybrid optimization algorithm. In: Sombattheera, C., Agarwal, A., Udgata, S.K., Lavangnananda, K. (eds.) MIWAI 2011. LNCS, vol. 7080, pp. 298–308. Springer, Heidelberg (2011)
12. Pardalos, P.M., Edwin, R.H.: Handbook of Global Optimization, vol. 2. Springer US (2002)
13. Harish, G., Sharma, S.P.: Reliability-Redundancy Allocation Problem of Pharmaceutical Plant. Journal of Engineering Science and Technology 8(2), 190–198 (2013)
14. Kirkpatrick, S., Gelatt Jr., C.D., Vecchi, M.P.: Optimization by Simulated Annealing. Science 220(4598), 671–680 (1983)
15. Wolpert, D.H., Macready, W.G.: No free lunch theorems for optimization. Evolutionary. EEE Transactions on Computation 1(1), 67–82 (1997)
16. Kennedy, J., Eberhart, R.: Particle swarm optimization. In: Proceedings of IEEE International Conference on Neural Networks IV, pp. 1942–1948
17. Glover, F.: Tabu Search - Part 1. ORSA Journal on Computing 1(2), 190–206 (1989)
18. Dorigo, M., Gambardella, L.M.: Ant Colony System: A Cooperative Learning Approach to the Traveling Salesman Problem. IEEE Transactions on Evolutionary Computation 1(1), 53–66 (1997)
19. Srinivas, M.: Rangaiah: Differential Evolution with Tabu list for Global Optimization and its Application to Phase Equilibrium and Parameter Estimation Problems. Industrial and Engineering Chemistry Research 46, 3410–3421 (2007)
20. Chauhan, N., Ravi, V.: Differential Evolution and Threshold Accepting Hybrid Algorithm for Unconstrained Optimization. International Journal of Bio-Inspired Computation 2, 169–182 (2010)
21. Li, H., Li, L.: A novel hybrid particle swarm optimization algorithm combined with harmony search for higher dimensional optimization problems. In: International Conference on Intelligent Pervasive Computing, Jeju Island, Korea (2007)
22. Mitchell, M.: An Introduction to Genetic Algorithms. MIT Press (1998)
23. Gao, X.Z., Wang, X., Ovaska, J.: Uni-Modal and Multi Modal optimization using modified harmony search methods. IJICIC 5(10(A)), 2985–2996 (2009)
24. Kaveh, A., Talatahari, S.: PSO, ant colony strategy and harmony search scheme hybridized for optimization of truss structures. Computers and Structures 87, 267–283 (2009)

25. Ravi, V., Reddy, P.J., Zimmermann, H.J.: Fuzzy global optimization of complex system reliability. IEEE Trans. Fuzzy Syst. **8**, 241–248 (2000)
26. Geem, Z.W. (ed.): Harmony Search Alg. for Structural Design Optimization. SCI, vol. 239. Springer, Heidelberg (2009)
27. Kusakci, A.O., Mehmet, C.: Constrained Optimization with Evolutionary Algorithms: A Comprehensive Review. Southeast Europe Journal of Soft Computing **1**(2) (2012)
28. Storn, R., Price, K.: Differential evolution - a simple and efficient heuristic for global optimization over continuous spaces. J. Global Optim. **11**, 341–359 (1997)

A New HMCR Parameter of Harmony Search for Better Exploration

Nur Farraliza Mansor, Zuraida Abal Abas,
Ahmad Fadzli Nizam Abdul Rahman,
Abdul Samad Shibghatullah and Safiah Sidek

Abstract As a meta-heuristic algorithm, Harmony Search (HS) algorithm is a population-based meta-heuristics approach that is superior in solving diversified large scale optimization problems. Several studies have pointed that Harmony Search (HS) is an efficient and flexible tool to resolve optimization problems in diversed areas of construction, engineering, robotics, telecommunication, health and energy. In this respect, the three main operators in HS, namely the Harmony Memory Consideration Rate (HMCR), Pitch Adjustment Rate (PAR) and Bandwidth (BW) play a vital role in balancing the local exploitation and the global exploration. These parameters influence the overall performance of HS algorithm, and therefore it is very crucial to fine turn them. However, when performing a local search, the harmony search algorithm can be easily trapped in the local optima. Therefore, there is a need to improve the fine tuning of the parameters.

N.F. Mansor(✉)
Faculty of Informatics and Computing, Universiti Sultan Zainal Abidin (Tembila Campus),
22200, Besut, Terengganu, Malaysia
e-mail: farralizamansor@unisza.edu.my

Z.A. Abas · A.F.N.A. Rahman · A.S. Shibghatullah
Optimization, Modelling, Analysis, Simulation and Scheduling (OptiMASS) Research
Group, Faculty of Information and Communication Technology, Universiti Teknikal
Malaysia Melaka (UTeM), Hang Tuah Jaya 76100,
Durian Tunggal, Melaka, Malaysia
e-mail: {zuraidaa,fadzli,samad}@utem.edu.my

S. Sidek
Center for Languages and Human Development, Universiti Teknikal Malaysia
Melaka (UTeM), Hang Tuah Jaya 76100, Durian Tunggal, Melaka, Malaysia
e-mail: safiahsidek@utem.edu.my

© Springer-Verlag Berlin Heidelberg 2016
J.H. Kim and Z.W. Geem (eds.), *Harmony Search Algorithm*,
Advances in Intelligent Systems and Computing 382,
DOI: 10.1007/978-3-662-47926-1_18

181

This research focuses on the HMCR parameter adjustment strategy using step function with combined Gaussian distribution function to enhance the global optimality of HS. The result of the study showed a better global optimum in comparison to the standard HS.

Keywords Meta-heuristic · HMCR · PAR · BW and gaussian distribution

1   Introduction

Optimization indicates the process of finding the best solution from various solutions defined according to the terms of mathematics or science. In this respect, every process has a chance to be optimised since it can be formulated or modeled in the form of optimization problems that focuses on achieving either a minimum or maximum objective function. The objective function are measured, particularly based on the efficiency, time, cost, profit and risk and these measurements are commonly applied when solving problems in engineering, business, economics and science. Relating to achieving the best solutions, solving real living problems are complicated as the solutions are usually fluid: The solution should be based on a specific approach applicable within a reasonable amount of time only. However, the conventional method of solving real living problems involve only a few steps; hence, leading to solutions that are subjected to inaccuracy and lack of feasibility. They also have limited assurance in reaching global solution. Nowadays, heuristic approach has been increasingly recognised as a suitable approach for finding solutions for complex optimization problems commonly found in the industry and services. Heuristic is categorized as an approximate algorithm based on trial and error approach. This approach either looks for good solutions (near optimal solutions) without aiming at optimal or feasible solutions, or claim the optimality of a particular feasible solution based on its closeness to the optimality in numerous cases [1]. Most meta-heuristics are inspired from the fields of physics, biology and ethology, in which the random variables and several parameters are used to achieve the objective function.

A variety of new meta-heuristics have emerged over the last four decades. They have demonstrated their respective strengths to solve critical optimization problems that exist in areas, such as resource allocation, industrial planning, scheduling, decision making, medical, engineering, computer engineering, and many others. Meta-heuristics can be classified based on the assessments of methodology presentation, and one of the main characteristics used to group meta-heuristics is based on the purpose of the designed algorithm. At present, the most inspired designs of an algorithm are the natural and the physical process as well as the animal social behavior, such as the Genetic Algorithm [2], Ant Colony Optimization [3], Particle Swarm Optimization [4], Bee Colony Optimization [5], Simulated Annealing, and Harmony Search Algorithm [6]. Among the latest algorithms classified into meta-heuristics is the Harmony Search Algorithm (HS). This design impersonates the concept of music improvisation and continues to polish their pitches in order to attain better harmony. The HS has several advantages in terms of simplicity, flexibility, adaptability and scalability [7][8]. It also has a novel stochastic derivative and imposes a less mathematical equation to produce new solutions at each iteration, particularly when an available solution is taken into consideration [9]. Since its development, HS has gained a foothold in multiple areas, such as fuzzy controller design [10], water management [11], broadcast scheduling packet [12] and environmental or economic dispatch [13]. Although it has many advantages, HS has a shortcoming as it has the tendency to be trapped in the local optima when numerical optimization problems are performed [9]. This situation also befalls on Particle Swarm Optimization and DE. In this regard, controlling the parameters becomes the main task when dealing with optimization performance.

Three parameters, namely the harmony memory consideration rate (HMCR), pitch adjusting rate (PAR) and distance bandwidth (BW) have motivated researchers to work in HS algorithm. Since the emergence of HS, the work has been largely focused on the adjustment of the parameters and its effects on the performance of HS algorithm. Each of these parameters has their own responsibilities in assisting HS to seek the best solutions. For example, the HMCR parameter is accountable for achieving faster convergence rate, PAR is responsible for the increase in the diversity of solutions and BW is used to ameliorate the diversity of final solutions at the end of the generation. Although the determination of parameters setting is crucial for all meta-heuristic algorithms, particularly the HS algorithm, it is a difficult task since the guidelines to decide the requirement to solve problem are still lacking. The common approach is to perform various experiments to get a good solution due to parameter setting value is dependent on the problem. Therefore, until now HS is still experiencing an enhancement phase, despite the standard HS has proven to be successfully implemented in different areas of optimization.

In this study, the value of HMCR is set at a small rate at the initial stage of generation in order to have an allowable range for the search of solutions. To improve the convergence efficiency and maintaining adequate exploitation capability of HS for this study, a Gaussian distribution function is employed to generate random numbers instead of applied uniform distribution function. This paper also combined with mechanism for adjusting dynamically of pitch adjustment rate and bandwidth parameters. The major purpose for this modification is to allow the behave strategy of algorithm explore at the initial of the searching and exploit toward end of searching process. This research is focus on regulating key parameters of HMCR using step function for the improvement of global search capability. The major contribution of this paper is to explore the impact of HMCR value to the HS algorithm altogether maintaining both the exploration ability and the diversity of population. This is our initial study on step function. Subsequently, the validation on the determination of range of interval for number of iteration will be presented. It is denoted that the value of HMCR parameter will be matched to the appropriate number of iteration. Our experimental results show that the proposed algorithm was better than the standard HS. Therefore, to prove the claims of our research, this paper has the following structure: After Section One that presents the introduction of this paper, Section 2 provides an overview of previous improvement on HS and brief standard HS algorithm, followed by Section 3 that describes the proposed method. The results and discussion is highlighted in Section 4 and lastly, Section 5 presents the conclusion.

2 Previous Improvement

These enhancements indicate the need for several possible actions, such as modification of the formulas for each parameter, integration with other algorithms, addition of some elements and introduction to new concepts. For example, a famous research of an extension in HS is Mahdavi et al. They developed Improved Harmony Search (IHS), which has been the first variant since the advent of HS. Their proposed algorithm is a modified setting parameter value of pitch adjustment rate (PAR) and a dynamic bandwidth. This is obviously different from the basic HS, where the parameter of PAR and bandwidth has fixed values. The PAR value is updated and increased linearly for each iteration and the bandwidth value is exponentially decreased.

Global-best HS algorithm (GHS) was initiated by Omran and Mahdavi [14]. Focusing on ways to overcome the limitation in HIS, they introduced a new improvisation mechanism by borrowing the concept of particle swarm optimization (PSO). This mechanism hybridizes the ideas of swarm intelligence with a conventional harmony search. According to [14], "each particle represents a candidate solution to the optimization problem. The position of a particle is influenced by the best position visited by itself (i.e. its own experience) and the position of the best particle in the swarm (i.e. the experience of swarm)." The idea of GHS is to eliminate the bandwidth with an introduction of the new pitch adjustment rule, where a new harmony will be influenced by the best harmony. Pan et al. [15] was the pioneer who introduced the self-adaptive global best HS (SGHS). They utilized a new improvisation mechanism and an adaptive parameter tuning method to improve the performance of GHS. [16] highlighted a mechanism to automatically modify the parameter setting according to self-consciousness, and they initialized harmony memory based on low-discrepancy sequence. [17] emphasized the enhancement of the PAR parameter setting. They suggested three types of PAR's strategy, known as the convex differential growth, convex differential changes and concave differential growth. These strategies could boost up a global searching and be applied to different types of problem because each problem has a different ratio of PAR.

In an analysis of selection methods, [18] revealed four methods of memory selections to replace the random selection method that has been widely adopted in harmony consideration recently. The methods are tournament, proportional, linear rank and exponential rank. These methods have been employed in an evolutionary algorithm to emulate the natural selection of the survival of the fittest. These four selection methods used selection probability to evaluate each HM vector. They claimed that a higher selection probability of HM vector would trigger memory consideration in choosing the value of the decision variable from the vector. A recent study by [19] made an adjustment to the parameter setting in HMCR and PAR. Their study recommended combinations of four different cases of HMCR and PAR with all of the parameters of PAR are linearly and exponentially decreased. At the same time, the HMCR is always increased linearly and exponentially.

2.1 The Fundamental of Standard Harmony Search Algorithm

Harmony Search (HS) algorithm is a new population-based solutions of meta-heuristics search technique that mimics the process of music improvisation. This process aims to achieve a perfect state of harmonies played simultaneously by more than one music instrument. This means that the pitch from every instrument is combined together to obtain the most harmonious rhythm and evaluated by aesthetic standards. HS is termed as a population-based solution because the improvisation of the pitch occurs iteratively using a set of solutions in Harmony Memory (HM): The harmony in music corresponds to a solution vector; each musical instrument corresponds to each decision variable; the pitch range of the musical instrument corresponds to the decision variable's value range; the audience's esthetic corresponds to the objective function and the musical improvisations correspond to the local and global search in optimization. Figure 1 illustrates the relationship between music improvisation and optimization.

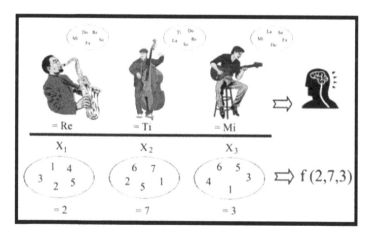

Fig. 1 Analogy between music improvisation and optimization [20]

A set of parameters is applied to HM to yield a new harmony vector so that a local optimum could be accomplished for each of the iterations. To execute the algorithm, there are five basic steps involved in HS, which are (i) initializing the problem and algorithm parameters, (ii) initializing the HM, (iii) improvising the new harmony, (iv) updating the HM with the exclusion of the worst and the best harmony, (v) checking the stopping criteria; if it is not satisfied, it goes back to step (iii). Figure 2 illustrates the flow diagram of the HS process. According to Geem et al. [11], the standard HS algorithm involves five basic parameters, which are the harmony memory size (HMS), the harmony memory consideration rate (HMCR), the pitch adjustment rate (PAR), the distance bandwidth (BW) and the number of improvisations (NI) [6]. HMS is the representation of the number of solution vector in harmony memory, while HMCR indicates the balancing between the value of exploration and exploitation, which ranges from zero to one. The PAR is a parameter which determines the requirements for further alteration, whether it is necessary or not, according to BW parameters. The termination criterion is NI, while BW is the step size of the PAR parameter.

To improvise a single pitch of each instrument, a musician has three choices, which are (1) playing any pitch from HM, (2) playing a pitch that is slightly different from HM with a PAR value, or (3) playing a totally random pitch from permissible pitch ranges. In a standard HS algorithm, the term "harmony" refers to each solution stored in HM in the form of n-dimensional real vector [15]. At the beginning of HS process, the initial population of harmony vectors is randomly generated. Next, new harmony vectors are generated from the three choices as mentioned previously. Then, the HM is updated by making an evaluation between the new harmony solutions with the existing one. If the existing HS is worse than the new harmony solutions, it will be excluded and replaced by a new HS in HM. The process continues and ends when the number of improvisations is met. The detail of the process is described below and Figure 3 summarizes the steps involved in the HS algorithm.

Step 1: Initialize the Problem and Algorithm Parameters
To execute the HS algorithm, the optimization problem must be specified as in Eq. (1) as provided by [15]:

$$\text{Minimize } f(x) \text{ s.t. } x(j) \epsilon \, [LB(j), UB(j)], \, j = 1,2,\ldots n,] \tag{1}$$

Where $f(x), x(j), n, LB(j)$ and $UB(j)$ denote the objective function, the decision variables, the number of design variables, and the lower and upper of the design variables respectively.

Step 2: Initialize the Harmony Memory (HM)
The "harmony memory" (HM) is a matrix of harmony solution vectors, where its size depends on the Harmony Memory Size (HMS). Each harmony is defined as one solution as shown in Eq. (2). The HM is occupied by many randomly generated solution vectors and sorted by the objective function value. Let $X_i = \{x_i(1), x_i(2), \ldots x_i(n)\}$ represents the ith harmony vector, which is randomly generated as follows: $x_i(j) = LB(j) + (UB(j) - LB(j)) \times r$ for $j = 1,2,\ldots n$ and $i = 1,2,\ldots, HMS$, where r is a uniform random number $[0,1]$.

$$HM = \begin{bmatrix} x_1^1 & x_2^1 & \ldots & x_N^1 & f(x^1) \\ x_1^2 & x_2^2 & \ldots & x_N^2 & f(x^2) \\ \vdots & \vdots & \vdots \vdots & \vdots \\ x_1^{HMS} & x_2^{HMS} & \ldots & x_N^{HMS} & f(x^{HMS}) \end{bmatrix} \tag{2}$$

Step 3: Improvise New Harmony

This step is crucial because it influences the overall performance of the HS algorithm. A new harmony vector, X_{new} is improvised using three selected operators, which are the memory considerations, pitch adjustment and randomization. This kind of improvisation enables HS to diversify the solutions that allow the algorithm to explore the variety of solutions, leading to a global optimum. The random numbers, r_1 and r_2 are ranged between 0 and 1. In each improvisation or iteration, if r_1 is less than HMCR, the solution vector, x_{new} (j) is generated from the memory consideration, otherwise, x_{new} (j) is selected through randomization. Next, if r_2 is less than PAR, the solution vectors x_{new} (j) from the memory consideration is further amended with the bandwidth distance using Eq. (3) as given by [15]. In this case, the value of x_{new} (j) will be evaluated using the probability of PAR.

$$x_{new} (j) = x_{new} (j) \pm r \times BW \tag{3}$$

where r is a uniform random number ranging from 0 to 1. The function of bw is an arbitrary distance bandwidth. The value of BW is determined by the changes in quantity of the new solution vector and its presence is enhanced by the performance of HS.

Step 4: Update Harmony Memory

To update the HM with a new solution vector, X_{new}, the function objective will be used to evaluate them. Then, comparative value of the objective function is made between the new vector solutions and the historical vector solution in HM. If the new vector solution is better than the worst historical vector solution, the worst historical is excluded and substituted with a new one. Otherwise, the new solution is neglected.

Begin
Define Objective Function, HMS, HMCR, PAR, BW and NI.
Initialize of HM with range of variables, X_i randomly.
While (currentIteration<maximumIteration)
  For (i<=number of variables)
    If (ran<HMCR), Choose a value from HM for the variable X_i
    If(ran2< pitch adjusting rate), Modify the value by adding or subtract of certain amount of BW
    End if
      Else Choose a random value
    End if
      Evaluate the newFitness (current objective function) and compared its value with worst function.
End while
  Sort the objective function by ascending order
  Satisfied number of iteration (NI)
  End

Fig. 2 The pseudo-code for HM Algorithm applied in the prototype

Step 5: Check The Stopping Criteria.
The process is repeated until the number of improvisation (NI) is satisfied, otherwise the process repeats step 3 and 4. Finally, the best solution is achieved and considered as the best result to the problem under investigation. The steps involved in the HS algorithm are summarized in Figure 3.

3   Proposed Method with Step Function

If the value of HMCR is kept constant during the improvisation procedure, there will be a delay in the convergence because it does not have the ability to achieve a proper balance in the main search strategy between exploitation (intensification) and exploration (diversification). It must be noted that the performance and efficiency of an algorithm relies on an accurate balance between the intensification and the diversification. If the diversification is too high, a large amount of areas in the solution space is not comprehensively examined, whereas if the diversification is low, the solution space is not investigated, leading it to be trapped in the local optima. In fact, the strategy for achieving a balance between these two elements is recognized as an optimization problem. Thus, to maintain the balance between intensification and diversification is not straightforward, but it requires a specific technique. In order to solve this issue, we proposed the use of different HMCR value at different group of iteration interval. By following the concept of step function, this method begins by setting the value of HMCR at a small rate and then increasing the value gradually. The setting of the HMCR value at a small rate at the beginning of the method allows the researchers to perform an exploration of a more diversed solution. In this study, the value of HMCR is set at a small rate at the initial stage of generation so that an allowable range for the search of the solutions can be performed. This action eventually leads to the achievement of a global optimum. Meanwhile, the value of HMCR is raised at the end of the optimization process as it is compulsory for the algorithm to search for the existing solutions in HM.

Following the step function (staircase function), a new Harmony Search algorithm is suggested in this study. In this method, the intensification and diversification are governed through the dynamic value of HMCR. The step function is one of the special types of function equation that can be depicted as a series of steps. It is a piecewise function that consists of pieces in constant value, in which the constant pieces function to specify the subsequent intervals. Further, the value of these pieces will be changed from one interval to another interval. Based on the description mentioned earlier, the step function is classified as a discrete function. In this experiment, y- axis and x-axis represent the HMCR value and interval of iteration respectively. To conduct this experiment, various rates of HMCR were used. To illustrate how the step function works, let say an experiment ranging from 0.1 until 0.9 with the increment of 0.1 throughout the iteration. The maximum number of iteration was fixed at 1000. Then, HMCR at 0.1 was run at the number of iteration from 1 to 100. This was followed by the value of HMCR at 0.2, which was executed at the number of iteration from 101 to 200. The next value of HMCR was at 0.3, which was carried out at the number of iteration from 201 to 300. This experiment continued until the value of HMCR at 0.9, which was run at the number of iteration from 801 to 1000. Figure 3 shows a corresponding variable applied translated for the previous explanation, in which Equation 4 represents the HMCR values with the interval range of NI.

The previous explanation is fundamental idea of this study and this concept is applied for my experiments, but the differences are the sequence of HMCR values and the increment. My study start at the initial value of HMCR 0.4001 and completed iterates at 0.99 with three steps increments for number of iteration. The purpose of this experiment was to find the minimum objective function at different values of HMCR so that the global optimum point can be accelerated. Herein, the objective function, f(x), illustrates the data from the driver scheduling problem seeking to find solution that minimizes the soft constraint violation [21][22]. To carry out this experiment, Table 1 is summarized the parameter setting.

$$
HMCR = \begin{cases}
0.1; & 0 < NI \le 100 \\
0.2; & 100 < NI \le 200 \\
0.3; & 200 < NI \le 300 \\
0.4; & 300 < NI \le 400 \\
0.5 & 400 < NI \le 500 \\
0.6 & 500 < NI \le 600 \\
0.7 & 600 < NI \le 700 \\
0.8 & 700 < NI \le 800 \\
0.9 & 800 < NI \le 1000
\end{cases} \tag{4}
$$

This experiment focused on the HMCR operator by observing the effect of adjustments to the results using different values. At the beginning of generalization, the harmony memory (HM) was fully occupied with the generated random solutions. Theoretically, the higher the HMCR, the stronger it forces the algorithm to search the existing solutions in HM. On the other hand, the small rate of HMCR leads to the search solutions in the permissible range, which is outside of HM. This method leads to a better exploration as it allows for a searching process that is beyond HM, but within the permissible range, hence leading towards global optimum When HMCR equals to 0.10, the number of iteration executed fall within the range of 0 and 100. This process was continued with HMCR at 0.20 with the number of improvisation at the range of 100 and 200. This process was terminated at HMCR 0.90 with number of iteration performed at the last range between 800 and 1000.

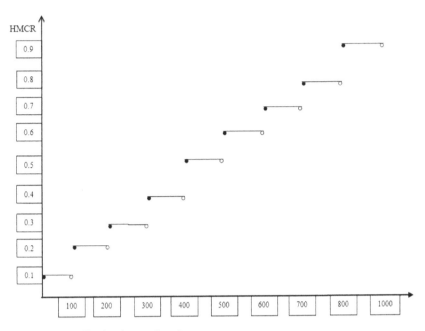

Fig. 3 HMCR and NI using in step function

3.1 Gaussian Distribution Function

The Gaussian probability density distribution is also known as normal probability distribution. The density function of Gaussian probability distribution used in this study is indicated as in Equation (5) where μ is mean value and σ is denoted as standard deviation. Other delegates for Gaussian distribution is N (μ, σ2).

$$p(x) = \frac{1}{\sqrt{2\pi}\sigma} \exp[-(z - \mu)^2/(2\sigma^2)] \tag{5}$$

The Gaussian distribution is a continuous distribution that random numbers are much more likely to be picked around the mean. Our target is to investigate the significance of this distribution to the performance of HS by introduced this Gaussian factor into the improvisation step particularly at PAR adjustment step. If satisfy the PAR condition, the algorithm will alter the vector solution which is taken from HM randomly and it is adjusted according to a Gaussian distribution as shown in Equation (6)

$$x_{new}(j) = x_{new}(j) \pm Gauss(-3,3) \times BW \tag{6}$$

The reason behind choosing the Gaussian random number manipulated with candidate solution is due to the searching around the vector solution is in wider region. On the contrary, the uniform distribution grant only small region of searching around the vector solution.

In this research, Gaussian probability function is adopted instead of uniform distribution function which normally implemented in meta-heuristic approach for generating random numbers. The outcome from this kind of distribution offers an enlargement of search efficiency and also ability to maintain a sufficient exploitation. Usually, random number is generated by uniform probability, but drawback of utilized this random generator are recognized such as [23][24] (1) they result in sequential correlation of successive cells and so the solution becomes sealed (2) Greater tendency to generate of high level random number are often compared to the lower random generator. Both disadvantages above guide the algorithm to escalate the run time as number of evaluations increased to get optimal solutions and the searching process is restricted to explore within the search space. Due to this constraints have lead other options to replace the uniform random number with better improvement in generating of random number. Aware of the weakness that occurred in uniform distribution, Khilwani et al, 2008[25] introduced Gaussian probability approach whereby appro-priate balance among exploration and exploitation is accomplished. Therefore, fast convergence and potential optimal solutions could be achieved.

3.2 The Computational Procedure of HMCR Parameter with Step Function

(1) Initialize the optimization problem and algorithm parameters. The parame-ters are HMS, set of each design variable, maximum number of improvisa-tion, HMCRmin, HMCRmax, PARmin, PARmax, BWmin and BWmax.
(2) Initialize harmony memory.
(3) Improvisation of new vector solution: A new vector solutions X_i = $\{x_i(1), x_i(2), ... x_i(n)\}$ is generated using procedure given below:

```
While(currentIteration<maximumIteration)
    HMCR=Generate through step function
    PAR= Equation 7 in [19]
    BW(m)=Equation 7 in [9]
    For (i<=number of variables)
        If (ran<HMCR), Choose a value from HM for the variable Xᵢ
        If(ran2< pitch adjusting rate), Modify the value by adding or subtract of
certain amount of BW
        End if
        Else Choose a random value
    End if
        Evaluate the newFitness (current objective function) and compared its value
with worst function.
    End while
    Sort the objective function by ascending order
    Satisfied number of iteration (NI)
    End
```

Fig. 4 The summarized proposed pseudocode for HMCR with step function

(4) Update the harmony memory: Selected the worst harmony vector x_{worst} in the current harmony memory and calculate $f(x_{new})$.
(5) Repeat Steps 3 and 4 until the termination criterion is satisfied. Figure 4 brief the algorithm for this proposed approach.

4   Result and Discussion

Table 2 and 3 delineates the experiment results obtained from the standard HS and step function respectively, which are referred as the minimum fitness function. It is important to note that justifications for mapping between HMCR value and interval of number improvisation are beyond the scope of this study. In other words, the rationalization for mapping each range of interval number of iteration to suitable value of HMCR will be provided in the subsequent experiment. Therefore, the framework will be equipped after the pattern of objective function is observed with numerous numbers of running experiments.

The result of the experiment was conducted in five runs for each experiment. Table 2 and 3 correspond to the results of the experiment that indicate the best objective function for respective technique. Based on the result, it reveals that the third run was an excellent objective function achieved by proposed approach at the value of HMCR 0.9416. This means that the global minimum is achieved at the value of HMCR 0.9416 at the 16248th number of iteration. While standard HS managed to reach the best objective function at 0.21 at the iteration of 9230th. This result proved theoretically that a low HMCR consider all permissible range of solutions. Table 2 shows the results of the experiment that used the traditional HS algorithm. Obviously, it shows that the traditional HS algorithm depends on the number of iterations to find an optimal solution. This shows a contradiction in the proposed algorithm that relies on combination of intensification and exploration. This contradiction is due to two reasons: Firstly, the use of fixed value of HMCR does not allow the algorithm to achieve the globally optimized solution and secondly, the way of adjusting HMCR is different from the standard HS.

Table 1 Summarized of parameter setting for HS's approach

	HMS	4
	HMCR	0.95
Standard HS	PAR	0.30
	BW	1
	NI	17700
	HMS	4
	$HMCR_{min}$	0.401
	$HMCR_{max}$	0.99
	PAR_{min}	0.01
Proposed Approach	PAR_{max}	0.99
	BW_{min}	0.01
	BW_{max}	0.02
	NI	17700

Table 2 The optimization results for standard HS

Number of experiment	Iteration	Best	Worst
1	9584	0.22	0.46
2	14584	0.22	0.44
3	9230	0.21	0.45
4	8193	0.22	0.45
5	2952	0.22	0.44

Table 3 The optimization results for proposed approach

Number of experiment	Iteration	Best	Worst
1	17322	0.15	0.45
2	17611	0.14	0.45
3	16248	0.13	0.44
4	16551	0.14	0.46
5	15833	0.15	0.45

5   Conclusion

In this paper, the effects on the adjustment strategy of HMCR using step function have been analyzed against the performance and efficiency of the HS algorithm. This investigation was compared to the standard harmony search (HS) algorithm, where the objective function was observed. It is found that the HMCR parameter was responsible to balance between the local exploitation and the global exploration in the HS algorithm. The step function introduced in this study has proven to successfully enhance the global minimum of the HS algorithm. The result also reported that low HMCR values should be applied in the early generation and ended with high values of HMCR in the final generation. This technique can avert the local optima, leading to the achievement of excellent global optimality. Further, the results from the experiments indicate that modification on the HMCR parameter results in some superior performances in comparison to the standard harmony search algorithm (HS). The adjustments of the HMCR parameter allows the parameter to obtain an excellent result of global minimum, whereas the fixed value of all parameters prevents us from achieving globally optimized solution. Therefore, fine tuning the HMCR parameters has a significant effect on the overall performance and is a good alternative in seeking excellent solutions. This framework is expected to be used in a variety of wide problems associated with optimization. For future work, it is suggested that the other two main operators involved in HS, which are the PAR and BW should be further manipulated because the judgment in determining the set of initial parameter values is still questionable. It requires further investigation in order to determine the initial value appropriately. Thus, this problem has motivated many researchers to further investigate this open problem in a wider discipline.

Acknowledgement This work was supported by research grants RAGS / 2013 / FTMK / ICT02 / 01 / B00039 under Ministry of Education of Malaysia and UniSZA for financial support.

References

1. Russell, S., Norvig, P.: Artificial intelligence: A modern approach, vol. 25. Prentice-Hall, Egnlewood Cliffs (1995)
2. Goldberg, D.E.: Genetic Algorithms in Search, Optimization and Machine Learning. Ad-dison-Wesley Longman Publishing Co., Inc., Boston (1989)
3. Dorigo, M., Maniezzo, V., Colorni, A.: Ant system: optimization by a colony of cooperating agents. IEEE Transactions on Systems, Man, and Cybernetics, Part B: Cybernetics **26**(1), 29–41 (1996)
4. Kennedy, J., Eberhart, R.C.: Particle swarm optimization. In: Proc. IEEE Intl. Conf. on Neural Networks. IEEE Service Center, Piscataway, pp. 1942–1948 (1995)
5. Walker, A., Hallam, J., Willshaw, D.: Bee-havior in a mobile robot: The construction of a self-organized cognitive map and its use in robot navigation within a complex, natural environment. In: IEEE International Conference on Neural Networks, pp. 1451–1456 (1993)
6. Geem, Z.W., Kim, J.H., Loganathan, G.V.: A new heuristic optimization algorithm: harmony search. Simulation **76**(2), 60–68 (2001)
7. Al-Betar, M.A., Doush, I.A., Khader, A.T., Awadallah, M.A.: Novel selection schemes for harmony search. Applied Mathematics and Computation **218**(10), 6095–6117 (2012)
8. Al-Betar, M.A., Khader, A.T., Zaman, M.: University course timetabling using a hybrid harmony search metaheuristic algorithm. IEEE Transactions on Systems, Man, and Cybernetics, Part C: Applications and Reviews **42**(5), 664–681 (2012)
9. Mahdavi, M., Fesanghary, M., Damangir, E.: An improved harmony search algorithm for solving optimization problems. Applied mathematics and computation **188**(2), 1567–1579 (2007)
10. Sharma, K.D., Chatterjee, A., Rakshit, A.: Design of a hybrid stable adaptive fuzzy controller employing Lyapunov theory and harmony search algorithm. IEEE Transactions on Control Systems Technology **18**(6), 1440–1447 (2010)
11. Ayvaz, M.T.: Application of harmony search algorithm to the solution of groundwater management models. Advances in Water Resources **32**(6), 916–924 (2009)
12. Ahmad, I., Mohammad, M.G., Salman, A.A., Hamdan, S.A.: Broadcast scheduling in packet radio networks using Harmony Search algorithm. Expert Systems with Applications **39**(1), 1526–1535 (2012)
13. Sivasubramani, S., Swarup, K.S.: Environmental/economic dispatch using multi-objective harmony search algorithm. Electric Power Systems Research **81**(9), 1778–1785 (2011)
14. Omran, M.G., Mahdavi, M.: Global-best harmony search. Applied Mathematics and Computation **198**(2), 643–656 (2008)
15. Pan, Q.K., Suganthan, P.N., Tasgetiren, M.F., Liang, J.J.: A self-adaptive global best harmony search algorithm for continuous optimization problems. Applied Mathematics and Computation **216**(3), 830–848 (2010)

16. Wang, C.M., Huang, Y.F.: Self-adaptive harmony search algorithm for optimization. Expert Systems with Applications 37(4), 2826–2837 (2010)
17. Chang-ming, X., Lin, Y.: Research on adjustment strategy of PAR in harmony search algorithm. In: International Conference on Automatic Control and Artificial Intelligence (ACAI 2012), pp. 1705–1708 (2012)
18. Al-Betar, M.A., Khader, A.T., Geem, Z.W., Doush, I.A., Awadallah, M.A.: An analysis of selection methods in memory consideration for harmony search. Applied Mathematics and Computation 219(22), 10753–10767 (2013)
19. Kumar, V., Chhabra, J.K., Kumar, D.: Parameter adaptive harmony search algorithm for unimodal and multimodal optimization problems. Journal of Computational Science 5(2), 144–155 (2014)
20. Ayvaz, M.T.: Identification of groundwater parameter structure using harmony search algorithm. Studies in Computational Intelligence 191, 129 (2009)
21. Using, B.U.S., Search, H., Shaffiei, Z.A., Abas, Z.A., Nizam, A.F., Rahman, A.: Optimization in Driver'S Scheduling For University, pp. 15–16 (2014)
22. Abas, Z.A., Binti Shaffiei, Z.A., Rahman, A.N.A., Shibghatullah, A.S.: Using Harmony Search For Optimising University Shuttle Bus Driver Scheduling For Better Operational Management
23. Caponetto, R., Fortuna, L., Fazzino, S., Xibilia, M.G.: Chaotic sequences to improve the performance of evolutionary algorithms. IEEE Transactions on Evolutionary Computation 7(3), 289–304 (2003)
24. Krohling, R.: Gaussian swarm: a novel particle swarm optimization algorithm. In: 2004 IEEE Conference on Cybernetics and Intelligent Systems, vol. 1, pp. 372–376 (2004)
25. Khilwani, N., Prakash, A., Shankar, R., Tiwari, M.K.: Fast clonal algorithm. Engineering Applications of Artificial Intelligence 21(1), 106–128 (2008)

KU Battle of Metaheuristic Optimization Algorithms 1: Development of Six New/Improved Algorithms

Joong Hoon Kim, Young Hwan Choi, Thi Thuy Ngo, Jiho Choi,
Ho Min Lee, Yeon Moon Choo, Eui Hoon Lee, Do Guen Yoo,
Ali Sadollah and Donghwi Jung

Abstract Each of six members of hydrosystem laboratory in Korea University
(KU) invented either a new metaheuristic optimization algorithm or an improved
version of some optimization methods as a class project for the fall semester 2014.
The objective of the project was to help students understand the characteristics of
metaheuristic optimization algorithms and invent an algorithm themselves focusing those regarding convergence, diversification, and intensification. Six newly
developed/improved metaheuristic algorithms are Cancer Treatment Algorithm
(CTA), Extraordinary Particle Swarm Optimization (EPSO), Improved Cluster HS
(ICHS), Multi-Layered HS (MLHS), Sheep Shepherding Algorithm (SSA), and
Vision Correction Algorithm (VCA). This paper describes the details of the six
developed/improved algorithms. In a follow-up companion paper, the six algorithms are demonstrated and compared through well-known benchmark functions
and a real-life engineering problem.

Keywords Cancer treatment algorithm · Extraordinary particle swarm
optimization · Improved cluster HS · Multi-layered HS · Sheep shepherding
algorithm · Vision correction algorithm

J.H. Kim(✉) · Y.H. Choi · T.T. Ngo · J. Choi · H.M. Lee · Y.M. Choo · E.H. Lee
School of Civil, Environmental and Architectural Engineering, Korea University,
Seoul 136-713, South Korea
e-mail: {jaykim,younghwan87,y999k,dlgh86}@korea.ac.kr, tide4586@yahoo.com,
{chooyean,hydrohydro}@naver.com

D.G. Yoo · A. Sadollah · D. Jung
Research Center for Disaster Prevention Science and Technology, Korea University,
Seoul 136-713, South Korea
e-mail: godqhr425@naver.com, sadollah@korea.ac.kr, donghwiku@gmail.com

© Springer-Verlag Berlin Heidelberg 2016
J.H. Kim and Z.W. Geem (eds.), *Harmony Search Algorithm*,
Advances in Intelligent Systems and Computing 382,
DOI: 10.1007/978-3-662-47926-1_19

197

1   Introduction

Each of six members of hydrosystem laboratory in Korea University (KU) invented either a new metaheuristic optimization algorithm or an improved version of Harmony Search (HS) [1,2] or ParticleSwarm Optimization (PSO) [3] as a class project for the fall semester 2014. The objective of the project was to help students understand the characteristics of metaheuristic optimization algorithms and invent an algorithm by themselves considering convergence, diversification, and intensification. Two new algorithms are Cancer Treatment Algorithm (CTA) and Vision Correction Algorithm and four algorithms are an improved version of existing algorithms: Extraordinary PSO (EPSO) and Sheep Shepherding Algorithm (SSA) are the variants of PSO, while Improved Cluster HS (ICHS) and Multi-Layered HS (MLHS) are the variants of HS. Through performance tests using well-know benchmark functions, the algorithms have been refined to enhance their global and local search ability and to minimize the number of the required parameters. This paper describes the details of the six optimization algorithms.

2   Six Metaheuristic Algorithms

Six new/improved metaheuristic optimization algorithms are CTA, EPSO, ICHS, MLHS, SSA, and VCA. ICHS and MLHS are the variants of HS while EPSO and SSA are those of PSO. CTA and VCA are newly developed in this study. The following subsections describe the details of the new/improved metaheuristic algorithms.

2.1   Cancer Treatment Algorithm (CTA)

CTA mimics the medical procedure of cancer treatment. Cancer is the result of cells that uncontrollably grow and do not die. If a person is diagnosed cancer from medical examinations such as magnetic resonance imaging and computed tomography, a type of treatment/surgery is then decided based on the tumor size and potential of transition. Generally, if the size is less than 2 cm, heat treatment is adopted. On the other hand, if the tumor is large and expected to have high potential of transition to another organ, it is removed by surgery (cancer removal surgery). This generic medical examination, treatment/surgery, and relevant decision processes help cure the disease (the best state), which is similar to the optimization process seeking the global optimum.

CTA, a population-based algorithm, generates new solutions for next generation based on three operators. By operation 1, k new solutions (out of total n new population) are generated considering the high fitness solutions stored in the memory called K (the memory size is K). While a single new solution is generated by operation 3, the remaining solutions $(n - (k + 1))$ of the next population are generated based on operation 2. Then, the combined $2n$ population (parent and children) is sorted according to fitness. The top n solutions become the parent population for the next generation and the top k solutions are stored in the memory K. This procedure is continued until stopping criteria are met. The following equations describe how new solutions are generated by three operations.

Operation 1:

$$x_{t+1}^{i,j} = x_t^{i,j} + rnd[-1,1] \times I^{-2\alpha} \qquad (1)$$

where $x_{t+1}^{i,j}$ is the ith component of the jth new solution ($j \in$ the memory K) in the $(t+1)^{th}$ generation, I is laser beam intensity and $I = \frac{\sum_{j=1}^{k} f_j}{k}$ where f_i is fitness value of the solution j in the memory K, and α is reduction factor and $\alpha = exp\left(\frac{f_j^{n-1}}{f_j^{n-2}}\right)$.

Operation 2:

$$x_{t+1}^i = x_t^i \quad if \ i \notin n_{remove} \qquad (2)$$

$$x_{t+1}^i = x_t^i \times \gamma \quad if \ i \in n_{remove} \qquad (3)$$

where n_{remove} is the set of solutions that will be generated by operation 2, and γ is revival rate and $1 - abs\left(\frac{f_{min}}{f_i^{t-1}}\right)$ where f_{min} is the smallest fitness in the population and f_i^{t-1} is the ith solution's fitness in the $(t-1)^{th}$ generation.

Operation 3:

$$x_{t+1}^i = x_t^i \quad if \ rnd > R \ or \ t = 1 \qquad (4)$$

$$x_{t+1}^i = x_t^i + rnd[-1,1] \times (x_t^i - x_{t-1}^i) \quad if \ rnd \le R \qquad (5)$$

where R is research rate.

2.2 Extraordinary Particle Swarm Optimization (EPSO)

PSO simulates the movement behaviors of animal swarm. Particles in a swarm have a tendency to share information on the current nearest location to food, which makes all particles move toward the current best particle. In addition, particle's past best location is also considered to determine the next location. However, the aforementioned strategies can result in being entrapped in local optimum because the search diversity can be limited.

Therefore, EPSO employed a different strategy. While high fitness particles have more probability to be selected as a target particle, each particle chooses a target particle which can be any particle in a swarm.

In EPSO, initial N_{pop} particles are randomly generated as original PSO. Each particle's location at the iteration $t+1$, $X_i(t + 1)$, is determined either by adding particle velocity to the current location or by randomly as

$$X_i(t + 1) = \begin{cases} X_i(t) + V_i(t + 1) & if\ T \in (0, T_{up}) \\ LB_i + rand \times (UB_i - LB_i) & otherwise \end{cases} \tag{6}$$

where $V(t + 1)$ is particle velocity for the component i and $C \times (X_{Ti}(t) - X_i(t))$ where C is movement parameter and $X_{Ti}(t)$ is the selected target particle's component i at the iteration t, T is a random integer generated in $[0, N_{pop}]$, T_{up} is selection parameter and round($\alpha \times N_{pop}$) where α is target range parameter, and LB_i and UB_i are the lower and upper bound of the decision variable i, respectively.

2.3 Improved Cluster Harmony Search (ICHS)

HS was inspired by the musical ensemble [1,2]. HS implements the harmony enhancement process and the sets of "good harmony" are saved to a solution space termed harmony memory (HM), which is a unique feature of HS compared to other evolutionary optimization algorithms.

HS generates new solutions either by random generation or by considering the good solutions in HM. For each component x_i of a new solution vector $\underline{x}$, one of the stored values of the component in HM is selected with harmony memory considering rate (HMCR) or a random value within the allowed range is chosen with the probability of 1-HMCR. Each stored value in HM is equally probable to be selected. After HM consideration, HS scans each decision variable in the new solution and changes it to the neighborhood values with pitch adjusting rate (PAR). If the new solution is better than the worst solution in HM, the latter is replaced with the former. The above process continues until stopping criteria are satisfied.

ICHS is a variant of HS. ICHS differs from other HS variants in the HM consideration. That is, ICHS assigns high probability to be selected for the component of the high fitness solutions. Note that in standard HS (SHS) the stored component values of the solutions in HM have the same probability to be selected regardless of the solutions' fitness. Other mechanisms such as pitch adjusting are same as SHS.

First, the solutions in HM are sorted according to fitness. Then, they are divided into subgroups by k-Means clustering. k-Means clustering partitions the solutions into k distinct clusters based on a Euclidean distance to the centroid of a cluster. A probability value of the solution j (P_{rj}) is calculated as

$$P_{rj} = e^{-\frac{\alpha f_j}{f_{max}}} \tag{7}$$

where α is selection pressure, f_j is fitness value of the solution j, and f_{max} is the maximum fitness value in HM.

Then, P_{rj} is normalized to have the cumulative probability of 1 as

$$P_{rni} = \frac{P_{rj}}{\sum_{j=1}^{nm} P_{rj}} \tag{8}$$

where P_{rnj} is the normalized probability of the solution j, and nm is the HM size.

Therefore, the kth cluster is selected with the probability $\sum_{j \in Cluster_k} P_{rnj}$ where $Cluster_k$ is the set of solutions classified as the kth cluster. In ICHS, any stored values of the component in the selected cluster can be selected with equal probability.

2.4   Multi-Layered Harmony Search (MLHS)

MLHS is another variant of HS. The special feature of MLHS is that HMCR and PAR operations are conducted based on multi-layered HMs. A layer is in the form of grid in which each cell has a sub-memory that stores m solutions (see Fig. 1.). There exists a hierarchical tournament between the two neighboring layers in which the best solutions in the lower layer are elevated to fill the sub-memories in the upper layer.

MLHS begins by initializing sub-memories in the first (bottom) layer (the largest grid in Fig. 1). Initial solutions are randomly generated same as SHS. For example, total $l^2 m$ solution are generated for the first layer with l by l grid and the sub-memory size of m ($8^2 \times 4 = 256$ where $l = 8$ and $m = 4$ as shown in Fig. 1). In the considered example in Fig. 1, the solutions (3, 4), (2,6), (6,2), and (9,6) are stored at the lower left corner cell of the grid in the first layer. Then, a new solution is generated for each sub-memory either by random generation or harmony memory consideration and pitch adjusting, as for SHS. If the new solution is better than the worst solution in sub-memory, the latter is replaced with the former.

The sub-memories of the second layer are filled with the best solutions from the sub-memories of the neighborhood cells in the first (lower) layer. For example, the best solutions from the four cells at the lower left corner of the first layer (the cells in the green box in Fig. 1), (2,6), (5,6), (4,4), and (5,4), are provided to the sub-memory of the lower left corner of the second layer. Similarly, a new solution is generated based on the SHS rules. This tournament between the two neighboring layers continues until the top layer where the grid dimension is 1 by 1.

In addition to the unique hierarchical structure, MLHS adopted dynamic parameters where HMCR and band width are changed over iterations.

Sub-memory 1~4 of 1st layer

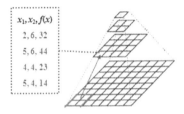

$x_1,x_2,f(x)$	$x_1,x_2,f(x)$	$x_1,x_2,f(x)$	$x_1,x_2,f(x)$
3, 4, 34	5, 6, 44	2, 4, 32	5, 4, 14
2, 6, 32	3, 6, 52	4, 4, 23	2, 2, 24
6, 2, 37	7, 3, 87	3, 1, 52	1, 2, 63
9, 6, 54	1, 5, 64	1, 1, 68	9, 3, 22

Sub-memory 1 of 2nd layer

$x_1,x_2,f(x)$
2, 6, 32
5, 6, 44
4, 4, 23
5, 4, 14

Fig. 1 Hierarchical tournament in MLHS

2.5   Sheep Shepherding Algorithm (SSA)

SSA is inspired by sheep shepherding which consists of two main operators: sheep gathering and shepherd herding (or guarding). SSA is similar with PSO in considering animal's flocking nature in the optimization. However, SSA adopts the shepherd operation which is intended to perturb worst solutions to enhance the probability of escaping from local optimum and finding global optimum. This unique feature of SSA and mimics the behavior of a shepherd which encourages the sheep which falls behind.

In SSA, a sheep represents a solution and initial sheep are randomly generated. In each iteration, a sheep can change its position or stay with the probability d and 1-d, respectively. If a sheep is determined to change its location, the component i of the jth sheep at the tth iteration $VecH_t(i,j)$ is determined as follows:

$$VecH_t(i,j) = H \times VecH_{t-1}(i,j) + C \times VecC_t(i,j) + P_s \times VecR_{s,t}(i,j) \quad (9)$$

where $VecC_t(i,j)$ is gathering vector and $CM_t(i) - sheep_t(i,j)$ where $CM_t(i)$ is the ith component of the center of sheep flock at t^{th} iteration and *sheep* represents the location of the sheep, $VecR_{s,t}(i,j)$ is herding vector and $sheep_t(i,j) - VecS_t(i)$ where $VecS_t(i)$ is the ith component of shepherd, and H, C, and P_s are user defined parameters.

If a sheep is determined to change its location in the next iteration, the new location is calculated as

$$sheep_t(i,j) = sheep_{t-1}(i,j) + VecH_t(i,j) \quad (10)$$

If a sheep stays at its previous location, minor movement is added to the previous position as

$$sheep_t(i,j) = (1 \pm m \times rnd) \times sheep_{t-1}(i,j) \qquad (11)$$

where m is movement parameter.

Finally, the component i of shepherd $VecS(i)$ is updated to keep perturbing the worst perform sheep.

$$VecS_t(i) = \begin{cases} VecS_{t-1}(i) + sheep_t(i,N) & if\ sheep(i,N) > H \\ VecS_{t-1}(i) + L & otherwise \end{cases} \qquad (12)$$

where $sheep_t(i,N)$ is the component i of the worst sheep (N^{th} sheep after sorting sheep), H is sheep criterion parameter, and L is a random term. The above updates of sheep and shepherd locations continues until stopping criteria are met.

2.6   Vision Correction Algorithm (VCA)

VCA simulates vision correction process. The repetitive processes for making a good lens (global optimum) are similar to optimization process. In VCA, a lens is a solution candidate. First, initial lens are generated within the allowed ranges and their fitness are calculated. For each iteration, a real random number r is generated and compared with the division rate $dr1$. If r is less than $dr1$, a new solution is generated randomly. Otherwise, roulette wheel selection is performed to select a solution from population. After the selection, only the last three operations (described later) are performed without processing myopia or hyperopia operations. A randomly generated solution is refined either by myopia or hyperopia operations. Myopia correction is carried out to treat near-sightedness (positive direction search) with the probability $dr2$. Hyperopia correction is performed to fix farsightedness (negative direction search) with the probability 1-$dr2$. Note that the division rates $dr1$ and $dr2$ are dynamic variables which change over iterations.

Myopia correction increases the focal length of lens and performs search in positive direction as given follows:

$$x(t,i) = x(t-1,i) + rnd \times (UB_i - x(t-1,i)) \qquad (13)$$

On the other hand, hyperopia correction decreases the focal length of lens and performs negative direction search given in the following equation:

$$x(t,i) = LB_i + rnd \times (x(t-1,i) - LB_i) \qquad (14)$$

Regardless of whether the above two operators are processed, the last three operators are conducted: brightness increase, compression, and astigmatism correction. Note that the frequency of processing each of the last three operators is controlled by a predefined probability. The brightness of lens is measured by modulation transfer function (MTF).

$$MTF = \sqrt{\sum_{i=1}^{n} ang_i} \tag{15}$$

where $ang_i = \sqrt{\dfrac{dx_i}{\sqrt{\sum_{i=1}^{n} dx_i^2}}}$ where dx_i is Euclidean distance in the ith axis (dimension) from the current best solution.

Lens is compressed to decrease its thickness. Compression factor (CF) is considered to control convergence speed. The bright increase and compression operation are embedded in the following equation:

$$x(t,i) = x(t,i) \times \left\{ 1 + MTF \times rnd[-1,1] \times (1 - \tfrac{t}{T})^{CF} \right\} \tag{16}$$

where T is maximum number of iteration allowed.

Finally, astigmatic correction is performed.

$$x(t,i) = x(t,i) \times \left\{ 1 + rnd[-1,1] \times (\sin\theta)^2 \right\} \tag{17}$$

where θ is axial parameter.. The whole operations continue until stopping criteria are met.

3 Summary

This paper introduced six new/improved metaheuristic optimization algorithms. CTA mimics generic cancer treatment process which consists of three differentoperators. All three operators are conducted to produce children population. On the other hand, VCA,inspired by vision correction procedures ,consists of myopia and hyperopia correction, brightness increase, compression,etc. The frequency of each operation is controlled by division rates. ICHS and MLHS are the variants of SHS. ICHS increases selection pressure for the solutions classified as high fitness clusters, while MLHS adopts hierarchical tournament for filling the HMs of the cells in the upper layers. EPSO and SSA are considered as the variants of PSO.

Table 1 indicates the number of parameters used in the six algorithms. The number of parameters is smallest in EPSO, while that is largest in MLHS. In a follow-up companion paper, the six algorithms are demonstrated and compared through well-known benchmark functions and a real-world engineering problem.

Table 1 Number of parameters (population size is included as a user parameter, while number of maximum iteration is excluded)

CTA	EPSO	ICHS	MLHS	SSA	VCA
5	3	6	7	6	7

Acknowledgement This work is supported by a National Research Foundation of Korea (NRF) grant funded by the Korean government (MSIP) (NRF- 2013R1A2A1A01013886).

References

1. Geem, Z.W., Kim, J.H., Loganathan, G.V.: A new heuristic optimization algorithm: harmony search. Simulation **76**(2), 60–68 (2001)
2. Kim, J.H., Geem, Z.W., Kim, E.S.: Parameter Estimation Of The Nonlinear Muskingum Model Using Harmony Search. JAWRA **37**(5), 1131–1138 (2001)
3. Kennedy, J., Eberhart, R.C.: Particle swarm optimization. In: Proc. IEEE International Conference on Neural Networks, Perth, Australia, pp. 1942–1948. IEEE Service Center, Piscataway (1995)

KU Battle of Metaheuristic Optimization Algorithms 2: Performance Test

Joong Hoon Kim, Young Hwan Choi, Thi Thuy Ngo, Jiho Choi, Ho Min Lee, Yeon Moon Choo, Eui Hoon Lee, Do Guen Yoo, Ali Sadollah and Donghwi Jung

Abstract In the previous companion paper, six new/improved metaheuristic optimization algorithms developed by members of Hydrosystem laboratory in Korea University (KU) are introduced. The six algorithms are Cancer Treatment Algorithm (CTA), Extraordinary Particle Swarm Optimization (EPSO), Improved Cluster HS (ICHS), Multi-Layered HS (MLHS), Sheep Shepherding Algorithm (SSA), and Vision Correction Algorithm (VCA). The six algorithms are tested and compared through six well-known unconstrained benchmark functions and a pipe sizing problem of water distribution network. Performance measures such as mean, best, and worst solutions (under given maximum number of function evaluations) are used for the comparison. Optimization results are obtained from thirty independent optimization trials. Obtained Results show that some of the newly developed/improved algorithms show superior performance with respect to mean, best, and worst solutions when compared to other existing algorithms.

Keywords Cancer treatment algorithm · Extraordinary particle swarm optimization · Improved cluster HS · Multi-Layered HS · Sheep shepherding algorithm · Vision correction algorithm

J.H. Kim(✉) · Y.H. Choi · T.T. Ngo · J. Choi · H.M. Lee · Y.M. Choo · E.H. Lee
School of Civil, Environmental and Architectural Engineering,
Korea University, Seoul 136-713, South Korea
e-mail: {jaykim, younghwan87,y999k, dlgh86}@korea.ac.kr, tide4586@yahoo.com,
{chooyean,hydrohydro}@naver.com

D.G. Yoo · A. Sadollah · D. Jung
Research Center for Disaster Prevention Science and Technology,
Korea University, Seoul 136-713, South Korea
e-mail: godqhr425@naver.com, sadollah@korea.ac.kr, donghwiku@gmail.com

© Springer-Verlag Berlin Heidelberg 2016
J.H. Kim and Z.W. Geem (eds.), *Harmony Search Algorithm*,
Advances in Intelligent Systems and Computing 382,
DOI: 10.1007/978-3-662-47926-1_20

1   Introduction

Each of six members of hydrosystem laboratory in Korea University (KU) invented either a new metaheuristic optimization algorithm or an improved version of Harmony Search (HS) [1,2] or Particle Swarm Optimization (PSO) [3] as a class project for the fall semester 2014. The objective of the project was to help students understand the characteristics of metaheuristic optimization algorithms and invent an algorithm by themselves considering convergence, diversification, and intensification. Two new algorithms are Cancer Treatment Algorithm (CTA) and Vision Correction Algorithm and four algorithms are improved versions of existing algorithms: Extraordinary PSO (EPSO) and Sheep Shepherding Algorithm (SSA) are the variants of PSO, while Improved Cluster HS (ICHS) and Multi-Layered HS (MLHS) are the variants of HS. The details of the six algorithms were described in the previous companion paper [4].

In this paper, the six algorithms were tested and compared through six well-known unconstrained benchmark functions and a pipe sizing problem of water distribution network. Performance measures such as mean, best, and worst solution (under given maximum number of function evaluations) were used for the comparison. This paper first describes the performance measures used and the details of test problems. Then, the optimization results of test problems were presented and conclusions were made at the end of the paper.

2   Performance Measures

2.1   Mean, Best, and Worst Error

Each test problem is solved by using the six metaheuristic algorithms. The predefined number of independent optimizations is performed with randomly generated initial solutions. Sufficiently large number of trials is considered to exclude the biased performance of better initial solutions. Mean, worst, and best errors obtained from the six algorithms are compared. Mean error is the averaged absolute deviation of the optimal solutions' objective function value (OFV) from the known optimal OFV and represents the expected solution quality obtained by an algorithm. Worst error indicates the fitness difference between the worst solution and the known optimal solution and the reliability of algorithm performance. Finally, best error is the best solution's absolute deviation. Theses error metrics can be measured only when given the known optimal OFV. Note that one of the three performance measures cannot solely used to compare algorithms' performances. Three measures should be simultaneously considered in the evaluation. For consistent comparison, the same predefined number of function evaluations (NFEs) is commonly allowed for the six algorithms as a stopping condition.

2.2   Success Rate

The success rate is defined as the ratio of the runs to total runs made in which the best solution's OFV found within the predefined NFEs falls within the acceptable limit (*Slimit*). This metric represents the robustness of algorithm performance.

An algorithm with high success rate has higher likelihood of finding near-optimal solution than the algorithm with low success rate.

2.3 Number of Function Evaluations

Mean number of functional evaluations (NFEs) is the expected NFEs at which the best solution's OFV falls less than *Slimit* and indicates search efficiency. Another NFEs measure was taken into account for the six algorithms comparison. Expected minimum NFEs (EMNFEs) is the NFEs at which the averaged evolution of the solutions falls less than *Slimit*. Therefore, in order to calculate EMNFEs, the OFVs of the best solutions in all trials are averaged at every iteration and checked whether less than *Slimit* or not.

3 Test Problems

3.1 Benchmark Functions

Six well-know benchmark problems are used to compare the six algorithms: three two-dimension (2D) and three 30D problems. Three 2D problems are Rosenbrock, Easom, and Goldstein functions. Three 30D problems are Rastrigin, Griewank, and Ackley functions. All problems are unconstrained, minimization problem that

Table 1 Benchmark Functions

Dimension	Functions	Equations
2D	Rosenbrock	$Minimize\ F = 100(x_2 - x_1^2)^2 + (x_1 - 1)^2$ $\min(F) = f(1,1) = 0$
	Easom	$Minimize\ F = -\cos(x_1)\cos(x_2)\,exp\big(-((x_1 - \pi)^2 + (x_2 - \pi)^2)\big)$ $\min(F) = f(\pi,\pi) = -1$
	Goldstein	$Minimize\ F = \{1 + (x_1 + x_2 + 1)^2(19 - 14x_1 + 3x_1^2 - 14x_2 + 6x_1x_2 + 3x_2^2)\} \times$ $\{30 + (2x_1 - 3x_2)^2(18 - 32x_1 + 12x_1^2 + 48x_2 - 36x_1x_2 + 27x_2^2)\}$ $\min(F) = f(0,-1) = 3$
30D	Rastrigin	$Minimize\ F = -\sum_{i=1}^{30}\big(x_i \sin(\sqrt{|x_i|})\big)$ $\min(F) = f(0,\dots,0) = 0$
	Griewank	$Minimize\ F = \frac{1}{4000}\sum_{i=1}^{30} x_i^2 - \prod_{i=1}^{30} \cos\left(\frac{x_i}{\sqrt{i}}\right) + 1$ $\min(F) = f(0,\dots,0) = 0$
	Ackley	$Minimize\ F = -20\exp\left(-0.2\sqrt{\frac{1}{30}\sum_{i=1}^{30} x_i^2}\right) - \exp\left(\frac{1}{30}\sum_{i=1}^{30}\cos(2\pi x_i)\right) + 20 + e$ $\min(F) = f(0,\dots,0) = 0$

has known global optimum. An optimization run is considered a success if the best solution's error is less than or equal to 1×10^{-10} ($Slimit = 1 \times 10^{-10}$) for 2D and $Slimit = 1 \times 10^{-5}$ for 30D problems. For 2D problem, the maximum NFEs was set to 50,000. On the other hand, the predefined NFEs of 100,000 was allowed for 30D problems. The success threshold and stopping criterion were consistently considered for each algorithm for having fair comparison.

3.2   Water Distribution Network Pipe Sizing Problem

The pipe sizing of water distribution network (WDN) is to determine optimal discrete pipe sizes for a system to minimize total economic cost, while satisfying constraints on the pressure constraints and system hydraulics. WDN pipe sizing problem is generally difficult to solve because the problem consists of a set of nonlinear equation based on the conservation laws of mass and energy and its dimension is large. Therefore, the complexity of the problem has been reduced to solve the problem using traditional deterministic optimization methods, linear and nonlinear programming [5], [6]. Genetic algorithm was also used for solving the problem [7,8].

Balerma network is a well-known benchmark network supplied by gravity flow from four fixed head sources. In the WDN research community, numerous optimization algorithms has been tested by solving the pipe sizing problem of Balerma network and the optimal results are compared with respect to the solution quality and computational speed [9-12].

Balerma network is a loop-dominated network with total 8 loops and consists of 443 nodes and 454 pipes. The pipe sizing problem is to minimize total economic cost with the constraints on the pressure requirements.

$$Minimize\ F = \sum_{i=1}^{n}[C(D_i) \times L_i] \qquad (1)$$
$$s.t.\ \ P_j \geq P_{min} \quad j = 1, \dots, m \qquad (2)$$

where $C()$ is unit pipe cost (Euro/m), D_i is pipe diameter in mm of the ith pipe, L_i is length of the pipe i, P_i is pressure at the jth node, and P_{min} is the minimum pressure requirement. The conservations of mass and energy were implicitly satisfied through a hydraulic simulation performed on EPANET [13].

For consistent comparison of the six algorithms' performance, the same constraint handing approach was used. The penalty cost ($PCost$) is added to total economic cost if the pressure constraint is not met ($P_j < P_{min}$).

$$PCost = \sum_{j=1}^{m}[\alpha(P_{min} - P_j) + \beta] \qquad (3)$$

where α is deviation constant and 10^{20}, and β is interceptor constant 10^7. Commercial pipe diameter set and unit pipe cost reported in [14] are used. Following [14], new pipes are assumed to have Darcy-Weisbach roughness coefficient $e = 0.0025$ mm. For this problem the maximum 45,400 NFEs were allowed.

4 Application Results

We ran each algorithm 100 times for each benchmark problems, while 30 optimization tasks were made for the Balerma network pipe sizing problem. Mean, worst, and best errors were obtained, and success rates and NFEs measures were calculated from the optimization results.

Table 1 shows average errors of the six algorithms for the six benchmark problems. VCA outperformed other algorithms finding the global optimums for all benchmark problems (success rate is 100%). Note that the algorithm that has the smallest mean NFEs and EMNFE (the values are not presented in this paper) was different in each problem. EPSO was the second best algorithms which found the global optimums in all the function except Ackley functions. EPSO and SSA both of which were stem from PSO showed similarly good performances. The quality of solution found by SSA in the Rosenbrock and Ackley was the best among other solutions. It can be concluded that perturbing worst solutions by shepherd and considering more target particles enhances the efficiency of swarm-based algorithms. MLHS was performed worst finding the known optimal solution except for the Easom and Goldstein functions. VCA's adopting various operators such as myopia and hyperopia operators and astigmatic correction resulted in robustly good performances on the benchmark problems.

Table 2 Mean errors of the six algorithms in benchmark problems

Problems	CTA	EPSO	ICHS	MLHS	SSA	VCA
Rosenbrock	3.93E-11	0	3.29E-21	6.13E-12	2.21E-22	0
Easom	2.46E-11	0	0	0	0	0
Goldstein	2.33E-11	0	0	0	0	0
Rastrigin	0	0	8.93E-05	7.29E-07	0	0
Griewank	0	0	0	2.36E-06	0	0
Ackley	4.44E-16	1.47E-14	1.11E-09	1.70E-06	8.70E-16	0

It is worth mentioning that for most methods different user parameter sets have been used for the benchmark functions. Talking about the second problem, for each algorithm, thirty optimization runs were performed to solve the Balerma pipe sizing problem. Mean, worst, and best costs of the six algorithms were obtained and compared. Again, VCA outperformed other algorithms, while SSA was the second best algorithm. The best cost found by VCA was about 84% of that found by EPSO which was the worst performed algorithm. On average, CTA required the smallest NFEs for finding a feasible solution. However, the quality of final solution was mostly determined by algorithm's exploitation efficiency. As can be seen in Fig. 1, VCA has the highest rate of improvement after the exploration phase (i.e., for NFEs > 10,000).

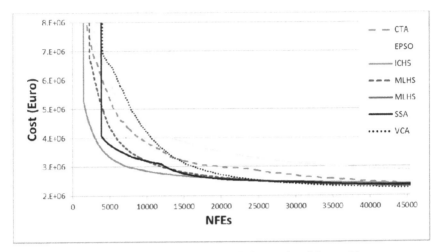

Fig. 1 Average evolution of solution costs for the Balerma pipe sizing problem

5 Summary and Conclusions

Six new/improved metaheuristic optimization algorithms were developed by members of Hydrosystem laboratory in Korea university (KU) in the previous companion paper. The six algorithms are CTA, EPSO, ICHS, MLHS, SSA, and VCA. In this study, the six algorithms were tested and compared through six well-known unconstrained benchmark functions and Balerma network pipe sizing problem. Mean, best, and worst solutions of the six algorithms were compared for the comparison. Regardless of the types of problems, VCA outperformed other algorithms, and overall, SSA was the second best algorithm.

However, this study has several limitations that future research should address. First, each of the six algorithms should be tuned more to make their efficiency comparable to recently published algorithms. Second, the number of parameters in each algorithm should be decreased to mitigate the difficulty in setting the parameters. Therefore, the sensitivity analysis on the existing parameters should be conducted to remove the parameters that have no significance influence on the algorithm's performance.

Acknowledgement This work is supported by a National Research Foundation of Korea (NRF) grant funded by the Korean government (MSIP) (NRF- 2013R1A2A1A01013886).

References

1. Geem, Z.W., Kim, J.H., Loganathan, G.V.: A new heuristic optimization algorithm: harmony search. Simulation **76**(2), 60–68 (2001)
2. Kim, J.H., Geem, Z.W., Kim, E.S.: Parameter Estimation Of The Nonlinear Muskingum Model Using Harmony Search. JAWRA **37**(5), 1131–1138 (2001)
3. Kennedy, J., Eberhart, R.C.: Particle swarm optimization. In: Proc. IEEE International Conference on Neural Networks (Perth, Australia), pp. 1942–1948. IEEE Service Center, Piscataway (1995)
4. Kim, J.H., Choi, Y.H., Choi, J.H., Lee, H.M, Choo, Y.M., Ngo, T.T., Lee, E.H., Yoo, D.G., Sadollah, A., Jung, D.H.: KU battle of metaheuristic optimization algorithms 1: development of six new algorithms. In: Proceedings of the 2015 ICHSA International Conference on Harmony Search Algorithm (Seoul, Korea) (2015)
5. Alperovits, E., Shamir, U.: Design of optimal water distribution systems. Water Resour. Res. **13**(6), 885–900 (1997)
6. Lansey, K.E., Mays, L.W.: Optimization model for water distribution system design. J. Hydraul. Eng. **115**(10), 1401–1418 (1989)
7. Simpson, A.R., Dandy, G.C., Murphy, L.J.: Genetic algorithms compared to other techniques for pipe optimization. J. Water Resour. Plann. Manage. **120**(4), 423–443 (1994)
8. Savic, D.A., Walters, G.A.: Genetic algorithms for least-cost design of water distribution networks. J. Water Resour. Plann. Manage. **123**(2), 67–77 (1997)
9. Bolognesi, A., Bragalli, C., Marchi, A., Artina, S.: Genetic heritage evolution by stochastic transmission in the optimal design of water distribution networks. Adv. Eng. Software **41**(5), 792–801 (2010)
10. Saldarriaga, J., Paez, D., Cuero, P., Leon, N.: Optimal design of water distribution networks using mock open tree topology. In: World Environmental and Water Resources Congress, pp. 869–880 (2013)
11. Saldarriaga, J., Paez, D., Cuero, P., Leon, N.: Optimal power use surface for design of water distribution systems. In: 14th Water Distribution Systems Analysis Conference, WDSA 2012, Adelaide, South Australia, p. 468, September 24–27, 2012
12. Zheng, F., Simpson, A.R., Zecchin, A.C.: A combined NLP differential evolution algorithm approach for the optimization of looped water distribution systems. Water Resources Research **47**(8) (2011)
13. Rossman, L.: EPANet2 user's manual. U.S. Environmental Protection Agency, Washington, DC (2000)
14. Reca, J., Martínez, J.: Genetic algorithms for the design of looped irrigation water distribution networks. Water Resour. Res. **42**(5) (2006)

Part IV
Other Nature-Inspired Algorithms

Modified Blended Migration and Polynomial Mutation in Biogeography-Based Optimization

Jagdish Chand Bansal

Abstract Biogeography-based optimization is a recent addition in the class of population based gradient free search algorithms. Due to its simplicity in implementation and presence of very few tuning parameters, it has become very popular in very short span of time. From its inception in 2008, it has seen many changes in different steps of the algorithms. This paper incorporates the modified blended migration and polynomial mutation in the basic version of BBO. The proposed BBO is named as BBO with modified blended crossover and polynomial mutation (BBO-MBLX-PM). The performance of proposed BBO is explored over 20 test problems and compared with basic BBO as well as blended BBO. Results show that BBO-MBLX-PM outperforms over BBO and other considered variants of BBO.

Keywords Biogeography based optimization · Meta-heuristics · Evolutionary algorithms

1 Introduction

Biogeography-Based Optimization (BBO) is a meta-heuristic algorithm for numerical optimization introduced by Dan Simon in 2008 [9]. The algorithm is based on the natural migration of species between habitats. Migration of species between habitats allows the information flow between them. The candidate solution movement in the search space is similar to that of Particle Swarm Optimization (PSO), i.e. the original population is not die out after each iteration but makes itself better [8]. This movement is not like that of Genetic Algorithms, where less fit individuals die out and in next iteration a new population is evolved. Therefore, in author's view, BBO should not be regarded as an evolutionary algorithm. Also it should be noted that in BBO, all the candidate solutions do not tend to cluster at a single point as in PSO. In this way, BBO process is significantly different from PSO also.

J.C. Bansal(✉)
South Asian University, New Delhi, India
e-mail: jcbansal@gmail.com

© Springer-Verlag Berlin Heidelberg 2016   217
J.H. Kim and Z.W. Geem (eds.), *Harmony Search Algorithm*,
Advances in Intelligent Systems and Computing 382,
DOI: 10.1007/978-3-662-47926-1_21

BBO has been modified in many ways from its inception in 2008. In [8] Lohokare et al. presented a memetic BBO named as aBBOmDE. In aBBOmDE, the performance of BBO is accelerated with the help of a modified mutation and clear duplicate operators. In order to make BBO more explorable for global optimization POLBBO is developed in [11]. In POLBBO, proposed polyphyletic migration operator utilizes four individuals' features to generate a new individual. Polyphyletic migration operator is introduced to increase the population diversity. Simultaneously, employing an orthogonal learning (OL) strategy based on orthogonal experimental design makes POLBBO quickly discover more useful information from the search experiences and efficiently utilize the information to construct a more promising solution. In [10], Simon et al. found that in BBO only one independent variable changes at a time and therefore BBO performs poorly on non-separable problems. In order to make BBO efficient to deal with non-separable problems they proposed a linearized version of BBO, namely LBBO. LBBO is combined with periodic re-initialization and local search operators to obtain an algorithm for global optimization in a continuous search space. Gong et al. proposed a real coded BBO (RCBBO) in [5] as an algorithm with better diversified population. For a better diversified population they implemented a mutation operator in BBO and did experiments with three mutation operators (i.e., Gaussian mutation, Cauchy mutation, and Levy mutation). In [6] Li et al. proposed a sinusoidal migration model called perturb migration in BBO and then Gaussian mutation is incorporated into perturbed BBO. The so obtained BBO is named as perturb biogeography based optimization (PBBO). Li et al. proposed two extensions to the basic BBO in [7]. They proposed multi-parent crossover and the Gaussian mutation operator in order to improve the exploration ability of BBO.

Migration and mutation are two important features of BBO algorithm. The performance of BBO can be enhanced significantly by revisiting original migration and mutation operators. In continuation of earlier modifications, this paper proposes a variant of BBO which incorporates two modifications in the basic BBO: a modified blended migration and polynomial mutation. The proposed variant of BBO is named as BBO with modified blended crossover and polynomial mutation (BBO-MBLX-PM). Experiments on test problems in the continuous search space show that there is a lot to do with migration and mutation operators of BBO. Rest of the paper is organized as follows: In section 2 original biogeography-based optimization is explained. In section 3 a new BBO with modified blended migration and polynomial mutation is proposed and analyzed. In section 4 BBO-MBLX-PM is tested over 20 test problems. Finally, in section 5 paper is concluded.

2   Biogeography-Based Optimization

Biogeography-based optimization is a population based global optimization algorithm which is inspired by the natural immigration and emigration of species within islands. Good geographical conditions of an island attract species to migrate from one island to other. Here each island is called habitat and its goodness is known as Habitat Suitability Index (HSI). It is to note that HSI is analogous to fitness of an individual like in any meta-heuristic algorithm. Here a habitat represents a candidate solution. Characteristics of a habitat like rainfall, land area and temperature represent the independent variables. These variables are called Suitability Index Variables (SIV). In short, a habitat is a n-tupple of n $SIVs$ whose fitness is denoted by HSI. In BBO, any solution improves based on the immigration and emigration of solution features within habitats. A high HSI habitat shares its good features with low HSI habitat and low HSI habitat accepts the new features of high HSI habitat. Repeated application of this process creates a better habitat.

   In BBO, the immigration rate λ and emigration rate μ balance the rate at which new species arrive and the rate at which the old species leave the habitat. These rates λ and μ are functions of the number of species on the habitat. If at a particular time, there are S species in a habitat, the immigration rate λ and emigration rate μ are calculated using the following two formulae:

$$\lambda = I \left(1 - \frac{S}{S_{max}} \right) \qquad (1)$$

$$\mu = E \left(\frac{S}{S_{max}} \right) \qquad (2)$$

where I is the maximum possible immigration rate, which occurs when there are zero species on the island and E is the maximum possible emigration rate, which occurs when the island contains the largest number of species S_{max} that it can support.

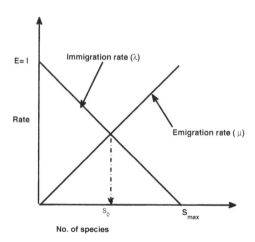

Fig. 1 Relationship among number of species, immigration rate and emigration rate

Fig. 1 shows the relationships between the number of species, λ and μ. Here S_0 is the equilibrium state of an island which is attained when the immigration rate and emigration rate are same. If number of species of a habitat is less than S_0 then the habitat is referred as low HSI habitat because they have high immigration and low emigration rate and if the number of species in a habitat is greater than S_0 then it is referred as high HSI habitat. More about immigration and emigration can be seen in [3]. The BBO algorithm primarily contains two operators migration and mutation. Algorithm 1 explains migration operator in BBO. Here $rand(0, 1)$ is a uniformly distributed random number between 0 and 1. In BBO, the migration operator is similar to the crossover operator of evolutionary strategies. SIV of a selected habitat is modified using this operator.

```
for i=1 to population size do
    Select habitat H_i with probability λ_i;
    if H_i is selected, i.e. if rand(0, 1) <λ_i then
        for j=1 to population size do
            Select habitat H_j with probability μ_j;
            if H_j is selected, i.e. if rand(0, 1) <μ_j then
                Randomly select an SIV from H_j;
                Replace a random SIV in H_i with randomly selected SIV of H_j in
                previous step, i.e. H_i(SIV) ← H_j(SIV)
            end
        end
    end
end
```

Algorithm 1. Migration in Biogeography-based Optimization

Mutation is analogous to the sudden changes in the habitat. This operator is responsible to maintain the diversity in population during BBO process. Mutation randomly modifies habitat SIVs based on the habitat's a priori probability of existence P_i obtained from the following differential equation:

$$\dot{P}_s = \begin{cases} -(\lambda_s + \mu_s)P_s + \mu_{s+1}P_{s+1} & \text{if } s = 0 \\ -(\lambda_s + \mu_s)P_s + \mu_{s+1}P_{s+1} + \lambda_{s-1}P_{s-1} & \text{if } 1 \leq s \leq s_{max} \\ -(\lambda_s + \mu_s)P_s + \lambda_{s-1}P_{s-1} & \text{if } s = s_{max} \end{cases} \quad (3)$$

Very high HSI solutions and very low HSI solutions are equally improbable. Medium HSI solutions are relatively probable [7]. Mutation is explained by algorithm 2. The mutation rate m_i is expressed as:

$$m_i = m_{max}\left(1 - \frac{P_i}{P_{max}}\right) \quad (4)$$

where m_{max} is a user defined maximum mutation probability. $P_{max} = Argmax P_i$; $i = 1, 2, ...,$ population size. For more details about migration and mutation operators see [9].

for *i=1 to population size* **do**
   Compute the probability P_i;
   for *j=1 to number of decision variables* **do**
      Select $H_j(SIV)$ with probability based on P_i;
      if $rand(0, 1) < m_j$ **then**
         | Replace $H_j(SIV)$ with a randomly generated SIV
      end
   end
end

Algorithm 2. Mutation in Biogeography-based Optimization

Create and Initialize a n-dimensional population;
Evaluate fitness (HSI) of each habitat;
while *termination condition met* **do**
   Sort the population from best to worst;
   Compute number of species S, λ and μ ;
   Apply migration algorithm 1;
   Update the probability for each habitat;
   Apply mutation algorithm 2;
   Evaluate fitness (HSI) of each habitat;
   Implement elitism for keeping best solutions;
end

Algorithm 3. Biogeography-based Optimization

3 Modified Biogeography-Based Optimization

Blended crossover, described below, is applied to BBO instead of its traditional migration operator in [8]:

$$H_i(SIV) \leftarrow \alpha H_i(SIV) + (1 - \alpha)H_j(SIV) \tag{5}$$

where α is a real number between 0 and 1. As mentioned in [8], α could be random or deterministic or based on the fitness of solutions H_i and H_j. This paper proposes a new setting of α in blended crossover as below:

$$\alpha = \alpha_{min} + (\alpha_{max} - \alpha_{min})k^t \tag{6}$$

Where α_{min} and α_{max} are minimum and maximum values of α. k is user defined parameter and should be less than 1. t is the generation counter. It is clear from (6) that due to $k < 1$, α decreases as generation increases and therefore the ratio of features (SIVs) of H_i and H_j in the new solution non-linearly changes over generations. The variation of α with respect to the generation counter is shown in Fig. 2. An inertia weight parameter for particle swarm optimization algorithm has also been defined using the formula (6) [2]. All of the mutation schemes that have been implemented for

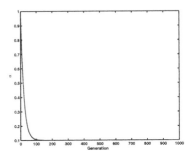

Fig. 2 Variation of α with respect to generation counter

GAs could also be implemented for BBO [9]. Polynomial mutation has already been successfully applied to GAs [4] and therefore this paper also explores the application of polynomial mutation in BBO algorithm. For a given solution $X \in [X_{lb}, X_{ub}]$, the mutated solution X' is defined as below [4]:

$$X' = \begin{cases} X + \delta_L(X - X_{lb}) & \text{if } u \leq 0.5 \\ X + \delta_R(X_{ub} - X) & \text{if } u > 0.5 \end{cases} \tag{7}$$

Here u is a random number belongs to [0, 1] and δ_L and δ_R are calculated as below:

$$\delta_L = (2u)^{\frac{1}{1+\eta m}} - 1 \ u \leq 0.5 \tag{8}$$

$$\delta_R = 1 - (2(1-u))^{\frac{1}{1+\eta m}} - 1 \ u > 0.5 \tag{9}$$

Where η_m is user defined index parameter in the range [20, 100].

In this paper, a comparative study among basic version of BBO, BBO with original mutation and modified blended migration, BBO with polynomial mutation and original migration and BBO with polynomial mutation and modified blended migration has been carried out. Results have also been compared with blended BBO of [8].

4 Numerical Experiments

In this section, the performance of BBO with polynomial mutation and modified blended migration is examined through 20 unconstrained test problems listed in Table 1. All the considered problems are of minimization type. Details of these problems can be seen in [1]. All the considered problems are scalable and the number of decision variables are fixed to be 30. For all experiments, parameters are set as follows:

– Maximum immigration rate: $I = 1$;
– Maximum emigration rate: $E = 1$;

- elitism size = 2;
- Mutation probability: $m_{max} = 0.005$;
- Number of habitats, i.e. population size =100;
- Total number of generations in each run = 1000;
- η_m (in polynomial mutation) = 20;
- $\alpha_{min} = 0.1$ and $\alpha_{max} = 1$ (in modified blended migration);
- $\lambda = 0.95$ (in modified blended migration);

Table 2 presents the average fitness of 30 runs for each problem. Second column entries are the results of basic BBO algorithm, while subsequent columns presents the results of blended BBO (BBO-BLX), BBO with original mutation and modified blended migration (BBO-MBLX), BBO with polynomial mutation and original migration (BBO-PM) and BBO with polynomial mutation and modified blended migration (BBO-MBLX-PM) respectively. It can be observed easily that BBO-BLX performs better than basic BBO as expected. BBO-MBLX modifies the BBO-BLX with a new setting of α in expectation of better results. As per expectation, BBO-MBLX performs better than BBO-BLX for 15 problems and equal for 4 problems out of 20. However, combination of original migration and polynomial mutation (BBO-PM) could not perform well but perform better than basic version of BBO and poor as compared to BBO-BLX. But the combination of modified blended migration and polynomial mutation in BBO did the job. BBO-MBLX-PM performed better than

Table 1 Test Problems

Sr. N0.	Problem Name
1	Sphere
2	De Jong's f4
3	Griewank
4	Rosenbrok
5	Rastrigin
6	Ackley
7	Alpine
8	Michalewicz
9	Cosine Mixture
10	Exponential
11	Zakharov's
12	Cigar
13	brown3
14	Schewel prob 3
15	Salomon Problem
16	Axis parallel hyperellipsoid
17	Pathological
18	Rotated hyper-ellipsoid function
19	step function
20	Quartic function

Table 2 Experimental Results

Sr. No.	BBO	BBO-BLX	BBO-MBLX	BBO-PM	BBO-MBLX-PM
1	5.3080E-03	2.6900E-05	1.1300E-11	7.3800E-05	1.6300E-13
2	5.4600E-05	3.1300E-08	0.0000E+00	1.2700E-08	0.0000E+00
3	9.4859E-01	1.5651E-02	1.1808E-02	5.8904E-02	4.9280E-03
4	7.6474E+01	2.7476E+01	2.6537E+01	5.4545E+01	2.5124E+01
5	7.5930E-01	6.2500E-03	3.3900E-09	1.6706E-02	2.3700E-10
6	5.6093E-01	1.8627E-02	2.3100E-05	3.5261E-02	2.1200E-06
7	2.6216E-02	7.0900E-04	1.5700E-07	4.7280E-03	3.7900E-08
8	-8.6536E+00	-8.6602E+00	-8.6602E+00	-8.5152E+00	-8.6325E+00
9	-2.9979E+00	-3.0000E+00	-3.0000E+00	-3.0000E+00	-3.0000E+00
10	-9.9992E-01	-1.0000E+00	-1.0000E+00	-1.0000E+00	-1.0000E+00
11	8.6806E+00	9.3443E-01	4.6111E-02	2.7731E+00	1.3717E-02
12	1.3916E+03	6.4074E+00	6.2300E-06	1.9825E+01	1.6000E-07
13	2.0170E-03	3.1300E-04	1.5400E-09	3.0000E-05	1.6800E-13
14	4.5027E-01	1.2898E-02	3.2000E-06	5.7692E-02	4.7400E-07
15	1.3695E+00	8.4546E-01	4.3011E-01	9.1708E-01	3.7006E-01
16	7.4550E-02	3.5800E-04	4.6700E-10	1.0330E-03	3.4400E-12
17	3.3626E+00	4.0155E+00	3.6173E-01	4.4856E+00	9.4797E-01
18	1.3245E+01	4.3322E-02	7.5255E-08	1.8473E-01	3.7777E-10
19	1.6000E+00	0.0000E+00	0.0000E+00	0.0000E+00	0.0000E+00
20	9.2439E+00	8.4372E+00	8.5114E+00	9.1217E+00	8.4506E+00

all considered versions of BBO over the considered test problems. Performance of the considered BBO versions can be summarized as below ($A > B \iff$ algorithm A is better than algorithm B):

$$BBO-MBLX-PM > BBO-MBLX > BBO-BLX > BBO-PM > BBO$$

5 Conclusion

This paper proposes a modification in the blended migration operator and use of polynomial mutation in BBO. Simultaneous application of proposed non-linear decreasing value of parameter α in blended migration operator and polynomial mutation makes better balance in exploration and exploitation in BBO. Experimental results over 20 unconstrained test problems exhibit that combination of these two operators in BBO makes BBO a promising continuous search algorithm. In the same way, studies can be carried out to check the performance of other combinations of different mutation and crossover operators already applied to the evolutionary algorithms.

References

1. Bansal, J.C., Sharma, H., Nagar, A., Arya, K.V.: Balanced artificial bee colony algorithm. International Journal of Artificial Intelligence and Soft Computing **3**(3), 222–243 (2013)
2. Bansal, J.C., Singh, P.K., Saraswat, M., Verma, A., Jadon, S.S., Abraham, A.: Inertia weight strategies in particle swarm optimization, pages 633–640 (2011)
3. Boussaïd, I., Chatterjee, A., Siarry, P., Ahmed-Nacer, M.: Biogeography-based optimization for constrained optimization problems. Computers & Operations Research **39**(12), 3293–3304 (2012)
4. Deb, K., Deb, D.: Analyzing mutation schemes for real-parameter genetic algorithms. KanGAL Report Number, 2012016 (2012)
5. Gong, W., Cai, Z., Ling, C.X., Li, H.: A real-coded biogeography-based optimization with mutation. Applied Mathematics and Computation **216**(9), 2749–2758 (2010)
6. Li, X., Wang, J., Zhou, J., Yin, M.: A perturb biogeography based optimization with mutation for global numerical optimization. Applied Mathematics and Computation **218**(2), 598–609 (2011)
7. Li, X., Yin, M.: Multi-operator based biogeography based optimization with mutation for global numerical optimization. Computers & Mathematics with Applications **64**(9), 2833–2844 (2012)
8. Ma, H., Simon, D.: Blended biogeography-based optimization for constrained optimization. Engineering Applications of Artificial Intelligence **24**(3), 517–525 (2011)
9. Simon, D.: Biogeography-based optimization. IEEE Transactions on Evolutionary Computation **12**(6), 702–713 (2008)
10. Simon, D., Omran, M.G.H., Cler, M.: Linearized biogeography-based optimization with re-initialization and local search. Information Sciences (2014)
11. Xiong, G., Shi, D., Duan, X.: Enhancing the performance of biogeography-based optimization using polyphyletic migration operator and orthogonal learning. Computers & Operations Research **41**, 125–139 (2014)

A Modified Biogeography Based Optimization

Pushpa Farswan, Jagdish Chand Bansal and Kusum Deep

Abstract Biogeography based optimization (BBO) has recently gain interest of researchers due to its efficiency and existence of very few parameters. The BBO is inspired by geographical distribution of species within islands. However, BBO has shown its wide applicability to various engineering optimization problems, the original version of BBO sometimes does not perform up to the mark. Poor balance of exploration and exploitation is the reason behind it. Migration, mutation and elitism are three operators in BBO. Migration operator is responsible for the information sharing among candidate solutions (islands). In this way, the migration operator plays an important role for the design of an efficient BBO. This paper proposes a new migration operator in BBO. The so obtained BBO shows better diversified search process and hence finds solutions more accurately with high convergence rate. The BBO with new migration operator is tested over 20 test problems. Results are compared with that of original BBO and Blended BBO. The comparison which is based on efficiency, reliability and accuracy shows that proposed migration operator is competitive to the present one.

Keywords Biogeography based optimization · Blended BBO · Migration operator

1 Introduction

The process of inspiring from nature many evolutionary algorithms (EAs) [1] and swarm intelligence (SI) algorithms [7] have been developed. Genetic algorithm (GA) [4], Genetic programming [3], Evolutionary programming [21], Differential evolution(DE) [20] and Neuroevolution algorithms [10] are in the category of EAs. SI algorithms such as Particle swarm optimization (PSO) [14], Ant colony optimiza-

P. Farswan · J.C. Bansal(✉)
South Asian University, New Delhi, India
e-mail: {pushpafarswan6,jcbansal}@gmail.com

K. Deep
Indian Institute of Technology Roorkee, Roorkee, India
e-mail: kusumfma@iitr.ernet.in

© Springer-Verlag Berlin Heidelberg 2016
J.H. Kim and Z.W. Geem (eds.), *Harmony Search Algorithm*,
Advances in Intelligent Systems and Computing 382,
DOI: 10.1007/978-3-662-47926-1_22

227

tion (ACO) [5] , Artificial bee colony (ABC) [13] and Spider monkey optimization (SMO) [2] etc. have been developed. Biogeography based optimization (BBO) algorithm falls down in the category of evolutionary algorithms because of some similar properties as evolutionary algorithm such as mutation and sharing the information within candidate solutions, admittedly. The origin of BBO algorithm started in the 19th century when the science of biogeography came in picture by Alfred Wallace and Charles Darwin. Then Robert Mac Arthur and Edward Wilson initiated work on biogeography theory and developed mathematical model of biogeography which stands for the mechanism, how species originate, how species dead and how species migrate among islands. Working process of BBO is motivated by this theory and improves the quality of solution by probabilistically sharing the information between population of candidate solutions. BBO has distinctive and effective capability to improve candidate solution using immigration rate (λ) and emigration rate (μ) of each island in all generations. These migration rates decide the immigrating habitat and emigrating habitat and responsible for updating solution by accepting information from promising solutions.

There are many developments in BBO algorithm by implementing and improving migration and mutation operators in original BBO algorithm. In [6] Du et al. proposed BBO with evolutionary strategy (ES) and immigration refusal (RE). In proposed BBO/ES/RE migration is based on immigration refusal and mutation is based on evolutionary strategy. In BBO/ES/RE immigrating island rejects the features from another islands which has low fitness than immigrating island and some threshold. In BBO/ES/RE, select only best n individuals among parent and child islands for next generation. In [17] Ma et al. proposed blended BBO. In blended BBO, migration operator combined the features of both immigrating and emigrating islands. In [19] Simon et al. proposed LBBO (linearized BBO) for improving solution of non seperable problems. LBBO combined with periodic re-initialization and local search operator and obtain algorithm for global optimization in a continuous search space. In [15] Lohakare et al. proposed a memetic BBO named as aBBOmDE, for improving convergence speed by modifying mutation operator and maintained exploitation by keeping original migration. In [12] Gong et al. proposed RCBBO (real coded BBO) in which each habitat was represented by real parameter. In RCBBO, to improving exploration ability and the diversity of population, some special mutation operators as gaussian mutation, cauchy mutation and levy mutation are incorporated into the habitat mutation. In [11] Gong et al. proposed a hybrid differential evolution with biogeography based optimization named as DE/BBO. In proposed algorithm exploration of DE combined with exploitation of BBO, generated the effective solution. In [16] Ma et al. presented BBMO for handling multiple objective with the help of BBO. In proposed algorithm, problem decomposed into sub problems and applied parallel BBO algorithm for optimizing each sub problems.

In this paper we introduced a modified migration operator. This operator is able to use the four individuals' information intelligently in essential step. This operator is modified for diversified search in promising area of search space. It can not be rejected that poor solution has good feature in some dimension as well as good solutions has possibility for bad feature in some dimension. In this way, poor solution may responsible for promising result. In basic migration operator of BBO and modified migration operator developed earlier can not make the best use of the search experiences. Therefore, acceptance of information for immigrating island from other candidate solutions is the important task. In this paper, modified migration operator is given for utilizing the best information of candidate solutions. The target of this paper is to enhance the performance of BBO by modifying migration operator in BBO algorithm. This paper is organized as follows: In section 2, description of BBO algorithm and its performance. In section 3, detail description to modified migration operator in BBO algorithm. In section 4, modified BBO is tested over 20 test problems. In section 5, paper is concluded.

2   Biogeography Based Optimization

An evolutionary algorithm, Biogeography based optimization(BBO) is recently developed by Dan Simon in 2008. BBO is inspired by geographical distribution of species within islands over period of time. The mathematical model of biogeography is based on speciation of species, extinction of species and migration of species within islands, but BBO is based on only the concept of migration of species within islands. So that speciation and extinction are not considered in BBO algorithm. In BBO, island represents the solution. The island which have large number of species corresponds to good solution and the island which have few number of species corresponds to bad solution. Good islands shares their features with bad islands. The features that characterize habitability are called SIVs (suitability index variables). SIVs are considered as independent variable. Island suitability index (ISI) represents the fitness. The island which is very friendly to life is said to have high island suitability index (ISI) and the island which is relatively less friendly to life is said to have low island suitability index (ISI) and the ISI can be considered as dependent variable. Low ISI island has high probability to accepts the new feature (good feature) from high ISI island. Emigration rate is decreases from high ISI to low ISI island so that highest ISI island has maximum emigration rate and immigration rate is increases from high ISI to low ISI island so that highest ISI island has minimum immigration rate. The immigration rate λ and emigration μ are calculated by two formulas.

$$\lambda_i = I\left(1 - \frac{k_i}{n}\right) \tag{1}$$

$$\mu_i = E\left(\frac{k_i}{n}\right) \tag{2}$$

λ_i stands for immigration rate of ith individual (island).
μ_i stands for emigration rate of ith individual (island).
I stands for maximum possible immigration rate.
E stands for maximum possible emigration rate.
n stands for maximum possible number of species that island can support.

K_i stands for fitness rank of ith island after sorting fitness of ith island, so that for worst solution K_i is taken as 1 and for best solution K_i is taken as n. It suffices to assume a linear relationship between number of species and migration rate for many application point of view. The relation between migration rate (λ and μ) and number of species is illustrated in Fig. 1. If there is zero species in the island then immigration rate is maximum, denoted by I. If there are maximum number of species (S_{max}) in the island then emigration rate is maximum, denoted by E. There is an equilibrium state where immigration rate and emigration rate are equal. The equilibrium number of species in this state is denoted by S_0. The island referred as high ISI island if the number of species is above than S_0 and the island referred as low ISI island if the number of species is less than S_0. Migration and mutation are two crucial operators in BBO. The migration operator is same as the crossover operator of evolutionary algorithm. Migration operator is responsible for sharing the feature among candidate solutions for modifying fitness. Mutation occur by sudden changes in island due to random event and is responsible for maintaining the diversity of island in BBO process. Algorithm [1, 2, 3] describes the Pseudo code of migration operator, mutation operator and BBO respectively.

Algorithm 1. Migration operator

Population size $\leftarrow n$;
for $i \leftarrow 1, n$ do
  Select the habitat H_i according to λ_i;
  if rand(0,1) $< \lambda_i$ then
    for $e \leftarrow 1, n$ do
      Select the habitat H_e according to μ_e;
      $H_i(SIV) \leftarrow H_e(SIV)$
    end for
  end if
end for

Algorithm 2. Mutation operator

Population size $\leftarrow n$;
for $i \leftarrow 1, n$ do
  Select the habitat H_i according to probability P_i;
  if rand(0,1) $< m_i$ then
    $H_i(SIV) \leftarrow$ randomly generated SIV
  end if
end for

Algorithm 3. Biogeography-based optimization

Population size $\leftarrow n$;
Define migration probabllity, mutation probability and elitism size ;
Evaluate ISI (fitness) value of each individual (island) ;
while Stoping condition not satisfied **do**
   Sorting the population according best fitness to least fitness;
   Apply migration :
   Update the ISI value of each island ;
   Apply mutation ;
   Keep elitism in population ;
end while

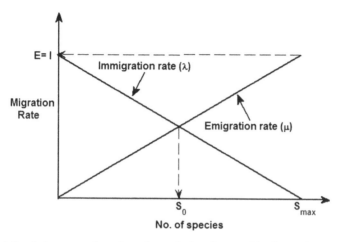

Fig. 1 Relation between number of species and migration rate Fig. from[18]

P_s is the probability when there are s species in the habitat is changes from t to $(t+\Delta t)$ as follows:

$$P_s(t + \Delta t) = P_s(t)(1 - \lambda_s \Delta t - \mu_s \Delta t) + P_{s-1}\lambda_{s-1}\Delta t + P_{s+1}\mu_{s+1}\Delta t \qquad (3)$$

Where λ_s is immigration rate when there are s species in the habitat. μ_s is emigration rate when there are s species in the habitat.
At time $t+\Delta t$ one of the following condition must hold for s species in the habitat.

1. If there were s species in the habitat at time t. Then no immigration and no emigration of species within time t and $t+\Delta t$.
2. If there were (s-1) species in the habitat at time t. Then one species immigrated within time t and $t+\Delta t$.
3. If there were (s+1) species in the habitat at time t. Then one species emigrated within time t and $t+\Delta t$.

For ignoring the probability of more than one immigration or emigration, we take Δt very small

Taking $\Delta t \longrightarrow 0$

$$\dot{P}_s = \begin{cases} -(\lambda_s + \mu_s)P_s + \mu_{s+1}P_{s+1}, & s = 0 \\ -(\lambda_s + \mu_s)P_s + \lambda_{s-1}P_{s-1} + \mu_{s+1}P_{s+1}, & 1 \leq s \leq s_{max} - 1 \\ -(\lambda_s + \mu_s)P_s + \lambda_{s-1}P_{s-1}, & s = s_{max} \end{cases} \quad (4)$$

3 Modified Migration in BBO

In biogeography based optimization process, migration operator plays a key role. The concept behind the migration operator is sharing the information within islands. Migration of solution feature within islands is motivated by the mathematical model of species migration in biogeography. Basic BBO algorithm suffers from lack of exploitation and stagnation at local minima. The possible way to enhance its exploitation capability is to improve the migration operator. Improved migration operator simultaneously adopts more information from other habitats and maintains population diversity as well as preserves exploitation ability and overcome stagnation at local minima. The immigrating habitat H_i and emigrating habitat H_j are selected according to the probability of immigration rate (λ_i) and probability of emigration rate (μ_i) respectively.

In basic BBO, migration process is taken as:

$$H_i(SIV) \leftarrow H_j(SIV) \quad (5)$$

In basic migration operator (5), immigrating island directly accepts the information from emigrating island only. In the modified migration operator, immigrating habitat accepts the information not only from emigrating habitat but also accepts the information from immigrating habitat, best habitat and random habitat (other than best habitat and immigration habitat). The working of new version of migration operator is given as:

$$H_i(SIV) \leftarrow \begin{cases} \frac{1}{G}(H_i(SIV)) + (1 - \frac{1}{G})(H_j(SIV)), & if \ G < G_{max} \\ \frac{1}{G}(H_r(SIV) - H_i(SIV)) + \\ (1 - \frac{1}{G})(H_{best}(SIV) - H_j(SIV)), & if \ G = G_{max} \end{cases} \quad (6)$$

Where G is the generation index. The core idea of the proposed modified migration operator is based on three considerations. First one is that if the immigrating habitat is selected and generation index is not met maximum. Immigrating habitat accepts the information only from immigrating and emigrating habitat. It is important to use the information from other habitat in suitable ratio to improve the population diversity. Here immigrating habitat use less information from itself and more information from emigrating habitat with increasing number of generation index. Here $\frac{1}{G}$ is decreasing

---

Algorithm 4. Modified migration operator

Population size $\leftarrow n$;
Generation index $\leftarrow G$;
For the selected habitat $H_i(SIV)$
if rand$(0,1) < \lambda_i$ **then**
   Initialize generation index G=1;
   if G $\leftarrow$ 1 to $G_{max} - 1$ **then**
      Select the random habitat within population other than best habitat and running index habitat;
      Update the current solution as
      $H_i(SIV) \leftarrow \frac{1}{G}(H_r(SIV) - H_i(SIV)) + (1 - \frac{1}{G})(H_{best(SIV)} - H_j(SIV))$
   else
      Update the current solution as
      $H_i(SIV) \leftarrow \frac{1}{G}(H_i(SIV)) + (1 - \frac{1}{G})(H_j(SIV))$
   end if
else
   Update the current solution as $H_i(SIV) \leftarrow H_{best}(SIV)$
end if

---

function of G and $1-\frac{1}{G}$ is increasing function of G. Second consideration is that if generation index is met maximum, then immigrating habitat uses information from first elite habitat, immigrating habitat, emigrating habitat and random habitat (except immigrating habitat and first elite habitat). Here immigrating habitat uses very less information from random habitat and immigrating habitat but uses relatively maximum information from first elite habitat and emigrating habitat. Third consideration is that if rand $(0,1) < \lambda_i$ is not met then immigrating habitat adopts the information of first elite habitat.

4   Numerical Experiments

In this section we compare modified BBO with other version of BBO algorithms and experiments performed on 20 unconstrained test problems. Parameters used in the algorithms are:
Maximum immigration rate:I=1
Maximum emigration rate :E=1
Mutation probability=0.01
Elitism size=2
Population size=50
Maximum no. of iteration=1000
No. of runs=100

Table 1 Test problems; TP: Test Problem, D: Dimensions, C: Characteristic, U: Unimodal, M: Multimodal, S: Separable, NS: Non-Separable, AE: Acceptable Error

TP	Objective function	Search Range	Optimum Value	D	C	AE
Alpine	$f_1(x) = \sum_{i=1}^{D} \mid (x_i, sin(x_i)) \mid + 0.1x_i$	[-10,10]	f(**0**) = 0	30	M, S	1.0E-05
Axis parallel hyper ellipsoid	$f_2(x) = \sum_{i=1}^{D} ix_i^2$	[-5.12,5.12]	f(**0**) = 0	30	U, S	1.0E-05
Cosine mixture	$f_3(x) = \sum_{i=1}^{D} x_i^2 - 0.1\sum_{i=1}^{D} \cos(5\pi x_i)$	[-1,1]	f(**0**) = -D*0.1	30	M, S	1.0E-05
De jong's f_4	$f_4(x) = \sum_{i=1}^{D} ix_i^4$	[-5.12,5.12]	f(**0**) = 0	30	U, S	1.0E-05
Exponential	$f_5(x) = -\exp\left(-0.5\sum_{i=1}^{D} x_i^2\right)$	[-1,1]	f(**0**) = -1	30	M, NS	1.0E-05
Griewank	$f_6(x) = \sum_{i=1}^{D} \frac{x_i^2}{4000} - \prod_{i=1}^{D} \cos(\frac{x_i}{\sqrt{i}}) + 1$	[-600,600]	f(**0**) = 0	30	M, NS	1.0E-05
Rosenbrock	$f_7(x) = \sum_{i=1}^{D}[100(x_i^2 - x_{i+1})^2 + (x_i - 1)^2]$	[-2.048,2.048]	f(**1**) = 0	30	U, NS	1.0E-02
Salomon prob 3	$f_8(x) = 1 - \cos\left(2\pi\sqrt{\sum_{i=1}^{D} x_i^2}\right) + 0.1\sqrt{\sum_{i=1}^{D} x_i^2}$	[-100,100]	f(**0**) = 0	30	M, S	1.0E-01
Schwefel	$f_9(x) = -\sum_{i=1}^{D} x_i sin(\mid x_i \mid^{1/2})$	[-512,512]	$f(\pm\mid\pi(0.5 + k)\mid^2]^2)$ $= -418.9829 * D$	30	M, S	1.0E-05
Schwefel221	$f_{10}(x) = \max\mid x_i \mid, 1 \leq i \leq D$	[-100,100]	f(**0**) = 0	30	U, S	1.0E-05
Schwefel222	$f_{11}(x) = \sum_{i=1}^{D} \mid x_i \mid + \prod_{i=1}^{D} \mid x_i \mid$	[-10,10]	f(**0**) = 0	30	U, NS	1.0E-05
Sphere	$f_{12}(x) = \sum_{i=1}^{D} x_i^2$	[-5.12,5.12]	f(**0**) = 0	30	U, S	1.0E-05
Step function	$f_{13}(x) = \sum_{i=1}^{D}([x_i + 0.5])^2$	[-100,100]	$f(-0.5 \leq x \leq 0.5)$ $= 0$	30	U, S	1.0E-05
Michalewicz	$f_{14}(x) = -\sum_{i=1}^{D} sin(x_i)\left[\frac{sin(ix_i^2)}{\pi}\right]^{20}$	[0,π]	$f_{min} = 9.66015$	10	M, S	1.0E-05
Zakharov's	$f_{15}(x) = \sum_{i=1}^{D} x_i^2 + \left(\sum_{i=1}^{D} \frac{i}{2}x_i\right)^2 + \left(\sum_{i=1}^{D} \frac{i}{2}x_i\right)^4$	[-5.12,5.12]	f(**0**) = 0	30	M, NS	1.0E-02
Cigar	$f_{16}(x) = x_0^2 + 100000\sum_{i=1}^{D} x_i^2$	[-10,10]	f(**0**) = 0	30	U, S	1.0E-05
Brown 3	$f_{17}(x) = \sum_{i=1}^{D-1}\left[(x_i^2)^{(x_{i+1}^2+1)} + (x_{i+1}^2)^{(x_i^2+1)}\right]$	[-1,4]	f(**0**) = 0	30	U, NS	1.0E-05
Easom	$f_{18}(x) = -cos x_1 cos x_2 e^{((-(x_1-\pi)^2-(x_2-\pi)^2))}$	[-100,100]	$f(-\pi,\pi) = -1$	2	U, N	1.0E-13
Ackley	$f_{19}(x) = -20\exp\left(-0.2\sqrt{\frac{1}{D}\sum_{i=1}^{D} x_i^2}\right) + 20 + e$ $\exp(\frac{1}{D}\sum_{i=1}^{D}\cos(2\pi x_i))$	[-30,30]	f(**0**) = 0	30	M, NS	1.0E-05
Rastrigin	$f_{20}(x) = \sum_{i=1}^{D}[x_i^2 - 10\cos(2\pi x_i)] + 10D$	[-5.12,5.12]	f(**0**) = 0	30	M, S	1.0E-05

Table 1 represents the list of benchmark functions used in the experiments. In order to see the effect of proposed modified BBO process success rate (SR), mean generation index (MGenIndex), mean error (ME), standard deviation (SD) and minimum error (Min E) are reported in Table 2. Table 2 shows that proposed modified BBO (MBBO) outperforms in terms of reliability, efficiency and accuracy as compare to basic BBO, blended BBO (BBBO) and previously modified BBO (m1_BBO) given in [8]. The proposed algorithm MBBO is compared with BBO, BBBO and M1_BBO based on SR, MGenIndex, ME and SD. The comparison of algorithms are based on this sequence as SR, MGenIndex, ME than SD . Firstly all algorithms are compared according to SR, if it is difficult to distinguish than compare based on MGenIndex, if still comparison is not possible than compare according to ME. Finally, if find difficulty to compare than compare according to SD. From Table 2 it is clearly shown that according to success rate, MBBO outperforms among all considered algorithms for the functions (f_1, f_2, f_3, f_5, f_6, f_8, f_{11}, f_{12}, f_{13}, f_{15}, f_{17}, f_{18}, f_{20}). Further comparison for remaining function is not possible by success rate than comparison according to mean generation index, MBBO is good for function f_4 among all considered algorithms. Still comparison is not possible by mean generation index. Then according to mean error, MBBO outperforms for functions (f_7, f_9, f_{10}, f_{16}, f_{19}) among all considered algorithms. Then finally according to SD, MBBO performance is better for the function f_{14} among all considered algorithms. From the above comparison the proposed modified BBO algorithm (MBBO) outperforms the considered algorithms. It is clearly says that MBBO is cost effective. All function given in Table 1 are high dimensional and include unimodal, multimodal, separable, non separable with different optimum solution.

Table 2 Comparison of results of BBO , BBBO, M1_BBO and MBBO

Test problem	Algorithm	SR	MGenIndex	ME	SD	Min E
f_1	BBO	0	1000.00	3.4172E-02	7.4295E-03	2.3251E-02
	BBBO	0	1000.00	2.3055E-02	1.8532E-02	7.6838E-03
	M1_BBO	0	1000.00	1.7875E-02	9.3983E-03	5.5819E-03
	MBBO	64	1000.00	1.0952E-05	1.2781E-05	4.2636E-07
f_2	BBO	0	1000.00	1.5005E-01	6.2059E-02	4.1119E-02
	BBBO	0	1000.00	1.1160E-02	4.8538E-03	3.7219E-03
	M1_BBO	0	1000.00	3.2510E-03	1.3871E-03	9.6735E-04
	MBBO	100	719.97	8.9174E-06	1.3975E-06	1.3044E-06
f_3	BBO	0	1000.00	4.5212E-03	1.7002E-03	1.4823E-03
	BBBO	0	1000.00	4.4780E-04	2.3971E-04	9.0809E-05
	M1_BBO	0	1000.00	1.1105E-04	4.8881E-05	4.0523E-05
	MBBO	100	392.67	9.2256E-06	9.1214E-07	5.2037E-06
f_4	BBO	0	1000.00	3.0398E-04	2.3429E-04	1.1133E-05
	BBBO	78	901.51	1.1061E-05	7.5301E-06	2.6671E-06
	M1_BBO	100	703.40	8.9391E-06	1.1892E-06	3.9210E-06
	MBBO	100	419.86	9.2644E-06	1.0104E-06	3.9390E-06

Table 2 (*Continued*)

Test problem	Algorithm	SR	MGenIndex	ME	SD	Min E
f_5	BBO	0	1000.00	1.9318E-04	6.0680E-05	7.5146E-05
	BBBO	10	994.36	1.9646E-05	8.0968E-06	7.8973E-06
	M1_BBO	94	824.13	9.5943E-06	1.3308E-06	7.0683E-06
	MBBO	100	337.17	8.6550E-06	1.4708E-06	1.9623E-06
f_6	BBO	0	1000.00	1.0345E+00	2.0308E-02	9.5733E-01
	BBBO	0	1000.00	4.2443E-01	1.3223E-01	1.6039E-01
	M1_BBO	0	1000.00	1.6664E-01	6.1318E-02	4.6993E-02
	MBBO	77	998.34	8.7389E-05	4.3441E-04	4.5035E-08
f_7	BBO	0	1000.00	5.8374E+01	2.9611E+01	1.3392E+01
	BBBO	0	1000.00	2.7988E+01	2.1686E-01	2.7148E+01
	M1_BBO	0	1000.00	3.5730E+01	2.1227E+01	7.9407E+00
	MBBO	0	1000.00	2.6829E+01	2.7721E+00	1.2812E+01
f_8	BBO	0	1000.00	1.3140E+00	2.2226E-01	7.9994E-01
	BBBO	0	1000.00	6.2195E-01	8.3602E-02	3.9987E-01
	M1_BBO	0	1000.00	5.5688E-01	7.5555E-02	3.9987E-01
	MBBO	9	1000.00	2.7669E-01	2.3649E-01	3.0275E-02
f_9	BBO	0	1000.00	9.4633E+00	3.3807E+00	3.7806E+00
	BBBO	0	1000.00	3.2176E+02	1.1911E+02	1.0410E+02
	M1_BBO	0	1000.00	1.7118E+00	1.0323E+00	4.7631E-01
	MBBO	0	1000.00	1.5174E+00	1.6345E+00	4.1500E-03
f_{10}	BBO	0	1000.00	5.1713E+00	1.1149E+00	2.2578E+00
	BBBO	0	1000.00	1.3480E+00	2.2576E-01	7.6256E-01
	M1_BBO	0	1000.00	1.1732E+00	2.0218E-01	6.9738E-01
	MBBO	0	1000.00	9.4184E-01	1.4169E+00	2.5628E-02
f_{11}	BBO	0	1000.00	6.8226E-01	1.1697E-01	4.2116E-01
	BBBO	0	1000.00	1.6261E-01	3.7529E-02	8.1126E-02
	M1_BBO	0	1000.00	6.7395E-02	1.4251E-02	3.4737E-02
	MBBO	8	1000.00	1.0901E-04	9.7789E-05	3.6625E-06
f_{12}	BBO	0	1000.00	1.0887E-02	3.9542E-03	3.3969E-03
	BBBO	0	1000.00	1.0088E-03	4.5187E-04	3.2521E-04
	M1_BBO	0	1000.00	2.8587E-04	1.2832E-04	4.5832E-05
	MBBO	100	470.41	9.3789E-06	6.5679E-07	7.0421E-06
f_{13}	BBO	0	1000.00	4.3800E+00	1.8683E+00	1.0000E+00
	BBBO	54	910.06	5.9000E-01	7.7973E-01	0.0000E+00
	M1_BBO	92	644.76	8.0000E-02	2.7266E-01	0.0000E+00
	MBBO	99	983.56	1.0000E-02	1.0000E-01	0.0000E+00

Table 2 (*Continued*)

Test problem	Algorithm	SR	MGenIndex	ME	SD	Min E
f_{14}	BBO	0	1000.00	9.6601E+00	1.4342E-12	9.6601E+00
	BBBO	0	1000.00	9.6601E+00	4.2958E-12	9.6601E+00
	M1_BBO	0	1000.00	9.6601E+00	2.5619E-12	9.6601E+00
	MBBO	0	1000.00	9.6601E+00	9.6772E-13	9.6601E+00
f_{15}	BBO	0	1000.00	2.5698E+01	1.1451E+01	9.0935E+00
	BBBO	0	1000.00	4.6674E-01	1.7943E-01	1.9471E-01
	M1_BBO	0	1000.00	1.1357E+00	6.9909E-01	2.8541E-01
	MBBO	50	1000.00	6.0615E+00	1.5010E+01	2.5766E-05
f_{16}	BBO	0	1000.00	3.1193E+03	1.1982E+03	9.9760E+02
	BBBO	0	1000.00	2.6013E+02	1.2657E+02	6.7269E+01
	M1_BBO	0	1000.00	7.1825E+01	2.6809E+01	2.4157E+01
	MBBO	0	1000.00	2.3801E-03	2.2788E-03	1.7076E-05
f_{17}	BBO	0	1000.00	5.0369E-03	1.8759E-03	1.5551E-03
	BBBO	0	1000.00	8.0826E-02	2.9152E-02	2.7447E-02
	M1_BBO	0	1000.00	1.9267E-04	9.8574E-05	5.1375E-05
	MBBO	100	785.24	5.9497E-06	3.1751E-06	3.6920E-09
f_{18}	BBO	0	1000.00	6.1028E-01	4.8976E-01	9.0018E-06
	BBBO	0	1000.00	5.0165E-06	7.3606E-06	2.0052E-08
	M1_BBO	23	894.04	1.2787E-12	2.6254E-12	1.2212E-15
	MBBO	85	441.13	1.4999E-01	3.5884E-01	1.1102E-16
f_{19}	BBO	0	1000.00	8.6347E-01	1.9081E-01	4.8123E-01
	BBBO	0	1000.00	3.6781E-01	3.2400E-01	1.0799E-01
	M1_BBO	0	1000.00	7.2645E-02	2.0529E-02	3.8511E-02
	MBBO	0	1000.00	1.0587E-03	1.7407E-03	3.6592E-05
f_{20}	BBO	0	1000.00	1.4866E+00	5.2716E-01	6.3522E-01
	BBBO	0	1000.00	7.5405E+00	2.4895E+00	2.3855E+00
	M1_BBO	0	1000.00	6.8402E+00	2.1345E+00	2.0802E+00
	MBBO	95	993.51	4.3060E-06	5.3902E-06	1.2352E-08

5 Conclusion

This paper proposes a novel modified migration operator for better diversified search in promising area of search space. The proposed modified BBO (MBBO) uses the information from selected candidate solutions to find global optima more accurately with high convergence rate. To verify the performance of MBBO, 20 test problems with different characteristics are employed. Basic comparison with original BBO and other variant of BBO are conducted. In terms of efficiency, reliability and accuracy, the comparison results shows that MBBO outperforms the all considered algorithms.

References

1. Bäck, T., Fogel, D.B., Michalewicz, Z.: Evolutionary computation 1: Basic algorithms and operators, vol. 1. CRC Press (2000)
2. Bansal, J.C., Sharma, H., Jadon, S.S., Clerc, M.: Spider monkey optimization algorithm for numerical optimization. Memetic Computing 6(1), 31–47 (2014)
3. Banzhaf, W., Nordin, P., Keller, R.E., Francone, F.D.: Genetic programming: an introduction, vol. 1. Morgan Kaufmann, San Francisco (1998)
4. Davis, L., et al.: Handbook of genetic algorithms, vol. 115. Van Nostrand Reinhold, New York (1991)
5. Dorigo, M., Stützle, T.: Ant colony optimization (2004)
6. Du, D., Simon, D., Ergezer, M.: Biogeography-based optimization combined with evolutionary strategy and immigration refusal. In: IEEE International Conference on Systems, Man and Cybernetics, SMC 2009, pp. 997–1002. IEEE (2009)
7. Eberhart, R.C., Shi, Y., Kennedy, J.: Swarm intelligence. Elsevier (2001)
8. Farswan, P., Bansal, J.C.: Migration in biogeography-based optimization. In: Das, K.N., Deep, K., Pant, M., Bansal, J.C., Nagar, (eds.) Proceedings of Fourth International Conference on Soft Computing for Problem Solving. Advances in Intelligent Systems and Computing, vol. 336, pp. 389–401. Springer, India (2015)
9. Geem, Z.W., Kim, J.H., Loganathan, G.V.: A new heuristic optimization algorithm: harmony search. Simulation 76(2), 60–68 (2001)
10. Gomez, F.J., Miikkulainen, R.: Robust non-linear control through neuroevolution. Computer Science Department, University of Texas at Austin (2003)
11. Gong, W., Cai, Z., Ling, C.X.: De/bbo: a hybrid differential evolution with biogeography-based optimization for global numerical optimization. Soft Computing 15(4), 645–665 (2010)
12. Gong, W., Cai, Z., Ling, C.X., Li, H.: A real-coded biogeography-based optimization with mutation. Applied Mathematics and Computation 216(9), 2749–2758 (2010)
13. Karaboga, D.: An idea based on honey bee swarm for numerical optimization. Technical report, Technical report-tr06, Erciyes university, engineering faculty, computer engineering department (2005)
14. Kennedy, J.: Particle swarm optimization. In: Encyclopedia of Machine Learning, pp. 760–766. Springer (2010)
15. Lohokare, M.R., Pattnaik, S.S., Panigrahi, B.K., Das, S.: Accelerated biogeography-based optimization with neighborhood search for optimization. Applied Soft Computing 13(5), 2318–2342 (2013)
16. Ma, H.-P., Ruan, X.-Y., Pan, Z.-X.: Handling multiple objectives with biogeography-based optimization. International Journal of Automation and Computing 9(1), 30–36 (2012)
17. Ma, H., Simon, D.: Blended biogeography-based optimization for constrained optimization. Engineering Applications of Artificial Intelligence 24(3), 517–525 (2011)
18. Simon, D.: Biogeography-based optimization. IEEE Transactions on Evolutionary Computation 12(6), 702–713 (2008)
19. Simon, D., Omran, M.G.H., Clerc, M.: Linearized biogeography-based optimization with re-initialization and local search. Information Sciences 267, 140–157 (2014)
20. Storn, R., Price, K.: Differential evolution-a simple and efficient heuristic for global optimization over continuous spaces. Journal of Global Optimization 11(4), 341–359 (1997)
21. Yao, X., Liu, Y., Lin, G.: Evolutionary programming made faster. IEEE Transactions on Evolutionary Computation 3(2), 82–102 (1999)

Tournament Selection Based Probability Scheme in Spider Monkey Optimization Algorithm

Kavita Gupta and Kusum Deep

Abstract In this paper, a modified version of Spider Monkey Optimization (SMO) algorithm is proposed. This modified version is named as Tournament selection based Spider Monkey Optimization (TS-SMO). TS-SMO replaces the fitness proportionate probability scheme of SMO with tournament selection based probability scheme with an objective to improve the exploration ability of SMO by avoiding premature convergence. The performance of the proposed variant is tested over a large benchmark set of 46 unconstrained benchmark problems of varying complexities broadly classified into two categories: scalable and non-scalable problems. The performance of TS-SO is compared with that of SMO. Results for scalable and non-scalable problems have been analysed separately. A statistical test is employed to access the significance of improvement in results. Numerical and statistical results show that the proposed modification has a positive impact on the performance of original SMO in terms of reliability, efficiency and accuracy.

Keywords Spider monkey optimization · Tournament selection · Unconstrained optimization · Swarm intelligent techniques

1 Introduction

Spider Monkey Optimization (SMO) technique [1] is a newly developed swarm intelligent technique for solving unconstrained real parameter optimization problems. It is a simple and easy to implement swarm intelligent technique with few control parameters. It is inspired from the food searching strategy of Spider monkeys. Spider monkeys belong to the class of fission-fusion social structure

K. Gupta(✉) · K. Deep
Department of Mathematics, Indian Institute of Technology Roorkee, Roorkee 247667, Uttarakhand, India
e-mail: {gupta.kavita3043,kusumdeep}@gmail.com

© Springer-Verlag Berlin Heidelberg 2016
J.H. Kim and Z.W. Geem (eds.), *Harmony Search Algorithm,*
Advances in Intelligent Systems and Computing 382,
DOI: 10.1007/978-3-662-47926-1_23

239

based animals which live in a group of 40-50 individuals [4]. The group leader divides the whole group into small subgroups and the members of these subgroups search for their food in different directions. There are mainly two manipulation phases in SMO in each iteration which are responsible for the updation of the swarm. In one of these phases, members of the swarm get chance to update their position based on their probability. This probability is fitness proportionate, which is similar to roulette wheel selection in Genetic Algorithm (GA) [2,3]. Due to the use of fitness proportionate probability scheme in SMO, members of the swarm having higher fitness have better chances of updating their position as compared to the ones having lower fitness value. But sometimes even less fit members may contain some important information which can be very useful, but being not given a chance due to their low fitness, this important information is lost. Aiming at this limitation of SMO, we propose a variation of original SMO in which tournament selection based probability scheme has been used instead of fitness proportionate probability scheme of SMO. Tournament selection is one of the famous operators used in selection phase of GA [5]. Tournament selection based scheme will provide a chance to even less fit individuals to update their position. This modification at improving the search ability and convergence speed of original SMO by favoring more exploration with the help of probability scheme based on tournament selection. To the best of our knowledge, this is the first attempt of using tournament selection for calculating probabilities in SMO. Tournament of size two has been used in the proposed method. The objective of this paper is to study the impact on the performance of SMO in terms of reliability and convergence speed after replacing fitness proportionate probability scheme with tournament selection based probability scheme which is meant to increase the exploration ability of SMO.

The paper is organized as follows. Section 2 gives a brief introduction to Spider monkey optimization technique. In section 3, proposed modification in the original SMO has been discussed. Section 4 deals with experimental settings followed by discussion of experimental results in section 5. Section 6 provides the conclusion and future scope of proposed work.

2   Spider Monkey Optimization Algorithm

Like all other population based algorithms, SMO follows some iterative steps in the process of improving the swarm of randomly generated solutions. In addition to the initialization of the swarm, SMO follows six iterative steps. These are: Local leader phase, global leader phase, local leader learning phase, global leader learning phase, local leader decision phase and global leader decision phase. Detailed description of each iterative step along with their purpose in the algorithm can be found in [1]. Local leader limit, global leader limit, perturbation rate and maximum number of groups are four control parameters of SMO. Pseudocode for SMO has been provided in Fig. 1.

```
begin:
    Initialize the swarm
    Initialize Local leader limit, global leader limit, perturbation rate, maxi-
mum number
    of groups
    Iteration=0
    Calculate fitness value of the position of each spider monkey in the swarm
    Select Global Leader and Local Leaders by applying Greedy Selection
    while (termination criterion is not satisfied) do
            //Local Leader Phase
            //Calculate Probabilities
            //Global Leader Phase
            //Global Leader Learning Phase
            //Local Leader Learning Phase
            //Local Leader Decision Phase
            //Global Leader Decision Phase
            Iteration = iteration +1
        end while
    end
```

Fig. 1 Pseudocode for SMO

3   Tournament Selection Based SMO

TS-SMO is just a variation of SMO replacing fitness proportionate probability
scheme used in SMO with tournament selection based probability scheme. The
parameter associated with the tournament selection operator is the size of the tournament. This size indicates the numbers of members which will participate in the
tournament. In this paper, tournament size is two. In global leader phase, members
of the swarm get a chance to update their position based on their probability. Fitness proportionate probability scheme used in SMO provides more chances to
highly fit members to make themselves better which sometimes may lead to premature convergence because of attraction of the swarm to highly fit individuals
only. Tournament selection based probability scheme facilitates diversity in the
population thus avoiding premature convergence. Also, it may happen that even
less fit individuals may contain some important information about the optimal
solution. But since they do not have high probability, they have very less chances
of updating their position. To avoid the loss of important information contained in
less fit members of the swarm, it has been decided to use tournament selection
based probability scheme in place of fitness proportionate probability scheme so
that even less fit individuals may get chance to update their position. Let prob[i]
and fit[i] be the probability and fitness respectively of the ith member of the
swarm. Pseudocode for calculating probability in original SMO and TS-SMO has
been provided in Fig. 2 and Fig. 3 respectively.

```
for i: 1 to n
prob[i] = 0.9 × (fit[i]/maxfit) + 0.1
end for
```

Fig. 2 Pseudocode for calculating probability in SMO

```
for i: 1 to n
a[i]=0
    for j:1 to n
        if (fit[i]>=fit[j])
        a[i]=a[i]+1
        end if
    end for
end for
for i: 1 to n
prob[i] = a[i]/∑_{i=1}^{n} a[i]
end for
```

Fig. 3 Pseudocode for calculating probability in TS-SMO

4   Experimental Setup

The performance of SMO and TS-SMO has been tested over a benchmark set of 46 (1-30 are scalable problems and 31-46 are non-scalable problems) unconstrained optimization problems. All the scalable problems are of dimension 30 and dimension of each non-scalable problem is mentioned in the list of test problems given in the appendix. Below is the parameter setting and termination criterion for the experiment.

Swarm size =150
Perturbation rate (pr) = linearly increasing ([0.1, 0.4])
Maximum number of groups (MG) =5
Local leader limit =100
Global leader limit=50
Total number of runs =100
Maximum number of iteration= 4000
acceptable error =1.0e-05
Stopping criterion = either maximum number of iterations are performed or acceptable error is achieved.

Here error is the absolute difference between the optimal solution and objective function value of the global leader. In order to make a fair comparison between the two algorithms, both the algorithms starts with the same initial swarm..

5   Experimental Results and Discussion

In order to compare the performance of both the algorithms, number of successful runs and the average number of function evaluations of successful runs have been recorded. A run is said to be successful if the error value is less than the acceptable error. Reliability of the algorithms has been measured from the number of successful runs and efficiency is measured with the number of function evaluations of successful runs. Comparison between the two algorithms have been made in the following manner:

First the number of successful runs have been checked and the algorithm with more number of successful runs will be the winner. If both the algorithms have same number of successful runs, then their average number of function evaluations for successful runs have been checked and the algorithm with less number of function evaluations will be the winner. Further, to see if there is really any significant difference between the function evaluations of successful runs, one tailed t-test at a significant level of 0.5 has been employed. "=" indicates there is no significant difference between the average of function evaluations of two algorithms and "+" and "-" indicates that TS-SMO performs significantly better and worse than SMO respectively. T-test is only applied to the problems where number of successful runs for both the algorithms are same. The results of scalable and non-scalable problems have been provided in table 1 and table 2 respectively.

From Table 1, it can be observed that out of 30 scalable problems, there are 5 problems where both the algorithms do not have even a single successful run. From the rest of 25 problems, there are 21 problems were TS-SMO performs better than SMO. Also, from the results of t-test, significant difference between the function evaluations can be observed.

From Table 2, it is observed that out of 16 non-scalable problems, there are 3 problems where both the algorithms have no successful runs. From the remaining 13 problems, there are 7 problems where SMO performs better than TS-SMO, while there are 6 problems where TS-SMO performs better. Also, t-test results show significant difference in the number of function evaluations.

From the results depicted in Table 1 and Table 2, it can be concluded that whereas TS-SMO performs better than SMO on most of the scalable problems, its performance is not so good on non-scalable problems.

In order to observe the effect of the variation in the objective function value as the iterations proceed Convergence graphs are plotted. Convergence graph of selected problems have been provided in Fig. 4, Fig. 5, Fig. 6 and Fig. 7. From these graphs it can be seen that TS-SMO convergence faster than SMO in most of the problems.

Table 1 No. of successful runs and average number of function evaluations of successful runs for SMO and TS-SMO (Scalable problems)

Fun_no.	Percentage of Success		Function evaluations of successful runs		Outcome of t-test
	SMO	TS-SMO	SMO	TS-SMO	
1	100	100	33192	**32128**	+
2	100	100	26566	**25923**	+
3	100	100	**103625**	108601	=
4	0	0	0	0	N.A.
5	100	100	**229495**	254951	-
6	100	100	63917	**61660**	+
7	100	100	189827	**149630**	+
8	100	100	24027	**19040**	+
9	100	100	37701	**34987**	+
10	100	100	25421	**24404**	+
11	47	**61**	**650570**	682431	N.A.
12	100	100	58502	**56807**	+
13	100	100	32220	**31231**	+
14	100	100	62664	**60888**	+
15	0	0	0	0	N.A.
16	100	100	38287	**37013**	+
17	0	0	0	0	N.A.
18	100	100	14002	**13294**	+
19	100	100	22628	**22437**	+
20	0	0	0	0	N.A.
21	7	2	**1051267**	1117825	N.A.
22	83	**85**	354061	**349233**	N.A.
23	100	100	48381	**47167**	+
24	100	100	30839	**29457**	+
25	100	100	30695	**29586**	+
26	100	100	39151	**37892**	+
27	100	100	44665	**43394**	+
28	0	0	0	0	N.A.
29	100	100	106373	**101093**	=
30	100	100	63980	**61938**	+

Table 2 No. of successful runs and average number of function evaluations of successful runs for SMO and TS-SMO for Non-Scalable problems

Fun_no.	Percentage of Success		Function evaluations of successful runs		outcome of t-test
	SMO	TS-SMO	SMO	TS-SMO	
31	100	100	3738	**3636**	=
32	**98**	93	**430014**	465326	N.A.
33	100	100	**18801**	19552	-
34	100	100	3585	**3415**	+
35	100	100	256584	**211382**	+
36	100	100	3304	**3199**	=
37	100	100	131394	**131171**	=
38	100	100	**26119**	33359	-
39	100	100	**2009**	2281	-
40	100	100	3071	**2793**	+
41	0	0	0	0	N.A.
42	100	100	**24508**	45553	-
43	0	0	0	0	N.A.
44	100	100	**13458**	13519	=
45	0	0	0	0	N.A.
46	100	100	**10737**	12525	-

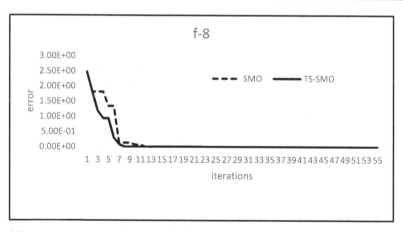

Fig. 4 Convergence graph of Function No. 8

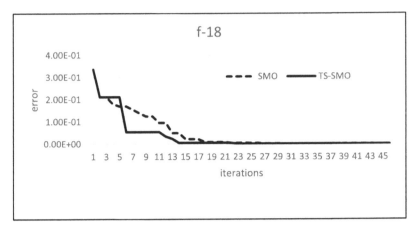

Fig. 5 Convergence graph of Function No. 18

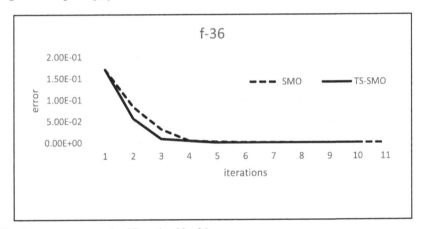

Fig. 6 convergence graph of Function No. 36

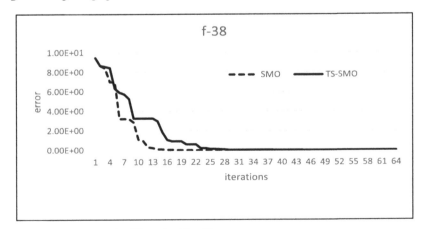

Fig. 7 Convergence graph of Function No. 38

6   Conclusion and Future Directions

In this paper, a new version of SMO abbreviated as TS-SMO has been proposed. The novelty of this new version lies in the use of tournament selection based probability scheme instead of fitness proportionate probability scheme which has been used in original SMO. Maintaining a high level of diversity while preserving convergence speed are two contradictory and necessary features of a metaheuristic technique and the numerical and statistical results show that whereas TS-SMO did a good job in balancing both the features by performing well on scalable problems, it could not perform so well on non-scalable problems. But since the experiments have been performed over a limited number of benchmark problems and no theoretical proof has been provided, making any concrete conclusion will not be justified. Further work needs to be done experimentally and theoretically to make any strong judgement about the superiority of TS-SMO over SMO.

Acknowledgement The first author would like to acknowledge the ministry MHRD, Government of India for financial support.

References

1. Bansal, J.C., Sharma, H., Jadon, S.S., Clerc, M.: Spider Monkey Optimization algorithm for numerical optimization. Memetic Computing **6**(1), 31–47 (2014)
2. Goldberg, D.E.: Genetic algorithms in search, optimization, and machine learning. Addison-Wesley, Reading (1989)
3. Holland, J.H.: Adaptation in Natural and Artificial Systems. University of Michigan press, AnnArbor (1975)
4. Symington, M.M.: Fission-fusion social organization inAteles and Pan. International Journal of Primatology **11**(1), 47–61 (1990)
5. Miller, B.L., Goldberg, D.E.: Genetic algorithms, tournament selection, and the effects of noise. Complex Systems **9**(3), 193–212 (1995)

Appendix

Table A1 Test problems

Test Problems	Objective function	Search range	Optimal value
Sphere Function	$f_1(x) = \sum_{i=1}^{D} x_i^2$	[-5.12,5.12]	0
De Jong's F4	$f_2(x) = \sum_{i=1}^{D} i x_i^4$	[-5.12,5.12]	0
Griewank	$f_3(x) = \sum_{i=1}^{D} \frac{x_i^2}{4000} - \prod_{i=1}^{D} \cos\left(\frac{x_i}{\sqrt{i}}\right) + 1$	[-600,600]	0
Rosenbrock	$f_4(x) = \sum_{i=1}^{D} [100(x_i^2 - x_{i+1})^2 + (x_i - 1)^2]$	[-100,100]	0
Rastrigin	$f_5(x) = \sum_{i=1}^{D} (x_i^2 - 10\cos(2\pi x_i) + 10)$	[-5.12,5.12]	0
Ackley	$f_6(x) = -20\exp\left(-0.2\sqrt{\frac{1}{D}\sum_{i=1}^{D} x_i^2}\right)$ $-\exp\left(\frac{1}{D}\sum_{i=1}^{D}\cos(2\pi x_i)\right) + 20 + e$	[-30,30]	0
Alpine	$f_7(x) = \sum_{i=1}^{D} |x_i\sin(x_i) + 0.1x_i|$	[-10,10]	0
Michalewicz	$f_8(x) = -\sum_{i=1}^{D} \sin(x_i)\left[\frac{\sin(ix_i^2)}{\pi}\right]^{20}$	[0,π]	-9.66015
Cosine Mixture	$f_9(x) = \sum_{i=1}^{D} x_i^2 - 0.1\sum_{i=1}^{D}\cos(5\pi x_i)$	[-1,1]	-D*0.1
Exponential	$f_{10}(x) = -\exp\left(-0.5\sum_{i=1}^{D} x_i^2\right)$	[-1,1]	-1
Zakharov	$f_{11}(x) = \sum_{i=1}^{D} x_i^2 + \left(\frac{1}{2}\sum_{i=1}^{D} i x_i\right)^2 + \left(\frac{1}{2}\sum_{i=1}^{D} i x_i\right)^4$	[-5.12,5.12]	0
Cigar	$f_{12}(x) = x_1^2 + 100000\sum_{i=2}^{D} x_i^2$	[-10,10]	0
Brown3	$f_{13}(x) = \sum_{i=1}^{D-1} \left[(x_i^2)^{(x_{i+1}^2+1)} + (x_{i+1}^2)^{(x_i^2+1)}\right]$	[-1,4]	0
Schewel Prob 3	$f_{14}(x) = \sum_{i=1}^{D} |x_i| + \prod_{i=1}^{D} |x_i|$	[-10,10]	0
Salomon Problem	$f_{15}(x) = 1 - \cos\left(2\pi\sqrt{\sum_{i=1}^{D} x_i^2}\right) + 0.1\sqrt{\sum_{i=1}^{D} x_i^2}$	[-100,100]	0
Axis Parallel Hyperellipsoid	$f_{16}(x) = \sum_{i=1}^{D} i x_i^2$	[-5.12,5.12]	0
Pathological	$f_{17}(x) = \sum_{i=1}^{D-1}\left[0.5 + \frac{\sin^2\sqrt{100x_i^2 + x_{i+1}^2} - 0.5}{1 + 0.001(x_i^2 + x_{i+1}^2 - 2x_i x_{i+1})^2}\right]$	[-100,100]	0
Sum Of Different Powers	$f_{18}(x) = \sum_{i=1}^{D} |x_i|^i$	[-1,1]	0

Table A1 (*continued*)

Step Function	$f_{19}(x) = \displaystyle\sum_{i=1}^{D} \left([x_i + 0.5]\right)^2$	[-100,100]	0
Quartic Function	$f_{20}(x) = \displaystyle\sum_{i=1}^{D} i x_i^4 + random[0,1)$	[-1.28,1.28]	0
Inverted Cosine Wave Function	$f_{21}(x) = -\displaystyle\sum_{i=1}^{D-1} exp\left(-\left(\dfrac{x_i^2 + x_{i+1}^2 + 0.5 x_i x_{i+1}}{8}\right)\right) *$ $cos\left(4\sqrt{x_i^2 + x_{i+1}^2 + 0.5 x_i x_{i+1}}\right)$	[-5,5]	-D+1
Neumaier3 Problem	$f_{22}(x) = \left\vert \displaystyle\sum_{i=1}^{D} (x_i - 1)^2 - \displaystyle\sum_{i=1}^{D} x_i x_{i+1} \right\vert$	[-900, 900]	0
Rotated Hyper Ellipsoid Function	$f_{23}(x) = \displaystyle\sum_{i=1}^{D} x_i^2$	[-65.536, 65.536]	0
Levi Montalvo 1	$f_{24}(x) = \dfrac{\pi}{D}[10 sin^2(\pi y_1) +$ $\sum_{i=1}^{D-1}(y_i - 1)^2\left(1 + 10 sin^2(\pi y_{i+1})\right) + (y_D - 1)^2],$ Where $y_i = 1 + \dfrac{1}{4}(x_i + 1)$	[-10,10]	0
Levi Montalvo 2	$f_{25}(x) = 0.1\left(sin^2(3\pi x_1)\right.$ $+ \displaystyle\sum_{i=1}^{D-1}\left[(x_i - 1)^2\left(1 + sin^2(3\pi x_{i+1})\right)\right]\bigg)$ $+(x_D - 1)^2\left(1 + sin^2(2\pi x_D)\right))$	[-5,5]	0
Ellipsoidal	$f_{26}(x) = \displaystyle\sum_{i=1}^{D} (x_i - i)^2$	[-D,D]	0
Shifted Parabola CEC 2005	$f_{27}(x) = \sum_{i=1}^{D} z_i^2 + f_{bias}$, $z=(x\text{-}o), \quad x=[x_1, x_2, ..., x_D], \quad O=[o_1, o_2, ..., o_D]$	[-100,100]	-450
Shifted Schwefel CEC 2005	$f_{28}(x) = \sum_{i=1}^{D}\left(\sum_{j=1}^{i} z_j\right)^2 + f_{bias}$, $z=(x\text{-}o), \quad x=[x_1, x_2, ..., x_D], \quad O=[o_1, o_2, ..., o_D]$	[-100,100]	-450
Shifted Greiwank CEC 2005	$f_{29}(x) = \sum_{i=1}^{D} \dfrac{z_i^2}{4000} - \prod_{i=1}^{D} cos\left(\dfrac{z_i}{\sqrt{i}}\right) + 1 + f_{bias}$, $z=(x\text{-}o), \quad x=[x_1, x_2, ..., x_D], \quad O=[o_1, o_2, ..., o_D]$	[-600,600]	-180
Shifted Ackley CEC 2005	$f_{30}(x) = -20 exp\left(-0.2\sqrt{\dfrac{1}{D}\sum_{i=1}^{D} x_i^2}\right)$ $-exp\left(\dfrac{1}{D}\sum_{i=1}^{D} cos(2\pi x_i)\right) + 20 + e + f_{bias}$, $z=(x\text{-}o), \quad x=[x_1, x_2, ..., x_D], \quad O=[o_1, o_2, ..., o_D]$	[-32,32]	-140

Table A1 (*continued*)

Step Function	$f_{19}(x) = \sum_{i=1}^{D} ([x_i + 0.5])^2$	[-100,100]	0
Quartic Function	$f_{20}(x) = \sum_{i=1}^{D} i x_i^4 + random[0,1)$	[-1.28,1.28]	0
Inverted Cosine Wave Function	$f_{21}(x) = -\sum_{i=1}^{D-1} exp\left(-\left(\frac{x_i^2 + x_{i+1}^2 + 0.5x_i x_{i+1}}{8}\right)\right) *$ $cos\left(4\sqrt{x_i^2 + x_{i+1}^2 + 0.5x_i x_{i+1}}\right)$	[-5,5]	-D+1
Neumaier3 Problem	$f_{22}(x) = \left\|\sum_{i=1}^{D} (x_i - 1)^2 - \sum_{i=1}^{D} x_i x_{i+1}\right\|$	[-900, 900]	0
Rotated Hyper Ellipsoid Function	$f_{23}(x) = \sum_{i=1}^{D} x_i^2$	[-65.536, 65.536]	0
Levi Montalvo 1	$f_{24}(x) = \frac{\pi}{D}[10sin^2(\pi y_1) +$ $\sum_{i=1}^{D-1}(y_i - 1)^2(1 + 10sin^2(\pi y_{i+1})) + (y_D - 1)^2],$ Where $y_i = 1 + \frac{1}{4}(x_i + 1)$	[-10,10]	0
Levi Montalvo 2	$f_{25}(x) = 0.1\left(sin^2(3\pi x_1)\right.$ $+ \sum_{i=1}^{D-1}\left[(x_i - 1)^2(1 + sin^2(3\pi x_{i+1}))\right]\bigg)$ $+(x_D - 1)^2(1 + sin^2(2\pi x_D))$	[-5,5]	0
Ellipsoidal	$f_{26}(x) = \sum_{i=1}^{D} (x_i - i)^2$	[-D,D]	0
Shifted Parabola CEC 2005	$f_{27}(x) = \sum_{i=1}^{D} z_i^2 + f_{bias}$, $z=(x-o), \quad x=[x_1, x_2, ..., x_D], \quad O=[o_1, o_2, ..., o_D]$	[-100,100]	-450
Shifted Schwefel CEC 2005	$f_{28}(x) = \sum_{i=1}^{D}\left(\sum_{j=1}^{i} z_j\right)^2 + f_{bias}$, $z=(x-o), \quad x=[x_1, x_2, ..., x_D], \quad O=[o_1, o_2, ..., o_D]$	[-100,100]	-450
Shifted Greiwank CEC 2005	$f_{29}(x) = \sum_{i=1}^{D} \frac{z_i^2}{4000} - \prod_{i=1}^{D} cos\left(\frac{z_i}{\sqrt{i}}\right) + 1 + f_{bias}$, $z=(x-o), \quad x=[x_1, x_2, ..., x_D], \quad O=[o_1, o_2, ..., o_D]$	[-600,600]	-180
Shifted Ackley CEC 2005	$f_{30}(x) = -20exp\left(-0.2\sqrt{\frac{1}{D}\sum_{i=1}^{D} x_i^2}\right)$ $-exp\left(\frac{1}{D}\sum_{i=1}^{D} cos(2\pi x_i)\right) + 20 + e + f_{bias}$, $z=(x-o), \quad x=[x_1, x_2, ..., x_D], \quad O=[o_1, o_2, ..., o_D]$	[-32,32]	-140
Shekel10	$f_{44}(x) = -\sum_{j=1}^{10}\left[\sum_{i=1}^{4}(x_i - C_{ij})^2 + \beta_j\right]^{-1}$	[0,10]	4	-10.5364
Dekkers and Aarts	$f_{45}(x) = 10^5 x_1^2 + x_2^2 - (x_1^2 + x_2^2)^2 + 10^{-5}(x_1^2 + x_2^2)^4$	[-20,20]	2	-24777
Shubert	$f_{46}(x) = -\sum_{i=1}^{5} icos((i+1)x_1 + 1)\sum_{i=1}^{5} icos((i+1)x_{2+1})$	[-10,10]	2	-186.7309

Optimal Extraction of Bioactive Compounds from Gardenia Using Laplacian Biogoegraphy Based Optimization

Vanita Garg and Kusum Deep

Abstract Bioactive compounds form different plant materials are used in a number of important pharmaceutical, food and chemical industries. Many conventional and unconventional methods are available to extract optimum yields of these bioactive compounds from various plant materials. This paper focuses on the extraction of bioactive compounds (crocin, geniposide and total phenolic compounds) from Gardenia *(Gardenia jasminoides Ellis)* by modeling the problem as a nonlinear optimization problem with multiple objectives. There are three objective functions each representing the maximizing of three bioactive compounds i.e. crocin, geniposide and total phenolic compounds. Each of the bioactive compounds are dependent on three factors namely: concentration of ethanol, extraction temperature and extraction time. The solution methodology is a recently proposed Laplacian Biofeographical Based Optimization. The results obtained are compared with previously reported results and show a significant improvement, thus exhibiting not only the superior performance of Laplacian Biogeographical Based Optimization, but also the complexity of the problem at hand.

Keywords Biogeography-Based optimization · Response surface methodology · RCGA · Extraction of compounds · Laplacian BBO

1 Introduction

The extraction optimization of bioactive compounds from plant material is an important issue in order to yield the maximum benefits. Extraction optimization

V. Garg(✉) · K. Deep
Department of Mathematics, Indian Institute of Technology Roorkee,
Roorkee 247-667, India
e-mail: {vanitagarg16,kusumdeep}@gmail.com

© Springer-Verlag Berlin Heidelberg 2016
J.H. Kim and Z.W. Geem (eds.), *Harmony Search Algorithm*,
Advances in Intelligent Systems and Computing 382,
DOI: 10.1007/978-3-662-47926-1_24

plays a crucial role in final outcome. Many authors have solved this type of optimization using response surface methodology. Many Meta-heuristic approaches like genetic algorithm, Particale swarm optimization, Ant colony optimization, Biogeography based optimization etc. are getting famous by each passing day. Use of these optimization methods for extracting the optimal amount of bioactive compounds could be useful. Nature inspired technique like Real Coded Genetic algorithm has already been applied to this problem in [1].To optimize different yields of three bioactive compounds from gardenia fruit i.e. crocin, geniposide and total phenolic compounds, we are applying weighted sum approach of multi-objective optimization method.

Gardenia is a plant having medicinal properties. Gardenia jasminoides Ellis is obtained from Gardenia fruits. These constituents of Gardenia fruits are used as herb medicines and natural dyes in many Asian countries. In real world scenario, using safe and natural colorants is a major concern as it can help in avoiding many harmful diseases [5].

Gardenia fruit is widely used in food industry for colored juice, candy, jelly and noodles. To serve this purpose, the extract of gardenia fruit which exhibit yellow, red and blue colors is used as a natural colorant because of their water solubility. Three main ingredients of gardenia fruit are considered in the present study i.e. crocin, geniposide and phenolic compounds.

In this paper, to maximize these yields, multi-objective optimization technique is used. After using weighted sum approach of multi-objective optimization technique, this problem is converted into single objective optimization problem and then is solved using a recently proposed Laplacian Biogeography Based optimization.

The paper is organized as follows: Section 2 gives the summary of the problem and its model. Section 3 describes the Algorithm used for solving the problem. Computational results are presented in section 4, response surface plots are given in section 5 and conclusions in section 6.

2 Formulation of the Problem of Extraction of Compounds from Gardenia

The problem is formulated in [1], where yields of three bioactive compounds namely crocin (Y 1), geniposide (Y 2) and total phenolic compounds (Y 3) obtained from Gardenia fruits which are affected by the three independent variables namely concentration of ethanol (X 1, extraction temperature (X 2) and extraction time (X 3). The modeling of the data is done using method of least square fitting. The data in [8] is used by [1] to formulate the problem.

Three different yields are formulated as a function of three independent variables. These functions are written in the form of second-order polynomial equation as follows:

$$Y_k = b_0 + \sum_{i=1}^{3} b_i X_i^2 + \sum_{i=1}^{3} b_{ij} X_i X_j \tag{1}$$

where Yk represents the yield, b0 is a constant, bi ,bii and bij are the linear, quadratic and interactive coefficients of the model, respectively. Xi and X j are the independent variables. The resultant equations for yieldsY1, Y2 and Y3 are as follows:

$Y_1 = 3.8384907903 + 0.0679672610\,X_1 + 0.0217802311 X_2 + .0376755412 X_3 - 0.0012103181 X_1^2 + 0.0000953785 X_2^2 - 0.0002819634 X_3^2 + 0.0005496524\,X_1 X_2 - 0.0009032316 X_2 X_3 + 0.0008033811 X_1 X_3$ \hfill (2)

$Y_2 = 46.6564201287 + 0.6726057655 X_1 + 0.4208752507 X_2 + 0.9999909858 X_3 - 0.0161053654 X_1^2 - 0.0034210643 X_2^2 - 0.0116458859 X_3^2 + 0.0122000907\,X_1 X_2 - 0.0095644212 X_2 X_3 + 0.0089464814 X_1 X_3$ \hfill (3)

$Y_3 = -6.3629169281 + 0.4060552042 X_1 + 0.3277005337 X_2 + 0.3411029105 X_3 - 0.0053585731 X_1^2 - 0.0020487593 X_2^2 - 0.0042291040 X_3^2 + 0.0017226318\,X_1 X_2 - 0.0011990977 X_2 X_3 + 0.0007814998 X_1 X_3$ \hfill (4)

Now, the problem represents a multi-objective optimization problem where we need to maximize all the three yields. In this paper, to optimize all the three yields Y1, Y2, Y3 simultaneously, LX-BBO is applied to solve the multi-objective optimization problem. A multi-objective problem is an optimization problem where a number of objective functions are to optimize simultaneously. The weighted method approach is an effective and simple technique to handle multi-objective optimization problem. In this technique different user defend weights are given to different objective functions and then the problem is treated as a single objective function .

In the present problem, equal weights are given to all the yield functions and then the problem is solved using LX-BBO [4].

Mathematically, for given yields Y1, Y2 and Y3, the objective function is to solve the following function:

$$\max g = w_1 Y_1 + w_2 Y_2 + w_3 Y_3 \tag{5}$$

Where w_1, w_2 and w_3 are user defined weights given to different yields.

3   Laplacian BBO (LX-BBO)

Biogeography-Based Optimization is a nature inspired technique which is inspired from the biogeography [7]. It deals with the immigration and emigration of habitats based on its suitability index. Dan Simon introduced BBO where a solution is considered as a habitat this solution or habitat is improved by using two operators: Migration and Mutation. These Operators are explained as follows:

Migration. The information sharing between the solutions (habitats) is termed as Migration. This information is shared probabilistically. Emigration and immigration rates decide which habitat is to be replaced and by which it should be replaced. These rates are based on Habitat Suitability index(HSI) usually taken as objective function.

Mutation. A sudden change in natural calamity or disease etc. can change the HSI of a habitat. This sudden change is termed as Mutation. [3]

Garg and Deep has proposed a new version of BBO named as Laplacian BBO [4]. In Laplacian BBO (LX-BBO), Migration operator of BBO is modified by integrating Laplace Crossover of Real coded Genetic Algorithm [2]. In migration operator of BBO, two habitats x_1 and x_2 are selected based upon immigration and emigration rates. These habitats are used to generate two new habitats y_1 and y_2 using Laplace crossover.

Two new habitats y_1 and y_2 are generated by the following equation:

$$y_1^i = x_1^i + \beta(x_1^i - x_2^i) \tag{6}$$

$$y_2^i = x_2^i + \beta(x_1^i - x_2^i) \tag{7}$$

Where random number which follows Laplace distribution is generated given by the equation:

$$\beta = \begin{cases} a - b * log(u), & u \le 1/2 \\ a + b * log(u), & u > 1/2 \end{cases} \tag{8}$$

$u_i \epsilon [0,1]$ is a uniform random number. $a \in R$ is called location parameter and b > 0 is called scale parameter.

These two new habitats further give rise to a new habitat z which is used in further generations.

$$z = \gamma\, y_1^i + (1 - \gamma)y_2^i \tag{9}$$

Pseudo Code of LX-BBO

```
Begin
        Initialize: Generate a random set of habitats (Islands)
        Compute HSI values of each habitat
        While (Stopping condition is not satisfied)
        For each habitat
        For each SIV
        Calculate the immigration rate and emigration rate based on HSI values
        Select habitat Hi with the help of immigration rate
        If Hi (i=1 to n, where n is the number of habitats) is selected then
            Select habitat Hj(j=1 to n)  with the help of emigration rate
            If Hj is selected then Find
```

$$H_i^1(SIV) = H_i(SIV) + \beta(H_i(SIV) - H_j(SIV))$$
$$H_i^2(SIV) = H_j(SIV) + \beta(H_i(SIV) - H_j(SIV))$$
$$H_i(SIV) \leftarrow \gamma H_i^1(SIV) + (1 - \gamma)H_i^2(SIV)$$

```
            End if
        End if
        Select Hi(SIV) based on mutation probability
        If Hi(SIV) is selected then
            Replace Hi(SIV) with a randomly generated SIV;
        End if
        End for
        End for
        For each habitat
        Re-compute HSI values
        End for
        End while
End
```

Fig. 1 Pseudo Code of Laplacian Biogeography- Based Optimization

4   Computational Results

In this section we present numerical results which are obtained by using Laplacian Biogeography based optimization method .The problem is to maximize the function g . All the three yields are of same importance. Thus, weights are set to equal for each objective function. In this paper, the weights w1, w2 and w3 are set equal to 0.33, 0.33 and 0.34 respectively. Population size is set equal to 30. The mutation rate in LX-BBO is set equal to 0.005. The maximum number of generation is set equal to 300.Differnt 30 independent runs are performed on the basis of time clock seed.

Extraction of compounds from gardenia is solved by response surface method [8] and RCGA [6]. The results obtained by LX-BBO are compared with the results obtained by these two algorithms.

Table 1 Yields obtained by different algorithms

Algo/yield	Y_1(Crocin) mg/g	Y_2(Genoposide) mg/g	Y_3(Total Phenolic compunds) CAE/g
LX-BBO	**9.2504**	**113.924**	**25.2279**
DDX-LLM	8.43	110.026	24.81
Yang et.al	8.36	108.5	24.5

In Table 1, we can clearly see that the yields obtained by LX-BBO are much more than those with DDX-LLM (which is a real coded GA proposed by [6]) and.[8] Out of these techniques, LX-BBO is emerged as a winner. Thus, in solving extraction of compounds problem, LX-BBO can be quite helpful.

5 Response Surface Plots

To check the behavior of the problem complexity, Response Surface plots are shown in following figures. Fig.2 (a) shows the response for (Y 1) yield of crocin when time is constant. Fig. 2(b) gives the response of (Y 1) yield of Crocin when temperature is constant. Fig. 3(a) shows the response for Y2 (Geniposide) when time is constant and Fig. 3(b) gives the response of (Y 2) yield of Geniposide when time is constant. Fig. 4(a) shows the response for Y3 (total Phenolic Compounds) when time is constant and Fig. 4(b) gives the response of (Y 3) yield of total Phenolic Compounds when time is constant.

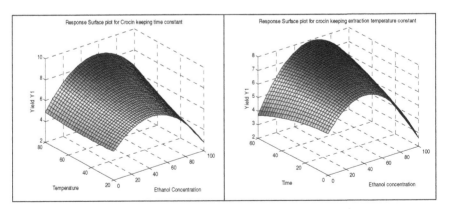

Fig. 2 Response surface plot for crocin keeping time(a) and temperature(b)constant

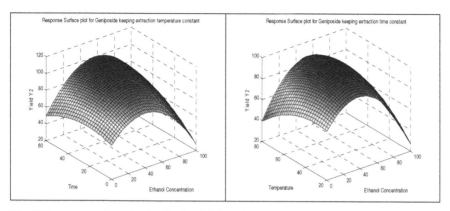

Fig. 3 Response surface plot for Geniposide keeping time(a) and temperature(b) constant

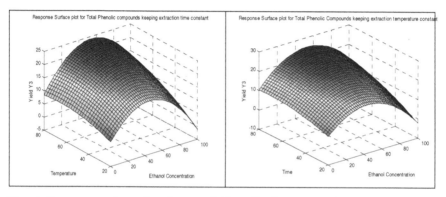

Fig. 4 Response surface plot for total Phenolic Compounds keeping time(a) and temperature(b) constant

6   Conclusion

In this paper the problem of extraction of bioactive compounds from a herb called gardenia, is modeled as a multi-objective optimization problem. The method of solution is a recently proposed hybrid version of Biogeography Based optimization called Laplacian Biogeography based optimization. The results are compared with earlier published results, which have been obtained to solve the problem at hand by a Real coded genetic algorithm namely LX-DDM. From the obtained results it is concluded that Laplacian Biogeography Based optimization is able to provide superior results as compared to the earlier published results.

References

1. Deep, K., Katiyar, V.K.: Multi objective extraction optimization of bioactive compounds from gardenia using real coded genetic algorithm. In: 6th World Congress of Biomechanics, vol. 31, pp. 1436-1466 (2010)
2. Deep, K., Thakur, M.: A new crossover operator for real coded genetic algorithms. Applied Mathematics and Computation, 895–912 (2007)
3. Garg, V., Deep, K.: A state of the art of biogeography based optimization. In: Das, K.N., Deep, K., Pant, M., Bansal, J.C., Nagar, A. (eds.) SocProS 2014. AISC, vol. 336, pp. 533–549. Springer, Heidelberg (2014)
4. Garg,V., Deep, K.: Performance of Laplacian Biogeography-Based Optimization Algorithm on CEC 2014 continuous optimization benchmarks and Camera Calibration Problem.(Resubmitted in Smarm and Evoulltionary Computation)
5. Owais, M., Sharad, K.S., Shehbaz, A., Saleemuddin, M.: Antibacterial efficacy of Withania somnifera (ashwagandha) an indigenous medicinal plant against experimental murine salmonellosis. Phytomedicine. 12, 229–235 (2005)
6. Shashi, B.: New Real Coded Genetic Algorithms and their Application to BioRelated Problem.PhD thesis, Indian Institute of Technology Roorkee. India (2011)
7. Simon, D.: Biogeography-Based Optimization. IEEE Transactions on Evolutionary Computation. 12(6), 702–713 (2008)
8. Yang, B., Liu, X., Yanxiang, G.: Extraction Optimization of Bioactive Compounds (crocin, geniposide and total phenolic compounds) from Gardenia (Gardenia jasminoides Ellis) Fruits with Response Surface Methodology. Innovative Food Sci Emerg Technol. 10, 610–615 (2009)

Physical Interpretation of River Stage Forecasting Using Soft Computing and Optimization Algorithms

Youngmin Seo, Sungwon Kim and Vijay P. Singh

Abstract This study develops river stage forecasting models combining Support Vector Regression (SVR) and optimization algorithms. The SVR is applied for forecasting river stage, and the optimization algorithms, including Grid Search (GS), Genetic Algorithm (GA), Particle Swarm Optimization (PSO), and Artificial Bee Colony (ABC), are applied for searching the optimal parameters of the SVR. For assessing the applicability of models combining SVR and optimization algorithms, the model performance is compared with ANN and ANFIS models. In terms of model efficiency, SVR-GS, SVR-GA, SVR-PSO and SVR-ABC models yield better results than ANN and ANFIS models. SVR-PSO and SVR-ABC models produce relatively better efficiency than SVR-GS and SVR-GA models. SVR-PSO and SVR-ABC yield the best performance in terms of model efficiency. Results indicate that river stage forecasting models combining SVR and optimization algorithms can be used as an effective tool for forecasting river stage accurately.

Keywords Support vector regression · Grid search · Genetic algorithm · Particle swarm optimization · Artificial bee colony

Y. Seo
Department of Constructional Disaster Prevention Engineering,
Kyungpook National University, Sangju 742-711, South Korea
e-mail: ymseo@knu.ac.kr

S. Kim(✉)
Department of Railroad and Civil Engineering, Dongyang University,
Yeongju 750-711, South Korea
e-mail: swkim1968@dyu.ac.kr

V.P. Singh
Department of Biological and Agricultural Engineering & Zachry Department of Civil
Engineering, Texas A & M University, College Station, Texas 77843-2117, USA
e-mail: vsingh@tamu.edu

© Springer-Verlag Berlin Heidelberg 2016
J.H. Kim and Z.W. Geem (eds.), *Harmony Search Algorithm,*
Advances in Intelligent Systems and Computing 382,
DOI: 10.1007/978-3-662-47926-1_25

259

1   Introduction

Accurate forecasting of river stage is significant for enhancing reservoir operation, water supply, flood prevention, water resources management, and decision support system [1]. Recently, support vector machines (SVMs), which are known as classification, regression and outlier detection methods in artificial intelligence or machine learning field, have gained the attention in hydrologic fields, including soil moisture, rainfall-runoff, streamflow and reservoir water level. The SVMs, which are a machine learning method based on the statistical learning theory, are based on the concept searching the hyperplane that has the largest distance to the nearest training data point of any class. When the SVMs are applied for regression problems, they are called support vector regression (SVR). Although the SVMs are known to be better than existing methods, such as artificial neural networks (ANNs) in terms of problems including over-fitting, local optimal solution and low convergence rate, the learning and generalization ability of the SVMs is very sensitive to the selection of their parameters. Therefore, it is essential to apply parameter optimization algorithms for selecting the optimal parameters effectively.

This study develops river stage forecasting models combining SVR and optimization algorithms and evaluates the model performance from a case study on Andong dam watershed. The SVR is applied for forecasting river stage and the optimization algorithms, including Grid Search (GS), Genetic Algorithm (GA), Particle Swarm Optimization (PSO) and Artificial Bee Colony (ABC), are applied for searching the optimal parameters of the SVR. For assessing the applicability of models combining SVR and optimization algorithms (SVR-GS, SVR-GA, SVR-PSO and SVR-ABC), the performance of models is compared with ANN and ANFIS models.

2   Material and Methods

2.1   Used Data

Daily river stage data of two streamflow gauging stations, Socheon and Dosan, were obtained from the observation archives of Water Management Information System (WAMIS), which is operated by the Ministry of Land, Infrastructure and Transport (MOLIT), South Korea. Fig. 1 shows the locations of streamflow gauging stations. The collected data were prepared for the period between 2002 and 2013. The data were divided into two parts, data of the first nine years for model training and the remaining three years data for model testing.

2.2   Support Vector Regression (SVR)

SVR is a machine learning method based on structural risk minimization (SRM) to solve nonlinear regression problems. The basic concept of SVR is that the original dataset is mapped into a high-dimensional feature space using a nonlinear

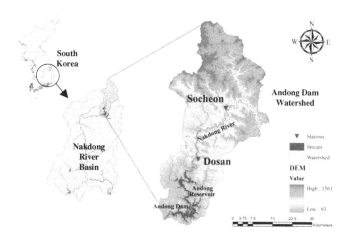

Fig. 1 Study region and locations of stage gauging stations

mapping function. Given a training data of n elements $\{\mathbf{x}_i, y_i\}_{i=1}^n$, where $\mathbf{x} \in R^m$ is the input vector of m components and $y \in R$ is the corresponding output value, the regression function of SVR is formulated as follows [2]:

$$f(\mathbf{x}) = \mathbf{w}^T \cdot \phi(\mathbf{x}) + b \tag{1}$$

where $\phi(\mathbf{x})$ is the nonlinear mapping function, $\mathbf{w}$ is the weight vector, and b is the bias term. By mapping the input vector into high dimensional feature space using the function $\phi(\mathbf{x})$, a simple linear regression of the feature space can cope with the complex nonlinear regression of the input space.

Coefficients $\mathbf{w}$ and b can be estimated by minimizing the regularized risk function R_{reg} based on ε-insensitive loss function L_ε as follows [2]:

$$L_\varepsilon(y, f(\mathbf{x}, \mathbf{w})) = \begin{cases} 0 & \text{if } |y - f(\mathbf{x}, \mathbf{w})| \le \varepsilon \\ |y - f(\mathbf{x}, \mathbf{w})| - \varepsilon & \text{otherwise} \end{cases} \tag{2}$$

$$R_{reg} = C \frac{1}{N} \sum_{i=1}^n L_\varepsilon(y_i, f(\mathbf{x}_i)) + \frac{1}{2} \|\mathbf{w}\|^2 \tag{3}$$

where ε is the insensitive loss function parameter that is the radius of insensitive tube, C is a regularization parameter that specifies the trade-off between an approximation error and the weight vector. Both ε and C should be determined beforehand by the user.

SVR performs linear regression in the high-dimensional feature space using ε-insensitive loss function and tries to reduce model complexity by introducing non-negative slack variables ξ_i, ξ_i^*, $i = 1, \cdots, n$, and minimizing $\|\mathbf{w}\|^2$. The slack variables represent the distance from actual values to the corresponding boundary values of the tube. The regularized risk function R_{reg} is transformed into the following constrained form [2]:

$$\text{minimize } \frac{1}{2}\|\mathbf{w}\|^2 + C\sum_{i=1}^{n}(\xi_i + \xi_i^*)$$

$$\text{subject to } \begin{cases} y_i - [\mathbf{w}^T \cdot \phi(\mathbf{x}_i) + b] \leq \varepsilon + \xi_i \\ [\mathbf{w}^T \cdot \phi(\mathbf{x}_i) + b] - y_i \leq \varepsilon + \xi_i^* \\ \xi_i \geq 0, \xi_i^* \geq 0, i = 1, \cdots, n \end{cases} \quad (4)$$

The optimization is a quadratic programming problem, which is solved by a dual set of Lagrange multipliers, α_i and α_i^*. Consequently, the general form of the SVR function can be expressed as follows [2]:

$$f(\mathbf{x}) = \sum_{k=1}^{m}(\alpha_k - \alpha_k^*)K(\mathbf{x}_k, \mathbf{x}) + b \quad (5)$$

where $\mathbf{x}_k$ is the support vector, m is the number of support vectors and $K(\mathbf{x}_k, \mathbf{x})$ is a kernel function to yield the inner products in the feature space $\phi(\mathbf{x}_k)$ and $\phi(\mathbf{x})$. There are several kernel functions used for SVM modeling, including linear function, radial basis function (RBF), polynomial function, hyperbolic tangent function, Bessel function of the first kind, Laplace RBF, ANOVA RBF, linear splines, etc. For example, RBF kernel can be written as follows:

$$K(x_i, x_j) = \exp\left(-\gamma\|x_i - x_j\|^2\right) \quad (6)$$

where $\gamma = 1/2p^2$ is the kernel parameter and p is the width parameter that reflects the input range of the training and testing data. Parameter γ controls the amplitude of the Gaussian function and thus, controls the generalization ability of the SVM model [3].

2.3 Parameter Optimization Algorithms

1) Grid Search (GS)
The grid search method involves setting up a suitable grid in the design space, evaluating an objective function at all the grid points, and finding the grid point corresponding to the lowest function value. The method can require prohibitively large number of function evaluations in most practical problems. However, for problems with a small number of design variables, the method can be used conveniently to find an approximate minimum. Also, the grid method can be used to find a good starting point for more efficient methods [4].

2) Genetic Algorithm (GA)

GA is a heuristic search technique that works on the principle of natural genetics and natural selection [5]. The main operations of GA are selection, crossover and mutation. The working procedure of GA usually starts with an initial population of randomly generated chromosomes which represent the values of SVR parameters. The selection operation is performed to select excellent chromosomes to reproduce. New offspring is created by the crossover and mutation operations. The crossover operation is performed randomly to exchange genes between two chromosomes. The mutation operation determines whether or not a chromosome should mutate to the next generation. Offspring replaces the old population and forms a new population in the next generation by the three operations. The evolutionary process proceeds until stop conditions are satisfied [6, 7].

3) Particle Swarm Optimization (PSO)

PSO is a population-based heuristic optimization algorithm that is inspired by the collective motion of biological organism, including bird flocking and fish schooling, to simulate the behavior seeking a food source [8]. The population, called the 'swarm,' consists of individuals called the 'particles.' A PSO algorithm is initialized with randomly produced population and velocity. To find the optimum solution, the velocity is dynamically adjusted according to the flying experience of itself and its companions, namely, *pbest* and *gbest*, respectively. Therefore, the particles fly through multidimensional search space with acceleration towards more optimum solutions during the search process.

4) Artificial Bee Colony (ABC)

ABC is an evolutionary optimization algorithm inspired by foraging and waggle dance behavior of honey bee colonies. In the ABC algorithm, the position of a food source represents a possible solution of the optimization problem, and the nectar amount of a food source corresponds to the fitness of the associated solution. The ABC algorithm consists of three phases: employed bee, onlooker bee and scout bee phases. The employed bee explores its food source. The employed bee produces a modification on the position (solution) in her memory, depending on the local information, and tests the nectar amount (fitness value) of the new source (new solution). The employed bee shares the nectar information of the food source with the onlooker bee. The onlooker bee evaluates the information and selects a food source to exploit, based on a probability related to its nectar amount. The employed bee, whose food source is abandoned, becomes the scout bee and then the scout bee is sent randomly to possible new food sources [9].

3   Results and Discussions

The determination of significant input variables is one of the most important steps in the development process of the SVR model. This study determined the input variables of the SVR model based on cross correlation function (CCF), autocorrelation function (ACF), and partial autocorrelation function (PACF) between the variables. Table 1 shows input and output variables for mode configuration.

Table 1 Input and output variables for model configuration

Input variables	Output variable
$WL_{SC}(t-2)$, $WL_{SC}(t-1)$, $WL_{SC}(t)$, $WL_{DS}(t-3)$, $WL_{DS}(t-2)$, $WL_{DS}(t-1)$	$WL_{DS}(t)$

WL: daily river stage, SC: Socheon, DS: Dosan.

Table 2 Comparison of model performance

Models	CE	d	r^2	RMSE (m)	MAE (m)	MSE (10^{-3} m²)	MSRE (10^{-7})	MS4E (10^{-4} m⁴)
ANN	0.965	0.991	0.967	0.079	0.053	6.195	2.154	11.909
ANFIS	0.980	0.995	0.981	0.060	0.036	3.600	1.262	2.729
SVR-GS	0.980	0.995	0.981	0.060	0.036	3.569	1.240	3.632
SVR-GA	0.981	0.995	0.975	0.058	0.034	3.373	1.171	3.947
SVR-PSO	0.981	0.995	0.982	0.058	0.034	3.352	1.164	3.833
SVR-ABC	0.981	0.995	0.982	0.058	0.034	3.352	1.164	3.787

The optimization algorithms, GS, GA, PSO and ABC, were applied for searching the optimal SVR parameters, including regularization parameter (C), RBF parameter (γ) and insensitive loss function parameter (ε). For assessing the applicability of models combining SVR and optimization algorithms (SVR-GS, SVR-GA, SVR-PSO and SVR-ABC), the model performance was compared with that of artificial neural network (ANN) and adaptive neuro-fuzzy inference system (ANFIS).

In this study, the performance of river stage forecasting models was evaluated using performance indexes, including the coefficient of efficiency (CE), the index of agreement (d), the coefficient of determination (r^2), the root mean squared error (RMSE), the mean absolute error (MAE), the mean squared error (MSE), the mean squared relative error (MSRE) and the mean higher order error (MS4E). Table 2 summarizes the values of performance measures for the models and Fig. 2 shows scatter plots for observed and forecasted values. It can be seen from Table 2 that ANFIS, SVR-GS, SVR-GA, SVR-PSO and SVR-ABC models have higher CE, d and r^2 values and lower RMSE, MAE, MSE, MSRE and MS4E values than those of the ANN model. ANFIS, SVR-GS, SVR-GA, SVR-PSO and SVR-ABC models have similar CE, and d values, whereas the ANFIS model has slightly higher RMSE, MAE, MSE and MSRE than those of SVR-GS, SVR-GA, SVR-PSO and SVR-ABC models. SVR-PSO and SVR-ABC models have slightly lower MSE and MSRE than those of SVR-GS and SVR-GA models. This indicates that in terms of model efficiency, SVR-GS, SVR-GA, SVR-PSO and SVR-ABC models are superior to ANN and ANFIS models, and SVR-PSO and SVR-ABC models are superior to SVR-GS and SVR-GA models. This also indicates that SVR-PSO and SVR-ABC models yield the best performance in terms of model efficiency.

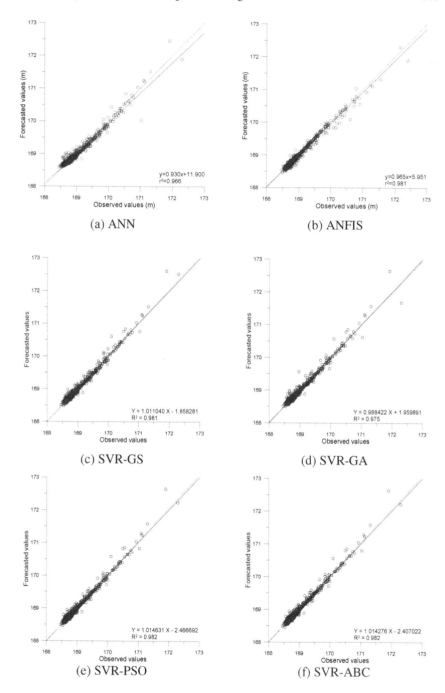

(a) ANN

(b) ANFIS

(c) SVR-GS

(d) SVR-GA

(e) SVR-PSO

(f) SVR-ABC

Fig. 2 Scatter plots for ANN, ANFIS and SVR models

4   Conclusions

This study develops river stage forecasting models using SVR and parameter optimization algorithms, including GS, GA, PSO and ABC, and evaluates the model performance from a case study on Andong dam watershed. In terms of model efficiency, SVR-GS, SVR-GA, SVR-PSO and SVR-ABC models yield better results than ANN and ANFIS models. SVR-PSO and SVR-ABC models produce relatively better efficiency than SVR-GS and SVR-GA models. SVR-PSO and SVR-ABC yield the best performance in terms of model efficiency. Therefore, it is found that the optimization algorithms applied in this study are very effective to search for the optimal parameters of SVR. Results indicate that river stage forecasting models combining SVR and optimization algorithms can be used as an effective tool for forecasting river stage accurately.

References

1. Seo, Y., Kim, S., Kisi, O., Singh, V.P.: Daily water level forecasting using wavelet decomposition and artificial intelligence techniques. J. Hydrol. **520**, 224–243 (2015)
2. Vapnik, V.N.: Statistical learning theory. Wiley, New York (1998)
3. Noori, R., Karbassi, A.R., Moghaddamnia, A., Han, D., Zokaei-Ashtiani, M.H., Farokhnia, A., Ghafari Gousheh, M.: Assessment of input variables determination on the SVM model performance using PCA, Gamma test and forward selection techniques for monthly stream flow prediction. J. Hydrol. **401**, 177–189 (2011)
4. Rao, S.S.: Engineering optimization: theory and practice, 4th edn. John Wiley & Sons Inc., Hoboken (2009)
5. Goldberg, D.: Genetic algorithms in search, optimization, and machine learning, 1st edn. Addison-Wesley, Boston (1989)
6. Yuan, F.C.: Parameters optimization using genetic algorithms in support vector regression for sales volume forecasting. Appl. Math. **3**(10A), 1480–1486 (2012)
7. Kaltech, A.M.: Wavelet genetic algorithm-support vector regression (Wavelet GA-SVR) for monthly flow forecasting. Water Resour. Manage. **29**(4), 1283–1293 (2015)
8. Bratton, D., Kennedy, J.: Defining a standard for particle swarm optimization. In: Proceedings of the 2007 IEEE Swarm Intelligence Symposium, Honolulu, H.I., USA, pp. 120–127 (2007)
9. Karaboga, D., Akay, B.: A comparative study of artificial bee colony algorithm. Appl. Math. Comput. **214**, 108–132 (2009)

Part V
Related Areas and Computational Intelligence

Online Support Vector Machine: A Survey

Xujun Zhou, Xianxia Zhang and Bing Wang

Abstract Support Vector Machine (SVM) is one of the fastest growing methods of machine learning due to its good generalization ability and good convergence performance; it has been successfully applied in various fields, such as text classification, statistics, pattern recognition, and image processing. However, for real-time data collection systems, the traditional SVM methods could not perform well. In particular, they cannot well cope with the increasing new samples. In this paper, we give a survey on online SVM. Firstly, the description of SVM is introduced, then the brief summary of online SVM is given, and finally the research and development of online SVM are presented.

Keywords Support vector machine · Machine learning · Generalization ability · Convergence

1 Introduction

Support Vector Machine (SVM) [1,2] is developed on the basis of VC dimension of statistical learning theory [3-5] and structural risk minimization principles, its success is due to its good generalization ability and has good convergence ability [6], as well as can show good performance in some difficult problems [7-8]. In the past decades, SVM as a new machine learning method has been successfully developed into one of the hot topic in the field of machine learning, and it was widely used in various fields, such as text classification, statistics, regression problems, pattern recognition, image processing, and so on.

Based on the development of optimal hyperplane in linear separable situation, SVM was initially proposed by Vapnik and his partners in 1995 [3]. It can solve

X. Zhou(✉) · X. Zhang(✉) · B. Wang
Shanghai Key Laboratory of Power Station Automation Technology,
College of Mechatronics Engineering and Automation,
Shanghai University, Shanghai 200072, China
e-mail: zhouxujun2013@163.com, {xianxia_zh,susanbwang}@shu.edu.cn

© Springer-Verlag Berlin Heidelberg 2016
J.H. Kim and Z.W. Geem (eds.), *Harmony Search Algorithm,*
Advances in Intelligent Systems and Computing 382,
DOI: 10.1007/978-3-662-47926-1_26

the problems in high dimension and local minima problems. SVM uses the large interval factor to control the training process of machine learning, and finds out the optimal hyperplane, which has the maximal margin classification. When dealing with the inseparable case, the slack variables are introduced to control the experience risk. This can not only satisfy the requirements, but also has good generalization ability. The process to find the optimal hyperplane can be transformed into solving a convex programming problem. Therefore, we can get global optimal solution theoretically. In view of the nonlinear classification problem, unlike traditional machine learning, SVM maps the input space to a high dimensional feature space by using nonlinear transformation, and still looks for optimal hyperplane in the high dimensional feature space. Consider the computational complexity, kernel functions that satisfy the Mercer condition should be used [9,10] to assure that the complicated nonlinear operations in original space can be transformed into inner product operations in the high dimensional feature space. The kernel function can cleverly solve the problem of complex calculation in high dimensional feature space, so that the "dimension disaster" can be avoided. For regression problems, it aims to find an optimal hyperplane that ensure the errors between all the training samples and the optimal hyperplane are smallest. However, for large-scare data sets, SVM usually takes a long training time and needs a lot of memory consumption. Especially, in real-time acquisition system, the traditional SVM will not be able to perform well when dealing with the new added samples. Therefore, in order to solve this problem, many online SVM algorithms are proposed by worldwide scholars, and have been successfully applied to many real-time acquisition systems, such as pedestrian detection and aircraft visual navigation system.

The rest of this paper is organized as follows. In Section 2, the preliminaries about SVM are introduced. The algorithms of online SVM are discussed in Section 3. Research and development trends of SVM are presented in Section 4. Finally, conclusions are given in Section 5.

2 Preliminaries of Support Vector Machine

Generally speaking, SVM can be divided into support vector classifier (SVC) and support vector regression (SVR), but the principle of them is similar; therefore, in this section we only present the basic idea and principle of ε - insensitive support vector regression (ε - SVR), which based on the ε - insensitive loss function, and then the nonlinear problem of ε - SVR and kernel function will be introduced.

Given the training data:

$$T = \left\{ (x_i, y_i) \mid x_i \in R^s, y_i \in R, i = 1,...,n \right\} \tag{1}$$

In ε - SVR, the purpose of regression problem is to find a function $f(x) \in F$ by using the training data sets, which makes the error between output value of each input points x_i and the corresponding target value y_i smaller than the given deviation ε, meanwhile the function should be as smooth as possible.

2.1 Linear ε - SVR

Consider the linear function:

$$f(x) = (w \cdot x) + b \quad w \in R^s, b \in R \tag{2}$$

For regression problems, it aims to find an optimal hyperplane that ensures the errors between all the training samples and the optimal hyperplane are smallest. Finding the optimal hyperplane is equivalent to solving a convex programming problem, which is shown as follows:

$$\min \quad \frac{1}{2}\|w\|^2$$

$$s.t. \quad \begin{cases} y_i - (w \cdot x_i) - b \leq \varepsilon \\ (w \cdot x_i) + b - y_i \leq \varepsilon \end{cases} \tag{3}$$

In order to prevent the formula (3) with no solution, Cortes&Vapnik [11] proposed a method to use "soft margin" loss function and introduce slack variable ξ_i, ξ_i^*. Therefore, according to the structural risk minimization principle, the primal problem can be transformed into a convex programming problem as follows:

$$\min_{w,b,\xi^{(*)}} \quad \frac{1}{2}\|w\|^2 + C \sum_{i=1}^{n}(\xi_i + \xi_i^*)$$

$$s.t. \quad \begin{cases} y_i - (w \cdot x_i) - b \leq \varepsilon + \xi_i \\ (w \cdot x_i) + b - y_i \leq \varepsilon + \xi_i^* \\ \xi_i, \xi_i^* \geq 0, \ i = 1, ..., n \end{cases} \tag{4}$$

where C is a penalty parameter, which is a constant that measure the complexity of model and the training error, and ε is a given error of regression function.

In order to solve the formula (4), Lagrange function has usually been introduced to transform the formula (4) into a dual form as follows:

$$L = \frac{1}{2}\|w\|^2 + C \sum_{i=1}^{n}(\xi_i + \xi_i^*) - \sum_{i=1}^{n}\alpha_i(\varepsilon + \xi_i - y_i + (w \cdot x_i) + b)$$

$$- \sum_{i=1}^{n}\alpha_i^*(\varepsilon + \xi_i^* + y_i - (w \cdot x_i) - b) - \sum_{i=1}^{n}(\eta_i \xi_i + \eta_i^* \xi_i^*) \tag{5}$$

subject to the following constraints:

$$\eta_i, \eta_i^*, \xi_i, \xi_i^* \geq 0, i = 1, \ldots, n$$

$$0 \leq \alpha_i, \alpha_i^* \leq C, \ i = 1, \ldots, n \tag{6}$$

where η_i, η_i^*, ξ_i, and ξ_i^* are Lagrange dual variables, α_i and α_i^* are Lagrange multiplier.

The dual formulation for ε - SVR is to find values for α_i, α_i^* that minimize the formula (5), and with first order derivative conditions for L, then we can have:

$$\begin{cases} \partial_b L = \sum_{i=1}^{n} (\alpha_i - \alpha_i^*) = 0 \\ \partial_w L = w - \sum_{i=1}^{n} (\alpha_i - \alpha_i^*) x_i = 0 \\ \partial_{\xi_i} L = C - \alpha_i - \eta_i = 0 \\ \partial_{\xi_i^*} L = C - \alpha_i^* - \eta_i^* = 0 \end{cases} \tag{7}$$

We can obtain the final decision function by using equation (5), (6) and (7):

$$f(x) = \sum_{i=1}^{n} (\alpha_i - \alpha_i^*)(x_i \cdot x) + b \tag{8}$$

where the parameters $(\alpha_i - \alpha_i^*)$ are nonzero, the corresponding sample point x_i is the support vector.

2.2 Nonlinear ε - SVR

Considering the nonlinear regression situation, we map the input space into a high dimensional feature space by using nonlinear transformation, and search for optimal hyperplane in the high dimensional feature space. Using kernel functions that satisfy the Mercer condition, the complicated nonlinear operation in original space can be transformed into inner product operations in a high dimensional feature space. It can not only solve the complex calculation problems in high dimensional feature space but also avoid "dimension disaster".

In high dimensional feature space, each input vector x_i is transformed to a high dimensional feature vector $\Phi(x_i)$. Similar to (8), the final decision function for a nonlinear ε - SVR is given as in (9)

$$f(x) = \sum_{i=1}^{n} (\alpha_i - \alpha_i^*)(\Phi(x_i) \cdot \Phi(x)) + b = \sum_{i=1}^{n} (\alpha_i - \alpha_i^*) K(x_i, x) + b \tag{9}$$

where $K(x_i, x) = (\Phi(x_i) \cdot \Phi(x))$ is a kernel function, $(\alpha_i - \alpha_i^*)$ is a nonzero parameter, and the corresponding sample point x_i is a support vector.

From the formula (9), we can conclude that the SVR has a three layer network structure similar to a neural network, as shown in Figure 1, where the output is a linear combination of several middle layer nodes and each middle layer node corresponds to the inner product, which is obtained from the input samples and support vectors.

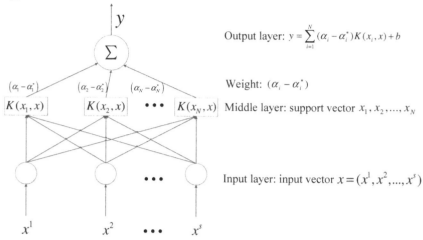

Output layer: $y = \sum_{i=1}^{N}(\alpha_i - \alpha_i^*)K(x_i, x) + b$

Weight: $(\alpha_i - \alpha_i^*)$

Middle layer: support vector $x_1, x_2, ..., x_N$

Input layer: input vector $x = (x^1, x^2, ..., x^s)$

Fig. 1 The network structure diagram of SVM (N is the number of support vectors)

2.3   Nonlinear $\varepsilon - SVR$

In the Statistical Learning Theory, the symmetric function $K(x, x') = (\Phi(x) \cdot \Phi(x'))$ that satisfies the limited semi-definite properties was called kernel function, that is to say the kernel function should satisfy the Mercer condition. In the symmetric function, $(\cdot)$ denotes the inner product in Hilbert space. Usually, the kernel function can be divided into two categories. One is called inner product kernel function and the other is called translation invariant kernel function. The common kernel functions are expressed as follows:

1) Polynomial kernel function: $K(x, x') = (x \cdot x')^d$, $K(x, x') = ((x \cdot x') + 1)^d$;

2) Gaussian radial basis kernel function: $K(x, x') = \exp(-\|x - x'\|^2 / \sigma^2)$;

3) Perceptron kernel function: $K(x, x') = \tanh(\beta x' + b)$;

4) Spline kernel function: $K(x, x') = B2n + 1(x - x')$.

Obviously, different kernel function will lead to different generalization ability. Thus, in specific situation, a proper kernel function should be chosen to obtain the best performance.

3   Online SVM Algorithms

As a new machine learning method, SVM has a good generalization performance and has various algorithms. However, its biggest disadvantage is that when training a SVM model with the large-scale data sets, it usually takes a lot of time and big memory consumption with the huge kernel matrix. Many traditional algorithms, such as Sequential Minimal Optimization (SMO) [12-14] and Chunking Algorithm [15] have been successfully applied to solve these problems; however, the online learning process is still not solved properly. Since the traditional methods usually need to retrain the SVM model when comes a new sample, these traditional algorithms are not suitable for real-time acquisition systems with the huge computational cost.

It is evident that online learning is very important in real-time pattern recognition system, and it can update the model in time as well as predict the trend of the system. Fortunately, some successful algorithms have been proposed to solve the online learning problem and successfully used in many fields. In general, the online algorithm of SVM can be mainly divided into two categories. The one is to solve the primal problem of SVM, and the other is to solve the dual problem of SVM. Both of them will be introduced in this section.

3.1   Online Algorithms in Primal Problem

In primal problem, many algorithms are using the stochastic gradient descent algorithm to handle the online learning problems [16]. The stochastic gradient descent method is to randomly select a sample to find gradient rather than select all the samples which can make the stochastic gradient descent method run faster than the common gradient descent method if the training samples are huge. Roberto [17] proposed an online algorithm based on stochastic gradient descent method. He put forward a new tube loss function called ρ-tube, which was similar to the ε-tube, updated its variables of current model according to some specific update rules by using stochastic gradient descent method, then the new model can obtained with the variables. It had been successfully applied to deal with orthogonal regression problems. In addition, Kivinen [18] proposed an online learning with kernel algorithm, by using the stochastic gradient descent within a feature space and the straightforward tricks, the online method can be applied to a wide range of problems, such as classification, regression, novelty detection, and so on.

3.2   Online Algorithms in Dual Problem

On the other hand, many online learning algorithms are proposed to solve the dual problem of SVM. Syed [19] proposed an algorithm, where the support vectors of current SVM model combined with the new added samples were taken as the new training set and used to retrain the model. In Ref. [20], Vatsa proposed a method called 2v-Online Granular Soft Support Vector Machine. The decision function was first obtained by the original training samples, and then the model would be retrained only by the newly added samples. The new support vectors were acquired at each time, and the existing support vectors that did not improve the model would be removed. Thus, it can ensure the training time would not increase significantly. Via using the weight, an online SVM method was proposed by Wang [21] for cell recognition. In this method, the old support vectors were removed, and weights were assigned to new training samples according to the importance of each training sample; therefore, it can adapt to the ever-changing experimental conditions.

Based on the Karush-Kuhn-Tucker (KKT) condition, Cauwenberghs [22] proposed an online SVM method, called IDSVM. The key was to retain the KKT conditions on all previously seen data, using the new samples and existing support vectors. When a new sample came, if it satisfied the KKT condition, then the current SVM model was updated until the whole samples satisfied the KKT condition. Inspired by IDSVM, Martin [23] put forward an online method which was an extension of the method developed by Cauwenberghs. His development opens the application of SVM regression to areas such as on-line prediction of temporal series or generalization of value functions in reinforcement learning. Furthermore, there are some other online learning methods. Bordes [24] proposed a simple and efficient online SVM algorithm, called Huller. When a new sample arrived, the convex hull was updated to retrain the model. This method tried to make the new sample become support vector, and deleted other support vectors. In Ref. [25], Wang proposed an online SVM based on convex hull vertices selection, which attempted to select the convex hull vertices of current training samples in the offline step that would be used for training. Via setting a threshold value, if a new coming sample lied in the given threshold, then we combined the new samples with the selected vertices of convex hull as the new training samples to update the current SVM model. This method can reduce the training time significantly, and can be applied to various online tasks, such as pedestrian and visual tracking.

4   Research and Development of Online SVM

Since its birth, SVM has attracted worldwide attentions in the field of machine learning because of its strong advantage in solving nonlinear, small samples and high dimensional problems. It integrates the optimal hyperplane, Mercer kernel, convex quadratic programming, sparse solution and slack variable technology, therefore, it obtains the best performance in some challenging applications. And SVM has been successfully applied to various fields, such as text classification, statistics, regression problems, pattern recognition, time series prediction, bioinformatics and image processing. In recent years, online SVM theory has made significant progress both in the algorithm implementation strategies and in practical applications. Although online SVM has been already used in several real-time acquisition systems, there are still some challenging problems when dealing with large-scare data sets, such as the training speed and computation complexity, therefore, this field needs further study and improvement. The research trend of online SVM can start from the following aspects. Computation time of kernel function takes most of the whole computation time, so computing and processing of kernel function is an important aspect; optimization of the existing online algorithm or research more useful and accurate algorithms are meaningful; the online algorithms are implemented in software is also a hot topic for researchers; last but not least, the online SVM must be applied in various practical application so that can embody its research value only.

5   Conclusions

The support vector machine (SVM) is based on statistical learning theory, by using structural risk minimization (SRM) to minimize the errors. For nonlinear problems, SVM maps the input space to a high dimensional feature space by using nonlinear transformation, which improves the generalization ability, and with no limit of data dimension. SVM as a new machine learning method, not only can solve the classification and pattern recognition problems, but also can solve the regression fit problem. Therefore, it is widely used in various fields.

In the aspect of SVM algorithm, many experts and scholars had put forward various algorithms. For small sample data sets, we can use the traditional methods, such as Newton method, conjugate gradient method, interior point methods; for large-scare sample data sets, we can use Block algorithm, Sequential minimal optimization (SMO), incremental learning algorithm. Incremental support vector machine learning algorithm can be applied in online training and learning, many experts proposed a lot of algorithms, and have been successfully used in time series prediction, weather forecasting and other areas. The training and achievement of SVM are necessary for further research, especially when considering the large-scare data sets. To get the faster and more accurate SVM algorithm is still need to further study. What's more, combined SVM with other various fields is also our research directions.

This work was supported by the project from the National Science Foundation of China under Grant 61273182.

References

1. Burges, J.C.: A tutorial on support vector machines for pattern recognition. Data Min. Knowl. Disc., 121-167 (1998)
2. Schölkopf, B.: Learning with Kernels: Support Vector Machines, Regularization, Optimization, and Beyond. MIT Press (2002)
3. Vapnik, N.V.: The Nature of Statistical Learning. Springer, New York (1995)
4. Vapnik, N.V., Lerner, A.: Pattern recognition using generalized portrait method. Automat. Rem. Contr+. **24**, 774–780 (1963)
5. Vapnik, N.V., Chervonenkis A.: A note on one class of perceptrons. Automat. Rem. Contr+. **25** (1964)
6. Cristianini, N., Shawe Taylor, J.: An Introduction to Support Vector Machines. Cambridge University Press (2000)
7. Dumais, S., Platt, J., Heckerman, D., Sahami, M.: Inductive learning algorithms and representations for text categorization. In: 7th International Conference on Information and Knowledge Management, ACM-CIKM 1998 (1998)
8. Osuna, E., Freund, R., Girosi, F.: Training support vector machines: an application to face detection. In: International Conference on Computer Vision and Pattern Recognition, CVPR 1997 (1997)
9. Schölkopf, B., Smola, A., Muller, K.R.: Kernel Principal Component Analysis. In: Schölkopf, B., Burges, C.J.C., Smola, A. (eds.) Advances in Kernel Methods-Support Vector Learning, pp. 327–352. MIT Press, Cambridge (1999)
10. Ahmad, A, Khalid, M., Yusof, R.: Kernel methods and support vector machines for handwriting recognition. In: IEEE Student Conference on Research and Development Proceedings (SCOReD 2002), pp. 309-312 (2002)
11. Cortes, C., Vapnik, N.V.: Support vector networks. Mach. Learn. **20**, 273–297 (1995)
12. Shevade, S.K., Keerthi, S.S., Bhattacharyya, C., Murthy, K.R.K.: Improvements to the SMO algorithm for SVM regression. IEEE Trans. Neural Netw. **11**(5), 1188–1193 (2000)
13. Chen, P., Fan, R., Lin, C.: A study on SMO-type decomposition methods for support vector machines. IEEE Trans. Neural Netw. **17**(4), 893–908 (2006)
14. Cai, F., Cherkassky, V.: Generalized SMO algorithm for SVM-based multitask learning. IEEE Trans. Neural Netw. Learn. Syst. **23**(6), 997–1003 (2012)
15. Cortes, C., Vapnik, V.: Support Vector Networks. Mach. Learn. **20**, 273–297 (1995)
16. Bordes, A., Bottou, L., Gallinari, P.: SGD-QN: Careful quasi-Newton stochastic gradient descent. J. Mach. Learn. **10**, 1737–1754 (2009)
17. Souza, R.C., Leite, S.C., Borges, C.C., Neto, R.F.: Online algorithm based on support vectors for orthogonal regression. Pattern Recognition Letters 34(12), 1394–1404 (2013)
18. Kivinen, J., Smola, A.J., Williamson, R.C.: Online learning with Kernels. IEEE Transactions on Signal Processing, 2165-2176 (2004)
19. Syed, N., Liu, H., Sung, K.: Handling concept drifts in incremental learning with support vector machines. In: 5th ACM SIGKDD International Conference Knowledge Discovery Data Mining, pp. 317–321 (1999)
20. Singh, R., Vatsa, M., Ross, A., Noore, A.: Biometric classifier update using online learning: A case study in near infrared face verification. Image Vis. Comput. **28**(7), 1098–1105 (2010)

21. Wang, M., Zhou, X., Li, F., Huckins, J., King, R., Wong, S.T.C.: Novel cell segmenta-
    tion and online SVM for cell cycle phase identification in automated microscopy.
    Bioinformatics. **24**(1), 94–101 (2008)
22. Cauwenberghs, G., Poggio, T.: Incremental and decremental support vector machine
    learning. In: Dietterich, T.G., Leen, T.K., Tresp, V. (eds.) Advances in Neural Infoma-
    tion Processing Systems, vol.13, pp. 409-415. MIT Press (2001)
23. Martin, M.: On-Line support vector machine regression. In: Elomaa, T., Mannila, H.,
    Toivonen, H. (eds.) ECML 2002. LNCS (LNAI), vol. 2430, pp. 282–294. Springer,
    Heidelberg (2002)
24. Bordes, A., Bottou, L.: The Huller: a simple and efficient online SVM. In: Gama, J.,
    Camacho, R., Brazdil, P.B., Jorge, A.M., Torgo, L. (eds.) ECML 2005. LNCS (LNAI),
    vol. 3720, pp. 505–512. Springer, Heidelberg (2005)
25. Wang, D.I.: Online Support Vector Machine Based on Convex Hull Vertices Selec-
    tion. IEEE Transactions on Neural Networks Learning Systems **24**(4), 593–608 (2013)

Computation of Daily Solar Radiation Using Wavelet and Support Vector Machines: A Case Study

Sungwon Kim, Youngmin Seo and Vijay P. Singh

Abstract The objective of this study is to apply a hybrid model for estimating solar radiation and investigate its accuracy. A hybrid model is wavelet-based support vector machines (WSVMs). Wavelet decomposition is employed to decompose the solar radiation time series components into approximation and detail components. These decomposed time series are then used as input of support vector machines (SVMs) modules in the WSVMs model. Based on statistical indexes, results indicate that WSVMs can successfully be used for the estimation of daily global solar radiation at Champaign and Springfield stations in Illinois.

Keywords Support vector machines · Wavelet decomposition · Solar radiation

1   Introduction

Solar radiation is the principal energy source for physical, biological and chemical processes, such as snow melt, plant photosynthesis, evaporation, and crop growth,

S. Kim(✉)
Department of Railroad and Civil Engineering, Dongyang University,
Yeongju 750-711, Republic of Korea
e-mail: swkim1968@dyu.ac.kr

Y. Seo
Department of Constructional Disaster Prevention Engineering,
Kyungpook National University, Sangju 742-711, Republic of Korea
e-mail: ymseo@knu.ac.kr

V.P. Singh
Department of Biological and Agricultural Engineering & Zachry Department of Civil
Engineering, Texas A & M University, College Station, TX 77843-2117, USA
e-mail: vsingh@tamu.edu

© Springer-Verlag Berlin Heidelberg 2016
J.H. Kim and Z.W. Geem (eds.), *Harmony Search Algorithm,*
Advances in Intelligent Systems and Computing 382,
DOI: 10.1007/978-3-662-47926-1_27

and is also a variable needed for biophysical models to evaluate risk of forest fires, hydrological simulation models and mathematical models of natural processes. Solar radiation plays an important role in the design and analysis of energy efficient buildings in different types of climate. In cold and severe cold regions, passive solar designs and active solar systems help lower the reliance on conventional heating means using fossil fuels. In tropical and subtropical climates, solar heat gain is a major cooling load component, especially in cooling dominated buildings. The effects of prevailing climate and local topography would determine the actual amount of solar radiation reaching a particular location. The objective of the present study is to develop support vector machines (SVMs) and wavelet-based support vector machines (WSVMs) models that can be used to estimate daily solar radiation at two locations (Champaign and Springfield stations) in Illinois.

2   Support Vector Machines (SVMs)

The SVMs model has found wide applications in several areas, including pattern recognition, regression, multimedia, bio-informatics, and artificial intelligence. The SVMs model is a new kind of classifier that is motivated by two concepts. First, transformation of data into a high-dimensional space can transform complex problems into simpler problems that can use linear discriminant functions. Second, the SVMs model is motivated by the concept of training and uses only those inputs that are near the decision surface[16~18]. The solution of traditional neural networks models may tend to fall into a local optimal solution, whereas global optimum solution is guaranteed for the SVMs model[6]. The current study uses an ε-support vector regression (ε-SVR) model. It has been successfully applied for modeling hydrological processes[8~10, 17]. During the ε-SVR model training performance, the purpose is to find a nonlinear function that minimizes a regularized risk function. This is achieved for the least value of the desired error criterion (e.g., RMSE) for various constant parameters C_C, and ε and various kernel functions with various constant σ values. Detailed information on the SVMs model can be found in[8~10, 16~18].

3   Wavelet Decomposition

Wavelet analysis is a multi-resolution analysis in time and frequency domains. The wavelet transform decomposes a time series signal into different resolutions by controlling scaling and shifting. It provides a good localization properties in both time and frequency domains [15]. It also has an advantage in that it has flexibility in choosing the mother wavelet, which is the transform function, according to the characteristics of time series.

A fast DWT algorithm, developed by Mallat (1989), is based on four filters, including decomposition low-pass and high-pass, reconstruction low-pass and high-pass filters. For practical implementation of Mallat's algorithm, low-pass and high-pass filters are used instead of father and mother wavelets, which are also called scaling and wavelet functions, respectively. The low-pass filter, associated with the scaling function, allows the analysis of low frequency components, while the high-pass filter, associated with the wavelet function, allows the analysis of high frequency components. These filters, used in Mallat's algorithm are determined according to the selection of mother wavelets [5]. Multiresolution analysis by Mallat's algorithm is a procedure to obtain 'approximations' and 'details' for a given time series signal. An approximation holds the general trend of the original signal, while a detail depicts high-frequency components of it. A multilevel decomposition process (Fig. 1) can be achieved, where the original signal is broken down into lower resolution components [3]. Detailed information for Mallat's algorithm can be found in [13].

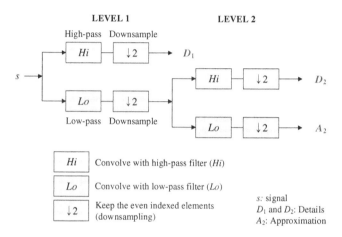

Fig. 1 Mallat's algorithm for two-level decomposition of a signal

4   Case Study

This study determined input variables. Daily weather data obtained from two weather stations (Fig. 2), Champaign (latitude, 40.0840° N; longitude, 88.2404° W; altitude, 219 m) and Springfield (latitude, 39.7273° N; longitude, 89.6106°W; altitude, 177 m) operated by the Illinois State Water Survey (ISWS), were used in this study (http://www.isws.illinois.edu/warm/). The ISWS is a division of the Prairie Research Institute of the University of Illinois at Urbana-Champaign and has flourished for more than a century by anticipating and responding to new challenges and opportunities to serve the citizens of Illinois. The weather data consisted of six years (January 2007 to December 2012, N=2,192 days) of daily records of air temperature (TEM), solar radiation (RAD), relative humidity (HUM), dew point temperature (DEW), wind speed (WIN), and potential evapotranspiration (ETO). Air temperature and relative humidity have been measured at 2 m above the ground, whereas wind speed has been measured at 10 m above the ground (prior to winter 2011/2012 measurement made at 9.1 m). Potential evapotranspiration has been calculated using the Food and Agricultural Organization (FAO) of the United Nations Penman–Monteith equation as outlined in FAO Irrigation and Drainage Paper No. 56 "Crop Evapotranspiration"([1]) since 1 December 2012 (Water and Atmospheric Resources Monitoring Program 2011). Prior to that time, the van Bavel method was used for calculating potential evapotranspiration [19].

For a data-driven model, data was split into training and testing data. The training data were used for optimizing the connection weights and bias of the data-driven model and the testing data were used to evaluate the chosen model against unseen data [4,7]. In all of these applications, the first 4 years data (2007–2010, N=1,461 days) was applied for training and 1 year (2012, N=366 days) for testing. The estimated solar radiation values were compared with observed ones using 5 performance evaluation criteria: the correlation coefficient (CC), root mean square error (RMSE), Nash-Sutcliffe coefficient (NS) [2,14], mean absolute error (MAE), and average performance error (APE). Although CC is one of the most widely used criteria for calibration and evaluation of hydrological models with observed data, it alone cannot discriminate which model is better than others. Since the standardization inherent in CC as well as its sensitivity to outliers yields high CC values, even when the model performance is not perfect. Legates and McCabe (1999)[11] suggested that various evaluation criteria (e.g., RMSE, MAE, NS, and APE) must be used to evaluate model performance. Fig. 3 shows a comparison of observed and estimated daily solar radiation values using the SVMs and WSVMs models (three input).

Fig. 2 Schematic map of two weather stations

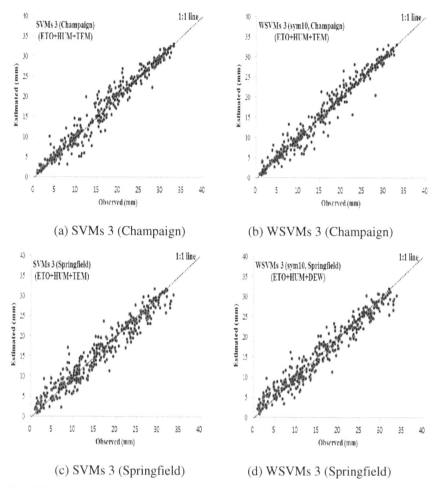

(a) SVMs 3 (Champaign)   (b) WSVMs 3 (Champaign)

(c) SVMs 3 (Springfield)   (d) WSVMs 3 (Springfield)

Fig. 3 Comparison of observed and estimated solar radiation values using the SVMs and WSVMs models

5   Conclusions

This study develops and evaluates data-driven models for estimating daily solar radiation at Champaign and Springfield stations in Illinois. The SVMs and WSVMs models are developed for three input combinations. Results indicate that the WSVMs model can successfully be used for the estimation of daily global solar radiation at Champaign and Springfield stations in Illinois. In this study, it can be found that the data-driven models can estimate daily solar radiation.

References

1. Allen, R.G., Pereira, L.S., Raes, D., Smith, M.: Crop evapotranspiration guidelines for computing crop water requirements. FAO Irrigation and Drainage. Paper No. 56. Food and Agriculture Organization of the United Nations, Rome (1998)
2. ASCE Task Committee: Criteria for evaluation of watershed models. J. Irrig. Drain. Eng. 119(3), 429–442 (1993)
3. Catalão, J.P.S., Pousinho, H.M.I., Mendes, V.M.F.: Hybrid wavelet-PSO-ANFIS approach for short-term electricity prices forecasting. IEEE Trans. Power Syst. 26(1), 137–144 (2011)
4. Dawson, C.W., Wilby, R.L.: Hydrological modelling using artificial neural networks. Prog. phys. Geog. 25(1), 80–108 (2001)
5. González-Audícana, M., Otazu, X., Fors, O., Seco, A.: Comparison between Mallat's and the 'à trous' discrete wavelet transform based algorithms for the fusion of multispectral and panchromatic images. Int. J. Remote Sens. 26(3), 595–614 (2005)
6. Haykin, S.: Neural networks and learning machines, 3rd edn. Prentice Hall, NJ (2009)
7. Izadifar, Z., Elshorbagy, A.: Prediction of hourly actual evapotranspiration using neural networks, genetic programming, and statistical models. Hydrol. Process. 24(23), 3413–3425 (2010)
8. Kim, S., Shiri, J., Kisi, O.: Pan evaporation modeling using neural computing approach for different climatic zones. Water Resour. Manag. 26(11), 3231–3249 (2012)
9. Kim, S., Seo, Y., Singh, V.P.: Assessment of pan evaporation modeling using bootstrap resampling and soft computing methods. J. Comput. Civ. Eng. (2013a). doi:10.1061/(ASCE)CP.1943-5487.0000367
10. Kim, S., Shiri, J., Kisi, O., Singh, V.P.: Estimating daily pan evaporation using different data-driven methods and lag-time patterns. Water Resour. Manag. 27(7), 2267–2286 (2013b)
11. Legates, D.R., McCabe, G.J.: Evaluating the use of "goodness-of-fit" measures in hydrologic and hydroclimatic model validation. Water Resour. Res. 35(1), 233–241 (1999)
12. Mallat, S.G.: A theory for multiresolution signal decomposition: The wavelet representation. IEEE Trans. Pattern. Anal. Mach. Intell. 11(7), 674–693 (1989)
13. Nason, G.: Wavelet methods in statistics with R. Springer, NY (2010)
14. Nash, J.E., Sutcliffe, J.V.: River flow forecasting through conceptual models, Part 1 – A discussion of principles. J. Hydrol. 10(3), 282–290 (1970)
15. Nejad, F.H., Nourani, V.: Elevation of wavelet denoising performance via an ANN-based streamflow forecasting model. Int. J. Comput. Sci. Manag. Res. 1(4), 764–770 (2012)
16. Principe, J.C., Euliano, N.R., Lefebvre, W.C.: Neural and adaptive systems: fundamentals through simulation. Wiley, John & Sons Inc., NY (2000)
17. Tripathi, S., Srinivas, V.V., Nanjundish, R.S.: Downscaling of precipitation for climate change scenarios: a support vector machine approach. J. Hydrol. 330(3–4), 621–640 (2006)
18. Vapnik, V.N.: The nature of statistical learning theory, 2nd edn. Springer, NY (2010)
19. van Bavel, C.H.M.: Estimating soil moisture conditions and time for irrigation with the evapotranspiration method. USDA, ARS 41–11, U.S. Dept. of Agric., Raleigh, NC, 1–16 (1956)

The Discovery of Financial Market Behavior Integrated Data Mining on ETF in Taiwan

Bo-Wen Yang, Mei-Chen Wu, Chiou-Hung Lin,
Chiung-Fen Huang and An-Pin Chen

Abstract In practice, many physics principles have been employed to derive various models of financial engineering. However, few studies have been done on the feature selection of finance on time series data. The purpose of this paper is to determine if the behavior of market participant can be detected from historical price. For this purpose, the proposed algorithm utilizes back propagation neural network (BPNN) and works with new feature selection approach in data mining, which is used to generate more information of market behavior. This study is design for exchange-traded fund (ETF) to develop the day-trade strategy with high profit. The results show that BPNN hybridized with financial physical feature, as compared with the traditional approaches such as random walk, typically result in better performance.

Keywords Data mining · Back-propagation neural network (BPNN) · Exchange-traded fund (ETF) · Financial physics · Behavior discovery

1 Introduction

The background of the study is for discovering abnormality in a financial market, which is more difficult than discovering the same in habitual behaviors.

B.-W. Yang(✉) · M.-C. Wu · C.-F. Huang · A.-P. Chen
Institute of Information Management, National Chiao Tung University,
Hsinchu 30010, Taiwan, R.O.C
e-mail: {nike4859,mcwu0715,amanda.huangcf,apc888888}@gmail.com

C.-H. Lin
Department of Information Management, Minghsin University of Science and Technology,
Hsinchu 30401, Taiwan, R.O.C
e-mail: lch@must.edu.tw

© Springer-Verlag Berlin Heidelberg 2016
J.H. Kim and Z.W. Geem (eds.), *Harmony Search Algorithm*,
Advances in Intelligent Systems and Computing 382,
DOI: 10.1007/978-3-662-47926-1_28

285

Since operating with the knowledge gleaned from the habitual behaviors leads to no significant profit or causes enormous loss on the part of investors, a single time series in no event generates desirable outcome for investment strategy formulation purpose. Additionally, exertion of external forces (i.e., invisible hands) to the market and variation thereof are omnipresent in the entire time series, complicating the analysis of the financial market and highlighting the flaws of the analysis on basis of the single time series. So long as the trends (or patterns of the behaviors) could be identified or discovered regardless of their corresponding locations in the time series, the analysis derived from the trends could be superior in forecasting/predicting the behaviors of the financial market.

The present study aims to identify the trends of ETF before formulating appropri-ate investment strategy (investment decision-making) so as to maximize profitability for any investments associated with ETF.

Global financial markets have been through extreme downturns in the wake of cri-sis associated with sub-prime mortgages, causing many investors to suffer from enormous loss of their investment. Despite the individual/general investors generally are not able to avoid any financial crisis of the similar scale from occurring again, they can reduce the risks with the aid of information technology. One primary reason for which the general investors were not able to reduce the risk associated with dynamics of the financial market is holding their investments too long while selling their investments too quick. ETF has been proposed to minimize the risk when the investors concentrate their investments in single market, or in stock variety. Many prior studies therefore focused on proposing effective/efficient models solely for the ETF investment. However, in view of the present study those prior studies still had some room to improve.

For example, an investment allocation method proposed by Harry Markowitz [1], a Nobel Prize Laureate in 1990 is widely perceived as having several unpractical as-sumptions therein including return and risk are normally distributed while at the same time concluding the investors are risk-averse, implicitly contradicting that particular assumption. In 1990, Dr. William Sharpe, another Nobel Prize Laureate, proposed a capital asset pricing model (CAPM) with the same assumption and conditioned upon existence of a perfect financial market. Those studies were associated with their corresponding restraints, which may distance the studies themselves away from the actual financial market. Further, since the financial market is at least driven by numerous factors even the investment strategies on basis of the studies with less number of the assumptions were still in the mode of searching for "transaction holy grail" in terms of their overall long-term performance. In other words, any currently-existing proposed investment studies could still go through a long period in which ups and downs in economy and even financial crises of much larger scale have occurred, without sustaining their profitability as succumbing to the fluctuations in the market. Therefore, the present study attempts to further improve the prior studies when having incorporating DM technique and weighted clustering to identify the trends of the market or the chosen stocks the ETF tracked so that the most profits associated with the investment of the ETF can be realized.

2 Literature Review

2.1 Exchange-Traded Fund (ETF)

Since the first investment trust fund that was first introduced in 1774 following the concept of "unity creates strength," the similar idea has been widely applied in many aspects of financial fields and exchange-traded fund (ETF) is just an variety of the mutual fund [2]. First index-based fund of a mutual fund type tracking SP 500 index was proposed in 1976. In 1993, first ETF (Standard Poor Depository Receipts, SPDR) tracking SP 500 index went public in U.S. In 1999, first HK-based index fund was issued. Next Track, a newly established market solely for transactions of in-dex/stock-based fund, was in place in 2001 in which ETF was referred to as "Tracker" in Europe. And many countries have introduced ETF of similar types since then.

ETF have been tracking a single country market, or global markets encompassing several countries' stock markets, designated industries/sectors, derivatives, commodities, currencies and even dividends [3]. More specifically, iShares also provides gold trust which was established on January 25th 2005 to reflect the price of gold owned by the trust at that time, less the expenses and liabilities of the trust, and such specific trust could fall into the category of commodity-index fund [4]. In other words, ETF has been a useful tool that extends beyond borders of the countries and provides additional opportunities for the investors to increase their investment returns while subjecting the investors to additional risks associated with the ETF investments as well.

2.2 Financial Physical

The application of physics to financial and economic problems has been considered in numerous prior studies. In practice, many physics principles have been employed to derive various models of financial engineering. For example, the widely applicable random walk theory illustrative of stock price fluctuation was derived from Brownian motion, while pricing model of options applies heat equation to closed-form solutions. Recently, quantum mechanics has been applied to market microstructure analysis to perform simulation [5] [6]. Further, statistical physics has been employed to simulate the probability and stochastic process in economic and financial issues [7] [8]. All of these have given rise to the study of physical phenomenon in economic and financial activities, which collectively is termed "econophysics" [9].

The background of philosophy of Classical Physics comes from fundamental Mathematics such as Taylor series. The Taylor series of a function $f(x)$ that is infinitely differentiable in a neighborhood of "a" could be represented in the follows:

$$f(a) + \frac{f'(a)}{1!}(x-a) + \frac{f^{(2)}(a)}{2!}(x-a)^2 + \frac{f^{(3)}(a)}{3!}(x-a)^3 + \cdots \quad (1)$$

And the same power series in a more compact form can be written as

$$\sum_{n=0}^{\infty} \frac{f^{(n)}(a)}{3!}(x-a)^n \quad (2)$$

where n! denotes the factorial of n and $f^{(n)}(a)$ denotes the nth derivative of f evaluated at a; while the zeroth derivative of f is defined to be f itself.

From the Taylor Series, both sides of the equation are representative of causes and results, respectively. For example, the derivatives on the right side (or in other words, the derivatives of $f(a)$) undoubtedly have impact on $f(x)$ on the left side. Put Taylor Series in the context of Classical Physics, a first order derivative may be indicative of momentum while a second order derivative may thus indicate an impulse. People have been trying to quantize cause-result analysis and utilizing result of the analysis to more accurately predict the result solely depending upon the cause. However, in the absence of absolute representation of the cause/effect relativity of the cause/effect instead becomes the focus of the analysis. In other words, as people would never succeed in matching the first order derivative and the second order derivative with any events in the dynamic financial market any corresponding "absoluteness" analysis presents little value.

Any dynamics in the financial market could be represented in terms of variation during a predetermined passage of time. For instance, when the variation in a stock index between yesterday and today is positive (e.g., the stock index is higher today) the first order derivative (which indicates the variation itself) of the stock index would remain positive. In terms of Classics Physics, the momentum in the stock market between yesterday and today is positive. And the second derivative of the stock index, which is the derivative of the first derivative in the passage of time, represents the "impulse" of the stock index. For example, when the variation (i.e., the first order derivative) between Day 1 and Day 2 is 1 and the variation between Day 3 and Day 4 is 2, the "impulse" between the pair of Day 1 and Day 2 and the pair of Day 3 and Day 4 is "1." Accordingly, when the impulse remains positive the stock index could be having larger potential of trending upwardly compared with the stock index having the positive first order derivative and the negative second order derivative.

Further, difference between the closing price of one specific stock in day N and the closing price at day N+1 may be indicative of difference between kinetic energy be-tween day N and day N+1. And such difference may result from an external force (F), which is the outcome of a volume of transaction, which is representative of mass, at day N+1 multiplying the "impulse" (or the second order derivative). The larger the volume of transaction could lead to larger external force when the impulse stays at the positive territory, indicating the entire market is going to trend upwardly. And since the larger external force also leads to larger kinetic energy the closing price at day N+1 should be larger than the closing price at day N, which is consistent with the presence of the larger external force that causes the upward trending.

Classical physics are widely perceived as guidance for developing growingly complicated financial decision-making strategy so as to ensure that developed models associated with that particular strategy could help enhance performance in profitability after processing enormous amount of data and learning from the processed data.

3   The BPNN Model for ETF

In order to enhance the performance of that particular model, multiple options have been explored. Since the financial physical is capable of helping improve the accuracy of the investment strategy, the present research not only considers the technical indicators but also the financial physical. More specifically, the present research further takes into account the financial physical of the technical indicators such as the first-order derivative and the second-order derivative of the technical indicators to be more accurately forecasting the trend of the ETF. As such, the input variables include the technical indicators themselves and their first-order derivatives for the purpose of enhancing the reliability of the improved model with its corresponding experiment discussed in sub-section as figure 1. Meanwhile, more specifically, the model proposed in the present research starts with :

1. Retrieving EWT-related information from iShare ETF database,
2. Calculating technical indicators utilized,
3. Applying financial physical to technical indicators for presenting variation in trend of physical force associated with technical indicators and difference in the variation in trend of physical force,
4. Inputting financial physical-related information as input variables of BPNN-based system for self-learning of BPNN-based system in order to predict trend in subsequent transaction day,
5. Calculating profitability of investment strategy.

3.1   Calculating Input Variables

The present research utilizes categories of price indicators, volume indicators, and theoretical indicators and picks one trend-type indicator and one swing-type indicator from each of these three categories to have MA (moving average), RSI (relative strength index), MV (market volume), VR (volatility ratio), MACD (moving average convergence divergence), and KD (stochastic oscillator) as the chosen technical indicators for the purpose of the present research. Please see the table 1 in the follows for the summary of the chosen technical indicators in the present research.

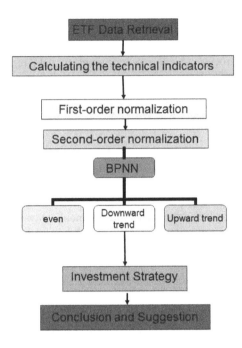

Fig. 1 The Proposed Model

Table 1 Technical Indicators parameters

Type	Technical Indicators	Time Unit
Price	MA	7
	RSI	7
Volume	MA	7
	VR	9
Theory	MACD	26, 15, and 9
	KD	9 and 3

4   Experiments

4.1   Data and Experiment Design

This study is design for trading index or derivatives of index. The empirical data are retrieve from iShares MSCI Taiwan Index Fund(EWT) . The index consists of stocks traded primarily on the Taiwan Stock Exchange. The trading period for this experiment is from June 23, 2000 to June 16, 2011. Eight and half years period of the fund data is for training data and the others for testing data. Table 1 includes attributes of the data such as opening price (Open), highest price (High), lowest price (Low), closing price (Close), volume of transactions (Volume).

Table 2 The portion of EWT price in 2007

Date	Open	High	Low	Close	Volume
2007/1/3	14.88	14.9	14.65	14.73	9268300
2007/1/4	14.6	14.81	14.58	14.78	4198700
2007/1/5	14.56	14.56	14.39	14.41	3826400
2007/1/8	14.46	14.46	14.27	14.36	2321900
⋮	⋮	⋮	⋮	⋮	⋮
2007/12/24	14.75	14.75	14.56	14.68	4518500
2007/12/26	14.68	14.72	14.58	14.64	6011600
2007/12/27	14.8	14.87	14.57	14.57	8728800
2007/12/28	14.95	15.06	14.85	14.85	5353700
2007/12/31	15.01	15.2	15	15.03	3244900

In order to verify proposed model, strategy of Random Walk Theory and buy-and-hold strategy are used as comparison model. The experimental results depend on the comparison between BPNN and the other two models during the test period. Training period uses to training neural network and determining parameters. The experiment design and the data intervals are shown in Table 3.

Table 3 Experiment design

Testing model	Training		Testing	
	Day-frames	Period	Day-frames	Period
(1) BPNN (2) Random (3) Buy and hold	2143 days	2000.06-2008.12	619 days	2009.01-2011.06

4.2 Performance Measures

We measure the accuracy of prediction and return respectively, and use them as indicators to evaluate the model. They are defined as follows:

$$\text{Accuracy} = \frac{\text{times of profit - making transactions}}{\text{times of transactions}} \tag{3}$$

$$\text{Total return} = \text{return of long and short position} \tag{4}$$

$$\text{Average return} = \frac{\text{return of long and short position}}{\text{times of transactions}} \quad (5)$$

Trading signals have two types which including long trade and short trade, and therefore the returns include long positions and short position. Long positions and short positions are demonstrated as follows:

$$\text{Return of long position} = \text{long price - short price - transaction cost} \quad (6)$$

$$\text{Return of short position} = \text{short price - long price - transaction cost} \quad (7)$$

Because of the buy and hold strategy has only one transaction, the comparisons of accuracy and average return only verifies BPNN and random model.

4.3   Comparative Results

The trading signals divide to two. When the signal above the top threshold, long EWT, and vice versa. In order to making higher quality decision, the trading thresh-old can filter the noises and improve the accuracy. In the training period, the different thresholds are tested for the BPNN. Table 4 shows the results. The return and accuracy are trade-off, so the thresholds setting 0.55 and 0.45 can make more profit and higher accuracy.

Table 4 Trading threshold

Trading threshold	Accuracy (%)	Average Return (USD)	Total Return (USD)
0.5/0.5	79.14	736	804,098
0.55/0.45	91.25	1,235	681,472
0.6/0.4	94.93	2,356	240,272

Table 5 is shown that proposed model has higher accuracy than Random model. The accuracy of random model isn't exceeding 50%. From Table 6, in terms of average return, proposed model has outperforms the Random model.

Table 5 Model comparison of accuracy

Model	Accuracy(%)
(1)BPNN	79.15
(2)Random	45.85

Table 6 Model comparison of average return

Model	Average Return(USD)
(1)BPNN	1,195
(2)Random	-398

From Table 7, it's shown that total return during the testing period, and the proposed model has better performance than other two models. Even though the buy and hold strategy buys EWT at the lower price in the testing period, it still only makes 72,700 USD. The results shown that proposed model can help investors make better decision in trading.

Table 7 Model comparison of total return

Model	Total Return(USD)
(1)BPNN	169,712
(2)Random	-72,100
(3)Buy and hold	72,700

5   Conclusions

The present research further capitalizes on characteristics including low cost and tax and trading flexibility associated with ETF to develop the day-trade strategy with high probability of positive investment return. Specifically, an approach proposed in the present study utilizes DM and back propagation neural network (BPNN). Experiment results show that when following the investment strategy prepared according to the present research the investment return is better and larger than following the traditional approaches such as Random Walk, notwithstanding the overall economy.

In future works, as the proposed model utilizes the daily financial data and the primary characteristic of ETF compared with mutual fund allows for the transactions at any point of the transaction day. Tick information of ETF could be used in the model offering additional opportunities of investment returns.

Acknowledgement The work was supported by the Ministry of Science and Technology, Taiwan, R.O.C. under Grant MOST103-2410-H009-023.

References

1. Markowitz, H.M.: Portfolio selection. J. Finance **7**(1), 77–91 (1952)
2. McWhinney, J.E.: A Brief History Of The Mutual Fund. Investopedia (2009). http://www.investopedia.com/articles/mutualfund/05/mfhistory.asp

3. Kennedy, M.: 14 types of ETFs and more Introducing Windows Azure. Aboutmoney (2010). http://etf.about.com/od/typesofetfs/tp/4_Types_of_ETFs.htm
4. iShare.: Fund Overview. iShare (2011). http://us.ishares.com/product_info/fund/overview/IAU.htm
5. Loffredo, M.I.: On the statistical physics contribution to quantitative finance. Int. J. Mod. Phys. A. **18**, 705–713 (2004)
6. Ausloos, M., Vandewalle, N., Boveroux, Ph, Minguet, A., Ivanova, K.: Applications of statistical physics to economic and financial topics. Physica A **274**(1–2), 229–240 (1999)
7. Jimenez, E., Moya, D.: Econophysics: from game theory and information theory to quantum mechanics. Physica A **348**, 505–543 (2005)
8. Stauffer, D.: Econophysics - A new area for computational statistical physics. Int. J. Mod. Phys. C **11**, 1081–1087 (2000)
9. Holland, J.H.: Escaping brittleness: The possibilities of general purpose learning algorithms applied to parallel rule-based systems. Computation & Intelligence, 593–624 (1986)

Data Mining Application to Financial Market to Discover the Behavior of Entry Point – A Case Study of Taiwan Index Futures Market

Mei-Chen Wu, Bo-Wen Yang, Chiou-Hung Lin, Ya-Hui Huang and An-Pin Chen

Abstract The value of the investment method is that investors who are anxious to pursue, there are many value investing methods have been proposed, but only a minority of the value investing method were proved to be effective. The study is based on messages generated defined value of the investment by Steidlmayer in 1984 proposed market profile theory. In order to extract trading behavior of dealer and product value by the huge financial trading information, the model used the trading data to capture feature patterns, and find the double distribution trend day generated by market profile. The experimental results show the single print as an entry point, and the reference to historical support and pressure line as an exit point. The results show the returns are 24.09 points and the accuracy achieved 57.45%.The results had shown the analysis model can find the investment goods real value from the huge trading information, and help investors obtain excess returns.

Keywords Data mining · Market profile theory · Double distribution trend day · TAIEX futures · Pressure support line

1 Introduction

Global financial markets as a whole, because the price of the financial markets are often subject to general economic, national policy, and even unpredictable factors

M.-C. Wu(✉) · B.-W. Yang · C.-H. Lin · Y.-H. Huang · A.-P. Chen
Institute of Information Management, National Chiao Tung University, Hsinchu 30010, Taiwan, R.O.C
e-mail: {mcwu0715,nike4859,sh9107272000,apc888888}@gmail.com

C.-H. Lin
Department of Information Management, Minghsin University of Science and Technology, Hsinchu 30401, Taiwan, R.O.C
e-mail: lch@must.edu.tw

© Springer-Verlag Berlin Heidelberg 2016
J.H. Kim and Z.W. Geem (eds.), *Harmony Search Algorithm*,
Advances in Intelligent Systems and Computing 382,
DOI: 10.1007/978-3-662-47926-1_29

all affect. Due to the ever-changing market prices, the market gradually, some people think the price is random and without regularity, and therefore cannot predict future price movements [1]. Efficiency market hypothesis is derived from random walk theory, Bachelier [2] in 1900 believes investors on the stock market is expected to be an isolated incident, sample collection will be close to normal distribution, represents about half of the people in the market that the share price will rise, and half think the stock price will fall, each act will cancel each other out, making the overall investor expectations of the market is equal to zero, in a very short period, price movements totally unpredictable. However, there is a blind spot above assumptions, when many people come together, decide between them affect each other, human behavior is no longer a stand-alone event, which is called "The Effect of Sheep Flock". Therefore, price movements are not random, but most investors will follow the view changes in one direction. Fama [3] in 1965 formally proposed efficiency market hypothesis, all the information on the price of the security can be immediate and adequate response to the market, he said. The current stock price will be past, present and even future can estimate of the impact of information. So investors regardless of any trading strategy, are unable to obtain excess returns, because everyone analyze the stock is independent, mutually affected, and the stock can quickly respond to all the information, so investors cannot use any strategy to predict the share price.

However, not all market can meet the above requirements, there will be some market information gap to some extent, but cannot reach full efficiency market, Fama [4] in 1970 reinforces his argument, he thought to be involved in the market both of rational participants to maximize profits for the termination, in an efficient market, prices reflect all information, any fundamental analysis or technical analysis cannot get excess returns from the market. Since the research methods, the study period, the researchers labeled different places, leading to the conclusion whether market efficiency are not the same, and therefore those who have different ideas on the market.

However, Steildmayer [5] in 1984, a book market profile theory mentioned market price is definitely not random development, there are different players on the market, according to their needs to enter or leave the market, and thus change the price, so the market has the logic sex. He believes that the market value = Price + time, with changes in the market price and time, participants from several different angles, resulting in changes in value. Wherein the range of values is 70% of participants agree with the price range while moving changes in their range of values, which represents most of the participants for the price produced a new view, we must immediately assess market trends, look for a relatively low risk trading position.

The study will outline the market theory, the use of Data Mining to dig out the number of days happen double distribution trend day, whichever form Single Prints range of values between the two, and the maximum and minimum day. Moving to judge the value range, whether it is High to Low or Low to High, on behalf of the whole range of values change, and whether long-term traders as the later transaction, which represents the recognition of this new range of values, thereby assess the market potential of the moving direction, and analyze hidden forces behind the market, to further assist investors to grasp the future trend changes, it proves that the market is not in line with the weak-form efficient market hypothesis.

2   Literature Review

Market profile theory is an outline can be used to observe and analyze changes in market prices J.Peter Steidlmayer new doctrine by the year of 1984 presented. The role of the free market is through a competitive auction, changing prices enough to attract buyers and sellers to conduct transactions. Outline the basic concepts of the theory of the market can be divided into the market structure, the transaction time and transaction logic. Market Structure That Market profile graph, but by its distinctive bell-shaped curve that the market structure, and then describe the process of changes in market prices and reflect the behavior of participants. Trading Time is represented by a bell-shaped curve, and the widest part of the figure shows the value of area participants consuming the most transactions, that is now recognized by the participants to find the range of values, understand the current market a variety of participants behavior. Trading Logic that is experienced by observing the characteristics of the market structure and trading hours are obtained, and then grasp the key factor in the cause of market operation, power, mode etc..

The major market participants into day traders and long-term traders. Day Traders also known Locals, on the one hand from the long hands of the buyer to buy the other hand, long-term and immediately sold to the buyer, if the role of market intermediaries frequently traded. The Other Timeframe Traders Trader is sustained for more than a day, and usually only when the price deviates from its news value or extrinsic trading decisions, will enter the market, the real key is to control market forces. Market controlled mainly by long offensive players, to influence changes in the market value of the interval, resulting in different forms of the trading day, the market profile theory there are seven common trading patterns, patterns used in this study for the double distribution trend day, at the beginning of the transaction have less trading activity, market participants are extremely non-confidence in the market, resulting in a narrow price range. After some time, because of changes or the occurrence of an event causing investors perceive profitable, speed to market, and thus the price into another grade. Moreover, the weak trend of the trading day will run for a period of time after a specified direction, stop and wait for further development of the market, and to assess the long-term traders in the new range of values is of a continuing nature, and then to find the best entry point.

3 Proposed Method

The subject is the TAIEX futures in this study. Experimental samples during the study from January 5, 2009 to December 31, 2013. Processing a total of 1,237 trading days of historical data, and it contains date, trade time and trade price. First, TAIEX futures tick data organized into one minute price and thirty minutes market profile graph. After TPO-processing, determine which day is the double distribution trend day. Double distribution trend day includes two value areas, and at least two consecutive TPO of price single letter between D letter and G letter (from an hour half after opening to an hour half before closing). That may have chance to complete a bell curve, and it's representing the value area to be accepted most of the investors and develop a new value area. A total of 152 double distribution trend days occurred between January 5, 2009 and 31 December 2013. Experimental groups determine long or short base on direction of double distribution trend days. Open mechanism is based on the maximum and minimum of Single Print, and evaluated the support capability. Stop mechanism is divided into two methods, first is stop-loss when loss ten points, and second is reference the single-letter support line. In the part of the control group, open on the maximum and minimum of Single Print by random long or short. Stop mechanism is stop-loss when loss ten points.even though the open should filtered by long term days, cause market profile theory point out when the price keep longer which means long-term traders strongly agree the value area. Therefore, when value area moves in the other direction, long-term response traders will have more power to defense the new direction. The experimental groups divided four kinds of filters of days, contains after 1 day, 3 days, 5 days and 7 days. Two evaluation methods to measure the model performance by simulated trading for accuracy and profitability.

In this study, experimental group A is focus on move direction of Double Distribution Trend Days and duration days, and stop-loss when loss 10 points. If value area from high to low, the support line which at the maximum of day will be continue average 14.5 days. If value area from low to high, the support line which at the minimum of the day will be continue average 27.6 days. Therefore, open position chose maximum or minimum price on direction of Double Distribution Trend Days to determine the support line, and evaluate the performance. Determine the support line and the duration of support line that could help to trace the long-term response traders.

Experimental group B is focus on support capability based on Single Print of Double Distribution Trend Days, and stop-loss when loss 10 points. If value area move from high to low and generate a new value area, the maximum of Single Print will be support line and recover when the price back to the old value area. If value area move from low to high and generate a new value area, the minimum of Single Print will be support line and recover when the price back to the old value area.

Therefore, open position chose maximum or minimum price on Single Print of Double Distribution Trend Days to determine the support line, and evaluate the performance. Determine the support line and the duration of support line that could help to trace the long-term response traders.

Experimental group C is focus on support capability based on Single Print of Double Distribution Trend Days, and stop-loss when reach support line by historical Single Print. If value area move from high to low and generate a new value area, the maximum of Single Print will be support line and recover when the price back to the old value area. If value area move from low to high and generate a new value area, the minimum of Single Print will be support line and recover when the price back to the old value area. Therefore, open position chose maximum or minimum price on Single Print of Double Distribution Trend Days to determine the support line; close position use support line of yesterday, and evaluate the performance. Determine the support line and the duration of support line that could help to trace the long-term response traders.

Control group A is based on Weak Form of Efficient Market that assumptions is the stock price completely reflect all the history or existing information. Hence investors couldn't use historical information to be analyzed and predict stock prices. Control group A trades by random walk model and compares with experimental group A. For this reason, the open timing is same as experimental group A., however short or long position is determining by random. The close is stop-loss when loss 10 points, i.e. when open by long position then close by short position; when open by short position then close by long position. Finally, determine the duration since occurred Double Distribution Trend Days is longer enough to support the price.

Control group B is based on Weak Form of Efficient Market that assumptions is the stock price completely reflect all the history or existing information. Hence investors couldn't use historical information to be analyzed and predict stock prices. Control group B trades by random walk model and compares with experimental group B. For this reason, the open timing is same as experimental group B., however short or long position is determining by random. The close is stop-loss when loss 10 points, i.e. when open by long position then close by short position; when open by short position then close by long position. Finally, determine the duration since occurred Double Distribution Trend Days is longer enough to support the price. Performance evaluation with experimental group B, and then test the Weak Form of Efficient Market in TAIEX market rationality.

4 Experiments

During the study period of January 5, 2009 to December 31, 2013, a total of 1237 days, which took the weak trend in the trading day, a total of 152 strokes. According to the weak trend in the value range of the moving direction of the trading day, the maximum or minimum value to obtain the date, entering as a transaction, a total of 138 strokes. According to the weak trend in the value range of the moving direction of the trading day, a single print) uppermost edge or do the lower edge, entering as a transaction, a total of 141 strokes.

This study is divided into the accuracy of performance evaluation assessment method and two points with average earnings, in accordance with the evaluation criteria in order to compare the performance of. When trading strategies finished, all the statistics of transactions, if the transaction results showing a profit, and its value is set to 1, zero otherwise. All items will be positioned to capitalize on the sum, divided by the total items, see equation (1).

$$\text{Accuracy rate} = \frac{\sum_{i=1}^{n} Correct_i}{N} \quad (1)$$

$Correct_i$: i-items profit of correct information; N: total items.

Then, with an average loss of points per port to measure. TAIEX futures simulated trading in goods, and to calculate profit or loss of points, each with a limited trading, calculated as follows:

$$\text{Average profit point} = \frac{\sum_{i=1}^{n} Profit_i}{N} \quad (2)$$

$Profit_i$: i-items trading remuneration, N: total items.

As can be seen from the table a day later in the experimental group A (≥ 1 day admission), three days later (≥ 3 days of arrival), five days later (≥ 5 days of admission) accuracy rate rising, the loss per port getting smaller and smaller, but five days later (≥ 5 days of admission) and seven days later (≥ 7 days of arrival), the precise rate to a minimum, but the highest loss per port, on behalf of the more entry points with time night, the more entry points is not necessarily the reference value. After five days (≥ 5 days admission) before entering the entry point, Accuracy rate is relatively high, the smaller loss in the experimental group A mean showing a loss per port, maximum weak trend trading day produced or minimum value, does not have the capability of defense.

Experimental group B after a day (≥ 1 day admission), three days later (≥ 3 days of admission), five days (≥ 5 days of admission) and after seven days (≥ 7 days of admission) accuracy rate rising , increasing earnings per port, on behalf of the later time as the entry point, which is the entry point with a reference value, a single letter (Single Print) weak trend in the trading day produced has defenses.

Experimental group C after a day (≥ 1 day admission), three days later (≥ 3 days admission), five days (≥ 5 days of admission) and after seven days (≥ 7 days of admission) the accuracy of declining but the surplus is growing every mouth, on behalf of the later time as the entry point, increasing earnings per port, in addition to the entry point of a reference value, its appearance historical reference pressure support line,but also it has a significant effect.

Table 1 Experiment results

	Number of days before arrival	Accuracy rate (%)	Average profit point (point)	Total losses (point)	Number of transactions (number)
Experimental group A	after 1 day (≥ 1)	44.93%	-1.01	-140	138
	after 3 day (≥ 3)	45.26%	-0.95	-90	95
	After5 day (≥ 5)	46.15%	-0.77	-60	78
	after 7 day (≥ 7)	42.43%	-1.52	-100	66
Experimental group B	after 1 day (≥ 1)	55.32%	1.06	150	141
	after 3 day (≥ 3)	61.11%	2.22	120	54
	After5 day (≥ 5)	70.00%	4.00	160	40
	after 7 day (≥ 7)	76.67%	5.33	160	30
Experimental group C	after 1 day (≥ 1)	57.45%	24.09	3396	141
	after 3 day (≥ 3)	46.30%	35.89	1938	54
	After5 day (≥ 5)	42.50%	41.78	1671	40
	after 7 day (≥ 7)	36.67%	50.80	1524	30
Control group A	after 1 day (≥ 1)	42.03%	-1.6	-220	138
	after 3 day (≥ 3)	41.05%	-1.79	-170	95
	After5 day (≥ 5)	41.03%	-1.79	-140	78
	after 7 day (≥ 7)	39.40%	-2.12	-140	66
Control group B	after 1 day (≥ 1)	41.84%	-1.63	-230	141
	after 3 day (≥ 3)	46.30%	-0.74	-40	54
	After5 day (≥ 5)	47.50%	-0.5	-20	40
	after 7 day (≥ 7)	50.00%	0	0	30

The whole, in addition to the day after arrival (≥ 1 day admission), Accuracy rate the experimental group B the highest, showing a single print to support the pressure line, and its ability to withstand stronger, is preferred entry points, combined with fixed stop-loss benefit stops playing, the results will be better than the experimental group a and experimental group C. In the highest part of the profit or loss per port, per port or loss Experiment C, representing the history of pressure support, has significant effects, but with the increase in transaction time, accurate rate is getting lower and lower, behind is more representative of the higher profit also it has a higher risk.

Group A and group B randomly trading models were experimental group A of the control group and the control group, the experimental group B, its selection of approach methods are the same, with the exception of whether to do more or vent, then take a random way trade. Stop playing the same to stop Lee ten points, to close out the opposite direction, such as the approach to do more, then vented appearances; to vent approach is to do more appearances. Finally still weak trend trading screening occurs to calculate the number of entry points the day before yesterday, whether the late arrival, the pressure support more valuable.

Accuracy rate of the control group B with each port and losses are higher than A, Accuracy rate the control group B with each loss the same port number as the day before entering the screening, the later approach the higher Accuracy rate, fewer losses. Overall, the group's forecast performance and profitability are poor.

5 Conclusions

In this study, the market profile theory, analysis of the double distribution trend day, the different parts of whichever parameter value trading, statistical analysis, can be hidden on the back of strength in the market, and then to find it hidden including the knowledge and behavior, creating high profit, and view Taiwan futures market is in line with the weak form efficient market hypothesis.

Its maximum range from the study results that, in a weak trend trading Kusakabe, made of a single print or set minimum entry point, and then take advantage of Accuracy rate and to calculate profit per port, the obtained the remuneration to 24.09 points and 57.45% accuracy rate per port, the result was better than the maximum or minimum day. Next, the same single-letter approach, only played each stop loss or stop profit in each ten o'clock, and the reference pressure history as a yardstick support line. Can be learned from the results, stop-loss or stop profit each ten relatively low risk for the low paid, historical support line pressure is high risk and high reward, thereby learned the history of pressure support line has the reference value. In the long-term market contours, it must be combined with effective reference index, it has the effect, as the later arrival time, if the value of the interval by most traders agree, the harder it will be a breakthrough.

Acknowledgement The work was supported by the Ministry of Science and Technology, Taiwan, R.O.C. under Grant MOST103-2410-H009-023.

References

1. Regnault, J.: Calcul des chances et philosophie de la bourse. Mallet-Bachelier (1863)
2. Bachelier, L.: Theory of speculation. In: Cootner, P.H. (Ed.) (1900)
3. Cootner, P.H.: The Random Character of Stock Market Prices, pp. 17–78. MIT Press, Cambridge (1964)
4. Fama, E.F.: The behavior of stock-market prices. J. Bus. 34-105 (1965)
5. Fama, E.F.: Efficient capital markets: A review of theory and empirical work*. J. Fin. **25**(2), 383–417 (1970)
6. Steidlmayer, J.P.: Markets and Market Logic. Porcupine Press, Chicago (1984)
7. Kendall, M.G., Hill, A.B.: The analysis of economic time-series-part i: Prices. J. Roy. Soc. Sta. **116**(1), 11–34 (1953)
8. James, F.D., Eric, T.J., Robert, B.D.: Mind Over Markets: Power Trading with Market Generated Information, 2th ed. Traders Press (1999)
9. Thomsett, M.C.: Support and resistance simplified. Marketplace Books (2003)
10. Murphy, J.J.: Technical Analysis of the Futures Market: A Comprehensive Guide to Trading Methods and Applications. Prentice Hall1, New York (1986)
11. Osler, C.: Support for Resistance: Technical Analysis and Intraday Exchange Rates. Economic Policy Review (2000)

Sensitivity of Sensors Built in Smartphones

Zoltan Horvath, Ildiko Jenak, Tianhang Wu and Cui Xuan

Abstract In our earlier researches we examined the reliability and accuracy of sensors used in outdoor positioning. Results have shown that they are not as reliable as many think. So the question rightly arises: what sensor could be the solution for indoor navigation. The sensitivity and resilience of smartphone sensors are currently unknown, and their reliability is also questionable because of the many distracting factors. However, studies also show that it is possible to decrease distraction-caused errors with the help of appropriate algorithms. Hence first we must define sensitivity of sensors and understand their operational principles to find the appropriate algorithms.

Keywords Indoor positioning · Accelerometer · Gyroscope · Sensors sensitivity · Smart phones

1 Introduction

We began our research by examining GPS and GLONASS sensors used in outdoor navigation [11] [12]. Since nearly every fourth person has used navigation software, we found examining how accurate these sensors are and what factors may interfere with them to be interesting. This is also important because indoor positioning is more complex compared to the outdoor. Suffice it to say that outdoor positioning technologies don't work indoor, because the sensors used are shielded. Our research, however, clearly revealed that outdoor positioning is not that simple either, as many might think [10]. Furthermore we didn't get the same

Z. Horvath(✉) · I. Jenak
Department of Information Technology, Faculty of Sciences, University of Pecs, Pecs 7624, Hungary
e-mail: hz@gamma.ttk.pte.hu, jenak@ttk.pte.hu

T. Wu · C. Xuan
Northwestern Polytechnical University, Xi'an 710072, Shaanxi, People's Republic of China
e-mail: tihaw@foxmail.com

© Springer-Verlag Berlin Heidelberg 2016
J.H. Kim and Z.W. Geem (eds.), *Harmony Search Algorithm*,
Advances in Intelligent Systems and Computing 382,
DOI: 10.1007/978-3-662-47926-1_30

result using different softwares, but the same device, operating system and sensors in the same weather conditions [3]. In this case it often occurred that we measured 27-32% discrepancy among our data; this can be seen on Figure 1.

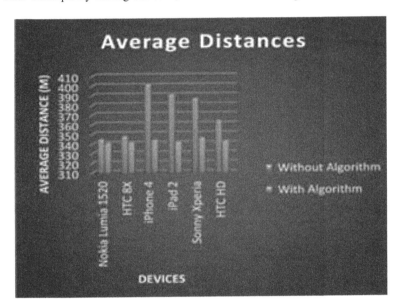

Fig. 1 Average Distance

I don't mention environmental factors, since it's evident that certain weather conditions can greatly distort the accuracy of results. Our experiments were complemented by examining how devices respond if different forces are applied on them. We did this for we would like to make use of the gyroscope and the accelerometer in indoor positioning. Our measurement results showed that G-force can distort the precision of data if it reaches 3G. The other part of our experiment was to examine indoor positioning. For this we used the software developed by Lanoga Kft, which we can see on Figure 2 [2].

The software can monitor sensors (accelerometer, gyroscope, etc.) in smartphones [6]. From the firs measures it turned out that the sensors are very sensitive during measurements [4]. The data detected during the test are discussed in chapter Results [7]. Many have attempted to use Wi-Fi networks for indoor positioning [1][8]. In this solution we meet with the same error as measuring inaccuracies caused by scattered and reflected signals observed in outdoor navigation.

Fig. 2 Monitoring software

2   Material and Method

2.1   Algorithms

Results provided by sensors are not enough to obtain evaluable data for indoor positioning. As we have already proven, it is necessary to use different algorithms to get evaluable and reliable data. First we examined how useful the Kalman filter is, since it is quite effective if we want to process data measured by the original signal from the scattered ones. During our research we recorded nearly 14,000 records, so it seemed obvious to examine the raw data with linear regression [5], since it gives a line; this is called a regression line. Its general formula is the following (1)[10]:

$$Y = \beta_0 + \beta_1 X \qquad (1)$$

However we should not forget that given this many data, measuring errors can occur, so do estimation errors in the regression. This can be described as follows (2):

$$\varepsilon_i = Y_i \left(\hat{\beta}_0 + \hat{\beta}_1 X_i \right) = Y_i - \hat{Y}_i \qquad (2)$$

Besides, we must mention the error of the regression line, the residual error, or the error variance. This can be expressed as follows (3):

$$Res = E\left[(Y_i - \hat{Y_l})^2\right] = E[\varepsilon_i^2] = E[(\varepsilon_i - 0)^2] = \sigma_{\varepsilon_i}^2 \qquad (3)$$

We encounter it especially when we measure too much reflected signals during recording. Of course, these errors can be filtered further, even with as mentioned above, or with the Bayesian Histogram (4).

$$f(y) = \sum_{h=1}^{k} 1\,(\xi h - 1 < y \leq \xi h)\frac{\pi_h}{(\xi_h - \xi_{h-1})}, y \in R \qquad (4)$$

Histogram filters have seen some use in robotics for localization due to their computational efficiency.
However, they do not offer an easy way to guarantee convergence due to information loss in the approximation step.

2.2 Seismograph

During tests we applied the software operating on the principles of a seismograph too. With this we wanted to examine whether smartphones suffer any resonance during measurements. This is an important factor during the research because if the software finds any displacement, a dispersion of a certain degree of measurement results could be explained by that. Otherwise we must find the factor inducing the scatter (5).

$$M = \frac{n(\mathrm{n}^2+1)}{2} \qquad (5)$$

3   Results

During test no displacement was measurable. The seismograph image confirms this, this can be seen on the next graph.
    It can be seen in the graph that none of the spatial axis suffered force. So we can state that dispersion described later and periodical errors couldn't be inflicted by resonance, or at least by measurable resonance.
    In this research of ours we examined mainly the magnetometer and the accelerometer. The following measurement results are originated from a measuring at a fixed point, so that we can illustrate the accuracy of sensors better. First we discuss the results measured with the accelerometer. We recorded nearly 14,000 records, thus we had a relatively large set of data to work with. WE measured in two dimensions, so it forms a plane; also the device was fixed to one point. The latter is important because the accelerometer resting on the surface of Earth measures 0 G with a good approximation.

Fig. 3 Seismograph

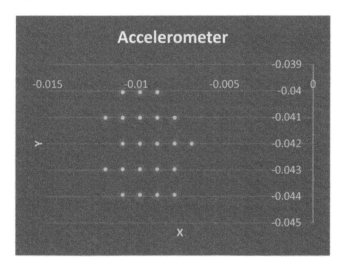

Fig. 4 The surface of Earth measures

This result is particularly interesting, since – as we have mentioned earlier – the measured results should be closer to 1, because the device conducting the measuring is fixed at the time of recording data. The results shown on the figure is similar to an event close to free fall, since the values converge to zero. We shouldn't ignore either that the values of the coordinate system are understood in the inertial system based on Newton's first law.

But it is an interesting result that nearly all measurement results show the same dispersion depicted in the inertial system.

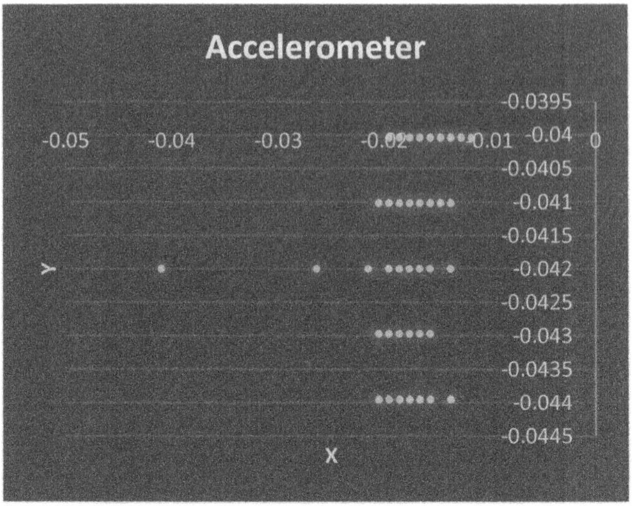

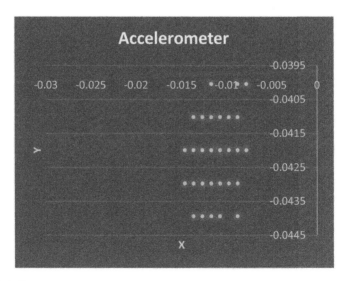

Fig. 5 The inertial system

Since we gathered more than 10,000 records, we could analyze measurement sequences separately. On the figure above we can see the results of records 3400-7200. Because we can see some value seeming to be errors located far from the central set, we examined their environment. As it is shown on the graph below, if we examine a smaller sequence (3400-4400), the salient results disappear. We observed during our experiments that these errors return periodically (in every 1209 records), and then 15-17 faulty values are stored. But of course, it could be the inaccuracy of the smartphone sensor too.

Fig. 6 Periodic errors

On the next figure we can see the values between 4700 and 6000, where the salient values outline spectacularly.

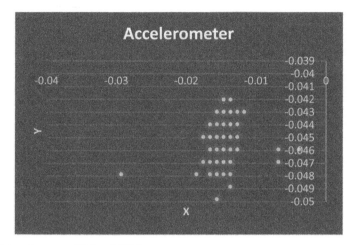

Fig. 7 Values between 4700 and 6000

Besides errors occurring periodically, using the magnetometer we noticed that other electric instruments (such as computers, refrigerator, etc.) can cause errors. The next graph will show an example to this interference.

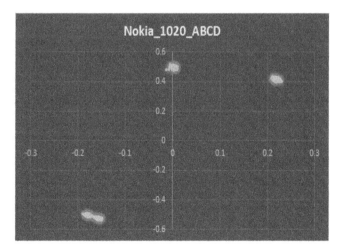

Fig. 8 Error of Interferences

During the experiment we performed measuring in the corners of the premises (950 per corner), then we imaged them in a coordinate system.

As we can see on the Figure 7, corner C and D became too close to each other, although they are further in reality. This error can be remedied in two different ways. Either we switch off the instrument in corner D which causes the interference, or we try to filter the signal. In the present case it can be seen that all measurement results are distorted, hence there will be no original signal to filter, and therefore the second solution is impossible. The next figure shows how measurement results change if the instrument is switched off, and if the results are filtered too.

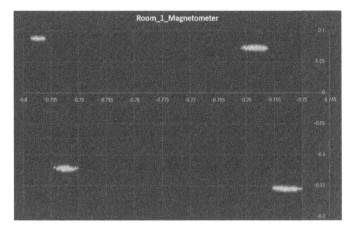

Fig. 9 The values without interferences

As we can see on the graph above, this way the four corners are palpable at the test premises. But we still experience some shift, so in this case we don't perfectly perceive the original testing area. We attempted to estimate the original size of the testing area based on the measurement results. According to theme the longer walls were estimated to be 450 cm, while the shorter ones to be 250 cm long. In reality the longer sides are 432 cm, and the shorter ones are 240 cm long. So we can assert that, even if the used method is not 100% accurate, it is suitable for approximate estimation.

4   Conclusions

At this stage of our research we mainly examined the magnetometer and the accelerometer. In our experiments we gathered and processed nearly 15,000 records. Our studies have shown that there are more interfering factors during indoor measurements than during outdoor measuring. In our opinion the interference causes the greatest problem, since our buildings are full of electric appliances; and the thickness of walls and cable insulation are not effective enough to eliminate it. Another problem is if the interfering appliance is positioned within a 10 cm radius of the device. In this case we cannot find the original signal, so we can't filter the faulty data. The smallest distance, where we could find signal with no disturbance, was 32 cm; but in this case we could use only 24 out of 1000 measurement data to filter.

References

1. Binghao, L., James, S., Andrew, G.D., Chris, R.: Indoor positioning techniques based on wireless LAN. In: First IEEE Iinternational Conference on Wireless Broadband and Ultra Wideband Communications, pp. 27–8 (2013). doi:10.1.1.72.1265
2. Brachmann, F.: A Multi-Platform Software Framework for the Analysis of Multiple Sensor Techniques in Hybrid Positioning Systems (2014)
3. Galvan-Tejada, C.E., Garcia-Vazquez, J.P., Brena, R.: Magnetic-field feature reduction for indoor location estimation applying multivariate models. In: 2013 12th Mexican International Conference on IEEE Artificial Intelligence (MICAI), pp. 128–132 (2013)
4. Galvan-Tejada, C.E., Garcia-Vazquez, J.P., García-Ceja, E., Carrasco-Jiménez, J.C., Brenaa, R.F.: Evaluation of Four Classifiers as Cost Function for Indoor Location Systems. Procedia Computer Science 32, 453–460 (2013)
5. Pearson, K.: Notes on regression and inheritance in the case of two parents. Proceedings of the Royal Society of London 58, 240–242 (1985)
6. Mobile phone sensors. http://www.baike.baidu.com/view/9077573.htm
7. Castro, P., Chiu, P., Kremenek, T., Muntz, R.: A probabilistic room location service for wireless networked environments. In: Abowd, G.D., Brumitt, B., Shafer, S. (eds.) UbiComp 2001. LNCS, vol. 2201, pp. 18–34. Springer, Heidelberg (2001)
8. Stella, M., Russo, M., Begušić, D.: Fingerprinting based localization in heterogeneous wireless networks. In: Expert Systems with Applications (2014). doi:10.1016/j.eswa.2014.05.016
9. Seber, G.A., Lee, A.J.: Linear regression analysis, vol. 936. John Wiley & Sons (2012)
10. Horvath, Z., Baranyi, A.: The problematics of indoor navigation. In: 10th International Miklos Ivanyi phd & dla Symposium Abstract Book, p. 53 (2014). ISBN 978-963-7298-56-1
11. Horvath, Z., Horvath, H.: More sensors or better algorithm? International Journal of Computers, Communications & Control (IJCCC) (2014)
12. Horvath, Z., Horvath, H.: The Measurement Preciseness of the GPS Built in Smartphones and Tablets. International Journal on Electronics and Communication Technology 1(5), 17–19 (2014)

Part VI
Optimization in Civil Engineering

Harmony Search Algorithm for High-Demand Facility Locations Considering Traffic Congestion and Greenhouse Gas Emission

Yoonseok Oh, Umji Park and Seungmo Kang

Abstract Large facilities in urban areas generate lots of traffic and cause congestion that waste social time and become a major source of greenhouse gas (GHG). To overcome a shortcoming of the fixed transportation cost in conventional facility models, the congestion effect by facility users as well as general drivers in networks, with increased GHG emission is considered. In this paper, several Harmony Search algorithms with local search are developed and compared to the existing Tabu Search algorithm in a variety of networks. The results demonstrate that the proposed approach and local search method can find better or comparable solution than other methods within a given time.

Keywords Facility location problem · Harmony search algorithm · Traffic congestion · Greenhouse gas emission

1 Introduction

Large facilities in urban areas, such as train station, bus terminals, department stores, or community centers typically generate lots of trips. These additional trips cause congestion and become major sources of greenhouse gas (GHG) emission. The conventional facility location models assume the transportation cost or time is fixed between demand and facilities, thus cannot analyze the network-wise impact of high demand facilities in urban transportation networks. The increased demand on the roadway and the rerouting of exiting drivers as well as increased GHG emission should be considered in finding the optimal location of facilities.

Y. Oh · U. Park · S. Kang(✉)
School of Civil, Environmental and Architectural Engineering,
Korea University, Seoul 136-713, South Korea
e-mail: {ysoh0223,puj2860,s_kang}@korea.ac.kr

© Springer-Verlag Berlin Heidelberg 2016
J.H. Kim and Z.W. Geem (eds.), *Harmony Search Algorithm*,
Advances in Intelligent Systems and Computing 382,
DOI: 10.1007/978-3-662-47926-1_31

Recent study by Hwang et al. [1] investigated this problem and showed that the suggested approach can significantly save the total cost than the existing location problems. They proposed metaheuristic algorithms for solving the mathematical model; Genetic Algorithm, Memetic Algorithm and Tabu Search(TS), and TS showed the best performances among those. The purpose of this paper is to develop new algorithms using Harmony Search Algorithm (HSA) to solve the capacitated facility location problem with traffic congestion and GHG emission and compare the effectiveness of the developed algorithms to existing TS results by Hwang et al.[1]

After Geem et al.[2] developed HSA, it has been applied to various optimization problems such as storage containers problem [3], music composition [4], web page clustering[5], structural design[6], water network design [7], dam scheduling [8],transportation energy modeling[9], etc.[1]Geem and Sim developed parameter-setting-free (PSF) harmony search algorithm[10], which can search good solutions without setting the fixed parameters for its best performance. Geem and Cho[11] solved optimal design of water distribution networks using PSF- HSA.

The following parts of the paper begin with the model formulation and introduce the suggested algorithms with flowcharts. Case study section presents the test results of algorithms with discussions. The paper ends with the conclusions and suggestions of the future research.

2   Model

In this section, a bi-level mixed-integer program is presented by combining fixed-charge facility-location model and the route-choice model. The method deals with facility location, traffic congestion induced by high-demand facilities, and emission gas.

Upper level:

Minimize
$$\sum_{j\in J} e_j Y_j + \sum_{a\in A}\left\{\alpha x_a t_a\left(x_a\right)+\beta x_a E_a\left(x_a\right)\right\}, \tag{1}$$

subject to
$$h_i=\sum_{l\in L^i} f_l^i, \ \forall i\in I^d, \tag{2}$$

$$\sum_{i\in I^d} h_i \leq \sum_{j\in J} C_j Y_j, \tag{3}$$

$$Y_j=\left\{0,1\right\}, \ \forall j\in J, \tag{4}$$

Lower level:

Minimize
$$\sum_{a\in A}\int_0^{x_a} t_a(\omega)d\omega, \tag{5}$$

subject to
$$x_a=\sum_{i\in I^d}\sum_{l\in L^i} f_l^i \delta_{a,l}^i + \sum_{i\in I^b}\sum_{k\in K}\sum_{m\in M^{i,k}} b_m^{i,k}\rho_{a,m}^{i,k}, \ \forall a\in A, \tag{6}$$

$$q^{i,k}=\sum_{m\in M^{i,k}} b_m^{i,k}, \ \forall i\in I^b, k\in K, \tag{7}$$

---

[1] Refer Milad and Pezhman(2013) for a detailed description of HSA including strengths and weaknesses.

$$v_j = \sum_{i \in I^d} \sum_{l \in L^i} f_l^i \theta_{j,l}^i, \quad \forall j \in J, \tag{8}$$

$$v_j \leq C_j Y_j, \quad \forall j \in J, \tag{9}$$

$$f_l^i \geq 0, \quad \forall i \in I^d, l \in L^i, \tag{10}$$

$$b_m^{i,k} \geq 0, \quad \forall i \in I^b, k \in K, m \in M^{i,k}. \tag{11}$$

Where,

$D(V, A)$: Directed graph which represents the roadway network

A, V : Set of directed links and nodes

j : Candidate nodes, $j \subseteq V$

C_j: Capacity of facilities when the facilities are located on the candidate node j

e_j: Fixed investment for the construction and operation of the facility

Y_j: Binary decision variable.

α, β: Parameter converting the time and amount of CO2 to monetary value

$I^b \subseteq V$, $K \subseteq V$: Set of origins and destinations of background traffic, respectively.

I^d: Set of origins of facility demand traffic

h_i: Facility demand at node i

$\sum_i h_i$: Total facility demand in the network

x_a: Flow on link a, $a \subseteq A$

$q^{i,k}$: Total background traffic flow from i ($i \in i^b$) to k ($k \in K$)

$b_m^{i,k}$: Background traffic flow from i to k on path m ($m \in M^{i,k}$)

$M^{i,k}$: Set of possible path from i to k

f_l^i: Facility demand flow from origin i to virtual sink node S

L^i: Set of possible paths from facility demand node i to the imaginary sink node S

$\rho_{a,m}^{i,k}$: Binary decision variable indicating whether link a is included background path m connecting from i to k or not

$\delta_{a,l}^i$: Binary decision variable indicating whether link a is included the facility demand flow l or not

$t_a(x_a)$, $E_a(x_a)$: Function of link cost and emission cost of traffic flow on link a, respectively.

v_j: Total assigned traffic flow on facility j

$\theta_{j,l}^i$: Binary decision variable indicating the facility j is used by demand flow start from i

An upper-level model was formulated to find the locations and number of facilities, while minimizing total cost including construction cost, transportation cost, and emission cost. The solution of upper-level model was derived by metaheuristic algorithms which we are tested in this paper then a lower-level model solves the traffic assignment problem of facility demand and background traffic.

This model adopts user equilibrium [12] as the rule for route choice with convex combination algorithm. Detail of the model is described in Hwang et al. [1]

3 Methodology

3.1 Tabu Search Algorithm (TS)

TS is a metaheuristic algorithm created by Glover in 1987 [13]. It uses a local search procedure to move from an initial solution to an improved solution in the neighborhood until certain criterion is satisfied, avoiding already visited area in the Tabu list. It is known that its computation time is relatively short and the solution process is effective. The detail procedure of TS is shown as follows (Fig. 1):

Step 1: In the first step, initial solution vector is built by using Drop Heuristic [14].

Step 2: Neighborhood solutions are generated by local search approach from the previous candidate solution, which is switching each element in current solution in turn. i.e. the cell value in the solution vector is 0, then replace it with 1, and vice versa. The best solution among the neighborhood solutions is selected as a candidate. If the candidate solution is in Tabu list, then the next best solution among neighborhood solutions becomes a candidate.

Step 3: The candidate becomes the new best solution and is included in the Tabu list, if it is better than prior best solution. If termination criterion is satisfied, the algorithm stopped, otherwise it moves to Step 2.

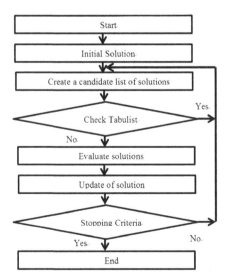

Fig. 1 Structure of Tabu Search Algorithm (TS)

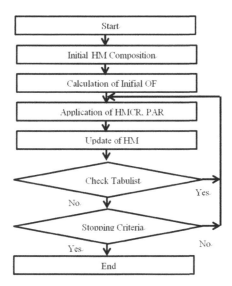

Fig. 2 Structure of Harmony Search Algorithm with Tabu list (T-HSA)

3.2 Harmony Search Algorithm with Tabu list(T-HSA)

HSA was inspired by the process of making a harmony by the musicians (Fig. 2). Its process has a memory system called Harmony Memory(HM), which contains the good solutions that have been found within its iterative process. In each iteration, it generates a new harmonies (a new solution) based on the 3 operators – Harmony Memory Considering Rate(HMCR), Pitch Adjust Rate(PAR) and Random Selecting (RS).

HMCR defines the probability that a specific part of the solution should be come from the existing HM. (Usual value of HMCR is greater than 0.5) PAR is the probability that a specific part of the solution is replaced by a random value. By RS, the search process finds the random solution and works as exploration component in HSA. (Usual value of PAR and RS is very small (less than 0.2)).

For the suggested problem in this paper, we have developed Harmony Search Algorithm with Tabu list(T-HSA). T-HSA constructs and update the Tabu list as used in TS within HSA process (Fig. 2). It checks the Tabu list while making a new solution, and ignores the solution and makes another one if it is in Tabu list. Because the computational bottleneck of the suggested model is to run convex combination algorithm and calculate the objective function, this process can save much of the time and lead to better and unexplored area.

3.3 Parameter-Setting-Free Harmony Search Algorithm with Tabu list (PSF-T-HSA)

PSF-HSA [10] is extended version of original HSA. The benefit of this algorithm is overcoming difficult parameter setting process, which was required in conventional HSA for better performance.

Not using fixed HMCR and PAR over all iterations, this algorithm uses Operation Type Memory (OTM). It iteratively calculates HMCR and PAR by checking how many solutions in HMCR are generated from each operator. HMCR and PAR are re-calculated every time that HMCR updated, but usually converged into certain values at the end of algorithm.

Fig.3 is the flowchart of this algorithm. The PSF-T-HSA also has a process of checking the Tabu list as in T-HSA.

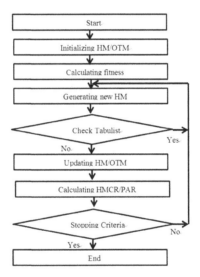

Fig. 3 Structure of Parameter-Setting-Free Harmony Search Algorithm with Tabu list (PSF-T-HSA)

3.4 Location-Based Local Search (LL)

In this paper, we applied another local search method for above algorithms. A local search method used in Hwang et al. [1] defines neighborhood solutions by flipping binary values in each candidate location in a solution vector. Location-based Local Search (LL) method (Fig. 4) uses the connection of each node in the transportation network, instead. It changes the locations of facility to another node directly connected to the existing one. This method can define the neighborhood in more reasonable way in the solution space than the previous method, which is more likely a random movement of facility location.

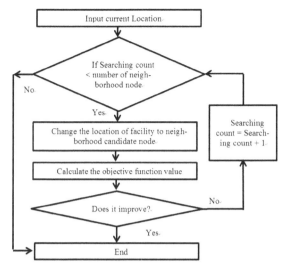

Fig. 4 Structure of Location-based Local Search (LL)

4 Case Study

In this section, the case study using aforementioned algorithms with specific local search method are tested on different sizes of network. The tested solution algorithms were TS, T-HSA, and PSF-T-HSA and its extensions using LL. Each algorithm was coded in VC++ and was tested using personal computers with 3.5GHZ CPU and 8 GB memory. In all cases, algorithms were terminated at 11,000 seconds. Parameters used for each algorithm are tabulated in Table 1.

Table 1 Parameters of solution algorithms

Algorithm	TS	T-HSA	PSF-T-HSA
Other parameters	Tabu list size 500	HMS 10 HMCR 0.99 PAR 0.02	HMS 10 Initial HMCR 0.9 Initial PAR 0.1 Calculate new parameters after replacing HM 50 times
Algorithm	**LL-TS**	**LL-T-HSA**	**LL-PSF-T-HSA**
Neighborhood node	3% of nodes of testing network		
Other parameters	Tabu list size 500	HMS 10 HMCR 0.99 PAR 0.02	HMS 10 Initial HMCR 0.9 Initial PAR 0.1 Calculate new parameters after replacing 50 times
End criteria	11,000 seconds		
Time value	$7.5/hour		
CO2 emission value	$150/ton·$CO_2$		

Here, we considered small- and large-scale test networks: the 24-node and 76-link Sioux Falls network [15] with 24 candidate locations, shown in Fig. 5, and the 408-node and 1,282-link Incheon network [16] with 72 candidate locations, shown in Fig. 6. The capacity of each facility was assumed to be 5,000vehicles/day. Facility building costs were converted into daily values with a 20-year planning horizon.

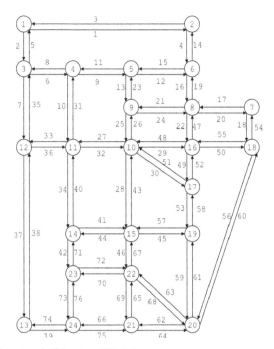

Fig. 5 Sioux-Falls network (24nodes, 76links)

Fig. 6 Incheon network (408 nodes, 1282 links)

Table 2 Result of small-size network (Sioux-Falls)

	TS(A)	T-HSA(B)	PSF-T-HSA(C)
Mean (I)	2,532,824	2,532,824	2,532,824
Min (J)	2,532,824	2,532,824	2,532,824
Time to best solution (sec)	3.4	1.3	2.8
	LL-TS(E)	LL-T-HSA(F)	LL-PSF-T-HSA(G)
Mean (K)	2,532,824	2,532,824	2,532,824
Min (L)	2,532,824	2,532,824	2,532,824
Time to best solution (sec)	2.5	0.8	33.5

4.1 Sioux-Falls Network : Small-Size

First we tested these algorithms into small size networks. As shown in Table 2, all algorithms could find the same solution within given time limit. This shows that all algorithms can get into the satisfactory level of accuracy in the small size network. LL method shows faster convergence than the original local search method in TS and conventional HSA.

4.2 Incheon Network : Large-Size

The facility building costs were categorized into three levels in the Incheon network depending on the land values of the candidate locations, including areas with high, medium, and low land value. This assumption makes this case more realistic and more complicated to solve, at the same time.

In this large-size network, all tested algorithms do not show a significant different in their performance. However, there is a slight improvement from original TS to LL-TS. LL-TS can find the same objective value in shorter time (9,406 sec vs 10,613 sec). T-HSA and LL-T-HSA could find comparable solutions, about 0.05~0.07% less than TS algorithms in the mean best objective values of multiple (30) runs. The best objective values of T-HSA and LL-T-HSA in 30 runs are about 0.3~0.6% less than the objective value of TS and LL-TS (Table 3). PSF is to reduce the effort the find a fixed parameter values in HSA, and is known that usually does not outperform the conventional HAS with optimal parameters. This can be confirmed by the test result of PSF-T-HSA and LL-PSF-T-HSA.

Table 3 Results of Large-size Network (Incheon)

	TS(A)	T-HSA (B)	B/A-1 (%)	PSF-T-HSA (C)	C/A - 1(%)
Mean (I)	260,808	260,628	-0.07	263,799	1.15
Min(J)		259,348	-0.56	262,758	0.75
	LL-TS(E)	LL-T-HSA(F)	F/E-1 (%)	LL-PSF-T-HSA(G)	G/E-1(%)
Mean (K)	260,808	260,679	-0.05	263,468	1.02
Min(L)		259,847	-0.37	262,315	0.58
(K/I)-1 (%)	0.00	0.02		-0.13	
(L/J)-1 (%)	0.00	0.19		-0.17	

5   Conclusion and Future Study

Defining the locations of high demand facilities are important task in urban areas since large amount of generated trips significantly affect the traffic conditions near the facilities. It can cause extra delays in nearby areas and these could be spread out to the other areas. In this paper, we applied several different versions of HAS and new local search method to solve the mathematical optimization model. The proposed algorithms have shown slight improvements in average and minimum objective function values in the large-size network, while all tested algorithms could reach the same objective value in the small-size network within given time.

This paper only considered the objective level and calculation time in comparison of algorithms. Future study may be extended to the analysis of location of facility and the traffic pattern of solutions from each algorithm.HSA is known to be more effective in the problem with general integer variables than binary variables. More complex location problems with multiple-size or discrete facility capacity can be a good examples that the proposed HSA can be applied.

References

1. Hwang, T., Lee, M., Lee, C., Kang, S.: Meta-heuristic approach for high-demand facility locations considering traffic congestion and greenhouse gas emission. Journal of Environmental Engineering and Landscape Management (2015). (under review)
2. Geem, Z.W., Kim, J.H., Loganathan, G.V.: A new heuristic optimization algorithm: harmony search. Simulation (2001) doi:10.1177/003754970107600201

3. Ayachi, I., Kammarti, R., Ksouri, M., Borne., P.: Harmony search algorithm for the container storage problem. In: 8th International Conference of Modeling and Simulation – MOSIM 2010 (2010)
4. Geem, Z.W., Choi, J.-Y.: Music composition using harmony search algorithm. In: Giacobini, M. (ed.) EvoWorkshops 2007. LNCS, vol. 4448, pp. 593–600. Springer, Heidelberg (2007)
5. Forsati, R., Mahdavi, M., Kangavari, M., Safarkhani, B.: Web page clustering using harmony search optimization. In: Canadian Conference On Electrical And Computer Engineering, pp. 1601–1604 (2008)
6. Geem, Z.W.: Harmony search algorithms for structural design optimization. Springer, Berlin (2009)
7. Geem, Z.W.: Particle-swarm harmony search for water network design. Eng. Optim. **41**, 297–311 (2009)
8. Geem, Z.W.: Harmony search algorithm for solving sudoku. In: Apolloni, B., Howlett, R.J., Jain, L. (eds.) KES 2007, Part I. LNCS (LNAI), vol. 4692, pp. 371–378. Springer, Heidelberg (2007)
9. Ayvaz, M.T., Kayhan, A.H., Ceylan, H., Gurarslan, G.: Hybridizing the harmony search algorithm with a spread sheet 'solver' for solving continuous engineering optimization problems. Eng. Optim. **41**, 1119–1144 (2009)
10. Geem, Z.W., Sim, K.B.: Parameter-setting-free harmony search algorithm. Appl. Mat. Comput. (2010). doi:10.1016/j.amc.2010.09.049
11. Geem, Z.W., Cho, Y.H.: Optimal design of water distribution networks using parameter-setting-free harmonysearch for two major parameters. J. Water Resour. Plann. Manage. (2011). doi:10.10161/(ASCE)WR.1943-452 .0000130
12. Wardrop, J.G.: Some theoretical aspects of road traffic research. ICE Proceedings: Engineering Divisions **1**, 325–362 (1952)
13. Glover, F.: Tabusearch, technical report center for applied artificial intelligence. University of Colorado, Boulde (1987)
14. Daskin, M.S.: Network and discrete location: model, algorithms, and applications. John Wiley and Sons, New York (1995)
15. http://www.bgu.ac.il/~bargera/tntp/ (2013)
16. http://www.ktdb.go.kr (2013)

Optimum Configuration of Helical Piles with Material Cost Minimized by Harmony Search Algorithm

Kyunguk Na, Dongseop Lee, Hyungi Lee, Kyoungsik Jung and Hangseok Choi

Abstract Helical piles are a manufactured steel foundation composed of one or multiple helix plates affixed to a central shaft. A helical pile is installed by rotating the central shaft with hydraulic torque motors. There are three representative theoretical predictions for the bearing capacity of helical piles: individual bearing method, cylindrical shear method, and torque correlation method. The bearing capacity of helical piles is governed by the helical pile's configuration, geologic conditions and penetration depth. The high variability of influence factors makes an optimum design for helical pile configuration difficult in practice. In this paper, the harmony search algorithm is adopted to minimize the material cost of helical piles by optimizing the components composing a helical pile based on the proposed bearing capacity prediction. The optimization process based on the combined prediction method with the aid of the harmony search algorithm leads to an economical design by saving about 27percent of the helical pile material cost.

Keywords Helical pile · Individual bearing method · Cylindrical shear method · Combined prediction method · Harmony search algorithm

1 Introduction

Helical piles are a manufactured foundation steel pile composed of one or multiple helix plates affixed to a central shaft. A helical pile can be driven to a designed

K. Na · D. Lee · H. Lee · H. Choi(✉)
School of Civil, Environmental and Architectural Engineering, Korea University,
Mail address Korea University Anam Campus, Anam-dong 5-ga,
Seongbuk-gu, Seoul, Korea
e-mail: {magicnapal,steallady,unin219,hchoi2}@korea.ac.kr

K. Jung
S-TECH Consulting group, Mail address 371-19, Sinsu-dong, Mapo-gu, Seoul, Korea
e-mail: Ksic2000@naver.com

© Springer-Verlag Berlin Heidelberg 2016   329
J.H. Kim and Z.W. Geem (eds.), *Harmony Search Algorithm*,
Advances in Intelligent Systems and Computing 382,
DOI: 10.1007/978-3-662-47926-1_32

depth by hydraulic torque motors rotating the central shaft. Helical piles have become widely used in the US and Europe, generating the booming market and also there have been many studies on the application of helical piles in the local soil conditions [1]. The rotary drilling construction method of a helical pile is suitable in fine and coarse grain soil layers. However, because of the limitation of the torque power generated by the torque motor that cannot penetrate the soil in full depth, they cannot be adopted in a bed-rock stratum.

There are three representative predictions for the bearing capacity of helical piles. The first is the torque correlation method predicted by the final installation torque measured in the construction [2]. The second is the individual bearing method assuming each of the helix plates has the end bearing capacity with the adhesion on the shaft above the top helix plate [3]. Finally the cylindrical shear method is composed of three components: the end bearing capacity from the bottom helix plate, the cylindrical friction surrounding the helix plates, and the shaft adhesion above the top helix plate [4]. These three theoretical prediction methods have been commonly used in many helical pile constructions.

In this study, by adopting the harmony search algorithm(HSA), one of the meta-heuristic optimization algorithms, the material cost of helical piles is minimized by optimizing the composing variables of helical pile configurations such as the shaft diameter, helix plate diameters or number, and the intervals between the plates based on a new prediction method for bearing capacity by combining the individual bearing method and the cylindrical shear method.

2   Input Function

2.1   Bearing Capacities of Helical Pile

The representative prediction methods adopted in this study are shown in Fig. 1. The individual bearing method is based on the failure mechanism that the each helix plate bears a uniform pressure distributed below the plate area with adhesion stresses along the shaft. The ultimate bearing capacity is given by

$$P_u = \sum_n q_{ult}A_n + \alpha H(\pi D) \tag{1}$$

Where
q_{ult}: Ultimate bearing pressure
A_n: Area of the helical bearing plate
α: Adhesion between the soil and the shaft
H: Length of the helical pile shaft above the top helix
D: Diameter of helix plate

By the cylindrical shear method, the spaces between the helix plates are considered as a uniformly cylindrical area. The bearing capacity composed of three components: the end bearing capacity from the bottom helix plate, the cylindrical friction surrounding the helix plates, and the shaft adhesion above the top helix plate is given by

$$P_u = q_{ult}A_1 + T(n - 1)s\pi D_{ave} + \alpha H(\pi D) \qquad (2)$$

Where
A_1: Area of the bottom helix
T : Soil shear strength
$(n - 1)s$: The length of a shaft above the top helix plate

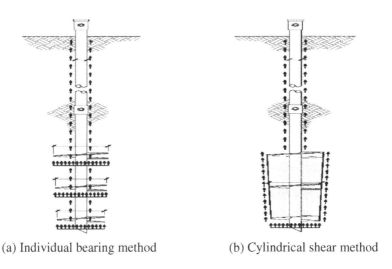

(a) Individual bearing method   (b) Cylindrical shear method

Fig. 1 Mechanisms of the helical pile bearing capacity

Referring the ratio between the bearing capacities calculated by the individual bearing method and the cylindrical shear method for the test site in Kimpo, South Korea, the combined prediction method, designed to consider with the local correction factor, can be defined as follows:

$$\text{Combined design method} = (0.85 \times P_i + 1.79 \times P_c) \times 0.5 \qquad (3)$$

Where
P_i: Bearing capacity from the individual bearing method
P_c: Bearing capacity from the cylindrical shear method

2.2 Site Investigation

The soil profile in Kimpo was investigated for the optimization of helical pile configuration. To validate the effect of the helical pile on its bearing capacity, the pile penetration depth was limited to a depth of 8.4m from the ground surface. The soil stratum is composed of coarse-grained soils with the dry unit weight of 1.76 t/m^3.

The ground water level was 3m below the ground surface. The N values in the SPT (shown in Fig. 2) were converted to the friction angle according to [5]

$$\phi = 0.3N + 27 \tag{4}$$

Where
ϕ: Friction angle
N: Blow count in SPT

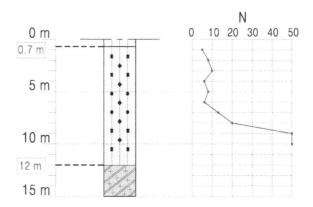

Fig. 2 Soil profile (N value) in test bed (Kimpo)

2.3   Decision Variables

With the combined prediction method, the total of 12 variables on the configuration of helical piles were considered for the specific optimization (shown in Table 1).

Table 1 Decision variables on helical pile configuration

Number of variable	1	2	3	4	5	6
Composing material	shaft diameter	plate number	plate1 diameter	Plate2 diameter	Plate3 diameter	Plate4 diameter
Number of variable	7	8	9	10	11	12
Composing material	plate5 diameter	Plate Space1	Plate Space2	Plate Space3	Plate Space4	Intrusion depth

2.4   Material Cost

Table 2 is the market price of a steel pipe advertised by POSCO (2015). The material cost of the helical pile can be calculated from Table 2.

Table 2 Market price of steel pipe (POSCO, 2015)

Number	Product name	Standard Size	Unit	Material price(won)
1	Steel pipe	J55 114.3 * 8.9T * 8M	PCS	185,040
2	Steel pipe	J55 139.8 * 10.5T * 6.6M	PCS	220,968
3	Steel pipe	API N80 88.9 * 10.6T * 12M	PCS	291,594

As helix plates commonly demand lower yield strength compared to shafts and couplers, it is assumed to use a steel of J55 for helix plate sand N80 for a shaft and couplers. The thickness of shaft and plate were assumed to be 0.65cm and 0.95cm, respectively. The material cost of the helical pile according to Table 2 can be expressed as follows:

$$\text{Cost(won)} = \pi[(0.5d)^2 - (0.5d - t_1)^2]z \times 9.319 +$$
$$\sum_{i=1}^{n} \pi[(0.5D_i)^2 - (0.5d)^2]t_2 \times 7.879 + \text{coupler} \qquad (5)$$

where
D_i: Helix plate diameter
d: Shaft diameter
t_1: Thickness of the shaft diameter+
t_2: Thickness of the helix plate diameter
n: The number of the helix plate
z: Driven depth

Considering constructability, the coupler was assumed to install at every 3m of the helical pile. The material cost for the coupler can be calculated as

$$\text{Coupelr(won)} = \pi[((0.5d + t_1)^2 - (0.5d)^2) \times 200$$
$$\pi[(0.5D_{ave})^2 - (0.5d)^2]nt_2 \times 7.065 \qquad (6)$$

Where
D_{ave}: Average helix plate diameter

3   Haromy Search Algorithm

Input parameters chosen in this study are shown in Table 3 such as HMS, the max iteration number, HMCR and PAR. The overall flowchart of HAS for this study is summarized in Fig. 3[6].

Table 3 Parameters adopted in HSA

Parameters in HS algorithm	
HMS (Harmony memory size)	50
Max iteration	5000
HMCR (Harmony memory considering rate)	0.85
PAR (Pitch adjusting rate)	0.3

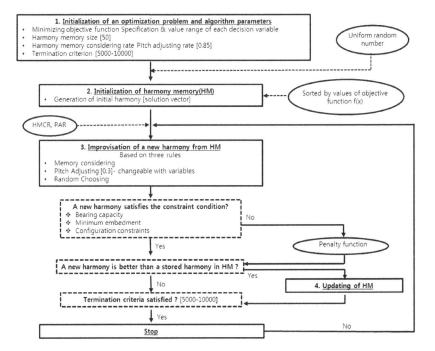

Fig. 3 Flow chart of optimization process of HSA

4   Constraint Conditions

The Criteria of helical pile configurations were considered in an aspect of empirical conformance with consideration of the minimum helical anchor embedment according to the soil conditions as shown in Table 4 and 5.

Table 4 Criteria of helical pile configuration [7]

Criteria
Round shaft with outside diameters between 73mm and 89mm
Helical plate diameters between 203mm and 356mm
Helical plates spaced along the shaft between 2.4 to 3.6 times the helix diameter

Table 5 Minimum helical anchor embedment [8]

Soil condition	Normalized Embedment Depth (H/D_t)
Fine-grain	5
Coarse-grain (loose)	7
Coarse-grain (medium)	9
Coarse-grain (dense)	11

The specific flow chart of the penalty function applied in this study is provided in Fig. 4 [9].

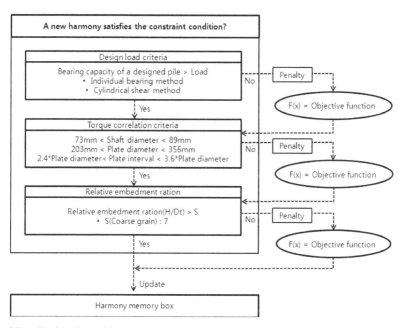

Fig. 4 Penalty function with constraint conditions

5 Optimization Results

The optimized configuration of helical piles based on the combined prediction method at each load level is presented in Table 6 and 7.

Table 6 Optimized configuration using combined method (load level of 5 to 45ton)

Load	(ton)	5	15	25	35	45
cost	(won)	34600	54100	79900	109300	120000
shaft dia.	(cm)	7.30	7.30	7.30	7.30	7.30
depth	(cm)	223	309	459	668	810
plate num.	(num.)	2	2	3	3	2
plate1 dia.	(cm)	20.91	31.36	32.43	33.16	31.15
plate2 dia.	(cm)	23.49	31.36	32.91	33.29	
plate3 dia.	(cm)			33.11	33.67	
plate4 dia.	(cm)					
plate5 dia.	(cm)					
interval 1	(cm)	50.42	75.35	115.92	102.65	75.05
interval 2	(cm)			79.58	85.53	
interval 3	(cm)					
interval 4	(cm)					

Table 7 Optimized configuration using combined method (load level of 55 to 97 ton)

Load	(ton)	55	65	75	85	97
Cost	(won)	122400	124500	131500	138900	147000
shaft dia.	(cm)	7.30	7.30	7.30	7.30	7.30
Depth	(cm)	836.37	839.70	839.74	839.67	839.61
plate num.	(num.)	2.00	2.00	3.00	4.00	5.00
plate1 dia.	(cm)	30.01	32.26	33.27	34.10	34.75
plate2 dia.	(cm)	30.01		33.27	34.10	34.82
plate3 dia.	(cm)			33.27	34.11	34.86
plate4 dia.	(cm)				34.15	34.99
plate5 dia.	(cm)					35.03
interval 1	(cm)	107.41	115.77	80.04	122.73	121.00
interval 2	(cm)			79.88	81.89	83.61
interval 3	(cm)				84.44	96.45
interval 4	(cm)					88.23

The optimized configurations in Tables 6 and 7 are schematically illustrated in Fig. 5 in detail.

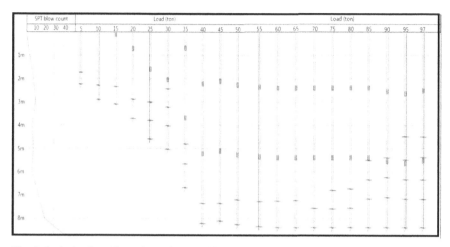

Fig. 5 Optimized configuration using combined method (load level of 5 to 97ton)

6 Cost Benefits and Discussion

Based on the bearing capacity measured from the pile load tests in Kimpo for the three helical pile configurations (shown in Fig. 6),the optimum configuration of the helical piles was evaluated along with the combined prediction method

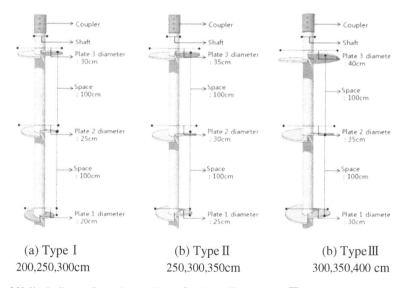

(a) Type I	(b) Type II	(b) Type III
200,250,300cm	250,300,350cm	300,350,400 cm

Fig. 6 Helical pile configurations of Type Ⅰ, Type Ⅱ, and TypeⅢ

Table 8 Costbenefitswith shaft diameters of 73, 89, and 114mm

Shaft dia. (mm)	Plate dia. (mm)	Driven depth (m)	Model number	Bearing capacity (ton)			
				Measured bearing capacity (ton)	Measured material cost (won)	Optimized material cost (won)	Cost-benefit (%)
73	Type III 400, 350, 300	4.5	1-1	21	81800	73600	10
		6	2-1	26.8	103700	84700	18
		7.5	3-1	32	122500	101500	17
		9	4-1	61	144400	123700	14
	Type III 350, 300, 250	4.5	1-2	16.7	76100	57200	25
		6	2-2	20	98000	70800	28
		7.5	3-2	38.9	116800	114800	2
		9	4-2	60	138700	123800	11
	Type III 300, 250, 200	4.5	1-3	15.5	71300	55400	22
		6	2-3	16	93100	55500	40
		7.5	3-3	20.5	112000	70200	37
		9	4-3	56	133800	122600	8
89	Type III 400, 350, 300	4.5	1-4	26	95700	84000	12
		6	2-4	27.5	122700	89400	27
		7.5	3-4	30	146100	90600	32
		9	4-4	52	173100	121400	30
	Type III 350, 300, 250	4.5	1-5	24.5	89900	80800	10
		6	2-5	30.5	117000	105500	10
		7.5	3-5	36	140400	112400	20
		9	4-5	50	167300	121300	28
	Type III 300, 250, 200	4.5	1-6	14.9	85100	54100	36
		6	2-6	29	112100	91500	18
		7.5	3-6	32	135500	101500	25
		9	4-6	51	162500	121300	25
89	Type III 400, 350, 300	4.5	1-7	22.6	117200	76200	35
		6	2-7	32	152200	101500	33
		7.5	3-7	35.5	182700	111200	39
		9	4-7	76	217700	131300	40
	Type III 350, 300, 250	4.5	1-8	20.4	111500	70200	37
		6	2-8	24.8	146500	80900	45
		7.5	3-8	32	166800	101500	39
		9	4-8	70	212000	126000	41
	Type III 300, 250, 200	4.5	1-9	20.5	106700	70200	34
		6	2-9	22	141700	74100	48
		7.5	3-9	32.2	172100	101600	41
		9	4-9	52	207200	121400	41

The cost benefits of the optimized configuration in comparison with the real helical piles constructed in the test site are summarized in Table8.The average cost-benefit is about 27.2 %. The cost benefits of the optimized configurations with different installation depths and model numbers are illustrated in Figs. 7 and 8.

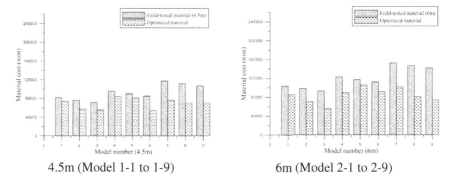

4.5m (Model 1-1 to 1-9)   6m (Model 2-1 to 2-9)

Fig. 7 Cost-benefits at depth of 4.5m and 6m

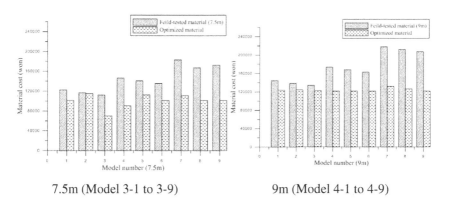

7.5m (Model 3-1 to 3-9)   9m (Model 4-1 to 4-9)

Fig. 8 Cost-benefits at depth of7.5m and 9m

7   Conclusions

The present paper investigates a supportive evidence for the applicability of HSA to optimizing helical pile configurations that leads to an economic feasibility of helical pile installation. According to the HAS results along with the new prediction method combining the individual bearing method and the cylindrical shear method, the optimized configuration of helical piles is suggested in accordance with a given loadlevel with the minimum material cost. The specific conclusions from the current study are as follows:

While the shaft diameter shows the lowest effect comparing to the other parameters, the pile penetration depth shows the strongest effect on the optimization process. As the penetration depth starts to converge to its constraint limitation, a change in the helix plate diameter at each load level does not occur. After the convergence of the penetration depth, the diameter of the helix plate increases gradually with an increase in the load level and eventually converges to its constraint condition. On the other hand, the number of helix plates increases with a slight decrease in diameter. These process repeats until the configuration meets the limitation of the load level.

When the required load level was 25ton, the helical pile with three helix plates-shows lower material cost than that with two helix plates. This tendency seems reasonable to conclude that these exceptional configurations suggest the significance of HSA application that enables to find the unpredictable optimized results.

Finally, In comparison with the material cost of the three helical piles constructed in the test site of Kimpo, the optimized configuration can reduce the material cost of helical piles about by 27.2% in the present study. Thus, it seems reasonable to conclude that the helical pile configuration optimized by HSA is economically feasible.

References

1. Perko, H.A.: Helical Piles: A Practical Guide to Design and Installation. John Wiley & Sons Inc., Hoboken (2009)
2. Hoyt, R.M., Clemence, S.P.: Uplift Capacity of Helical Anchors in Soil. Soil Mechanics and Foundation Engineering **2**, 1019–1022 (1989)
3. Meyerhof, G.G.: The Ultimate Bearing Capacity of Foundations. Geotechnique **2**(4), 301–332 (1951)
4. Mooney, J.S., Adamczak Jr., S., Clemence, S.P.: Uplift capacity of helix anchors in clay and silt. In: Uplift Behavior of Anchor Foundations in Soil, pp. 48–72. ASCE, Detroit (1985)
5. Peck, R.B., Hanson, W.E., Thornburn, T.H.: Foundation Engineering. John Wiley & Sons, Inc., New York (1974)
6. Geem, Z.W., Kim, J.H.: Loganathan. G.V.: A new heuristic optimization algorithm: harmony search. Simulation **76**(2), 60–68 (2001)
7. ICC-Evaluation Services: AC358 Acceptance Criteria for Helical pile Foundations and Devices (2007). http://www.icc-es.org
8. Ghaly, A.M., Hanna, A.M.: Ultimate pullout resistance of single vertical anchors. Canadian Geotechnical **31**(5), 661–672 (1994)
9. Darin Willis, P.E.: How to Design Helical Piles per the 2009 International Building Code (2010). http://www.foundationrepairnetwork.com/design-helical-piles.pdf

A Preliminary Study for Dynamic Construction Site Layout Planning Using Harmony Search Algorithm

Dongmin Lee, Hyunsu Lim, Myungdo Lee, Hunhee Cho and Kyung-In Kang

Abstract Construction site layout planning is a dynamic multi-objective optimization problem since there are various temporary facilities (TFs) employed in the different construction phase. This paper proposes the use of harmony search algorithm (HSA) to solve the problem that assigning TFs to inside of the building. The suggested algorithm shows a rapid convergence to an optimal solution in a short time. In addition, comparative analysis with Genetic Algorithm (GA) is conducted to prove the efficiency of the proposed algorithm quantitatively.

Keywords Site layout planning · Optimization · Harmony search

1 Introduction

Site layout planning is an important task that involves identifying the temporary facilities(TFs) needed to support construction operations, determining their size and shape, and appropriately positioning them within the limited construction space[1]. Such TFs include temporary restaurant, site offices, storage yard, formwork storage yard, storeroom, labor residence, restrooms, utility control room and equipment(e.g., cranes). The site layout problem can be formulated as a assigning of facilities to suitable position over the course of a construction project.

D. Lee · M. Lee · H. Cho(✉) · K.-I. Kang
School of Civil, Environmental and Architectural Engineering, Korea University,
Seoul 136-713, South Korea
e-mail: {ldm1230,iroze00,hhcho,kikang}@korea.ac.kr

H. Lim
Research and Development Center, Yunwoo Technology Co. Ltd.,
Seoul 135-814, South Korea
e-mail: md.lee@yunwoo.co.kr

© Springer-Verlag Berlin Heidelberg 2016
J.H. Kim and Z.W. Geem (eds.), *Harmony Search Algorithm*,
Advances in Intelligent Systems and Computing 382,
DOI: 10.1007/978-3-662-47926-1_33

Furthermore, the problems associated with the planning of a construction site layout with the consideration of changing site facilities and site space in different time intervals are termed as dynamic site layout problem[2]. In a dynamic site layout problem, finding the optimal time when the temporary facilities are installed and dismantled is a significant factor to improve construction productivity.

On the other hands, according to the recent increase of high-rise building project in the downtown area where the space is very limited for construction, inside space of the building has become a possible allocation area for TFs. This study suggests a new method to solve the optimization problem with harmony search algorithm(HSA) which is one of the most powerful optimization algorithm [3].

Literature reviews show that many models have already been developed using various methodologies such as Genetic algorithm(GA), Ant colony optimization(ACO), Artificial intelligence(AI), computer-aided design(CAD). However, most of the previous researches more focus on static conditions. In fact, the main challenge in developing optimized layouts is in reflecting the dynamic nature of the site over the course of a construction project. Construction activities change as the project progresses, and accordingly, the number and nature of associated objects are subject to change as well. Several TFs enter the site at different times, occupy space on the site for different periods of time, and leave the site when they are no longer required. Furthermore, previously conducted researches.[1, 4-10] only consider horizontal space when generating dynamic layouts even though there is not enough space for the planning in a downtown construction site.

This paper presents an innovative approach based on allowable principles, for the first time, considering the vertical layout planning. In fact, in practically lots of facilities are vertically arranged already on construction site. For example, office is located on 15th floor, storeroom is located on 20th floor, utility control systems are installed every twenty floor in a tall building project in Seoul. This paper recommends an optimization model for vertical layout planning of TFs especially proper for tall building construction site where the space is not enough. At the beginning of the construction, the TFs are allocated or installed on the floor level, and as the buildings are higher, TFs will dynamically be moved to inside of the building. The purpose of the model is to find which floor is best location for each TFs and, when is the best time for movement.

2   Dynamic Construction Site Layout Planning Model

Construction productivity is one of the significant interests in a project. The productivity is related with construction time, cost, and they are all related with site layout plan. The more efficient layout planning, the higher construction productivity according to reduce in working distances. Main movement of laborers are drawn in fig. 1.

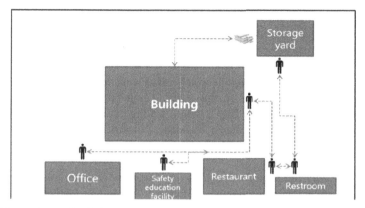

Fig. 1 Main moving path of laborers in construction site

To make a model for vertical layout model for TFs, decision variables are primarily defined. Facilities which can be or should be relocated, and the characteristics of usage patterns for those facilities were identified. According to the facilities, their main roles are different, consequently objective functions are also different each other. Also, even a safety factor is one of the most critical consideration factor when deciding TFs layout planning, by arranging TFs inside of the building, it is no more required to consider the safety factor between facilities since there is no interfere area or hazardous task between them. Therefore, only by minimizing the working distance, construction productivity could be increased until Pareto optimal point.

There are lots of TFs in the construction site and all of them have an intimate relation each other. Furthermore, working distance should be calculated based on the laborer's traffic line or materials transference line, but this paper only takes into account of traffic line of laborers as a preliminary study. Also, the horizontal movement of laborers are not included in the model since it has nothing to do with layout plan. In short, several assumptions are needed in this paper, and they are shown in below.

1. *Working distance is calculated based on the worker's traffic distance.*
2. *Movement of workers can be formulated as an objective function*
3. *Horizontal working distance is not considered when deciding the location of TFs.*

In the proposed model, its dynamic nature originated from the fact that the needed TFs inside of building change as the schedule. To determine the needed TFs in a specific time duration(between any two different tasks, it may could be a cycle time for floor), a three steps approach is used: (1) necessary TFs must be identified and their size should be decided first; (2) a schedule for the construction tasks should be confirmed; (3) each tasks requirements of the TFs are defined, similar to the requirements of labor, equipment etc.

STEP 1. Necessary TFs which need vertical arrangement in the construction site
There are a lot of TFs, but in this study, Temporary Restaurant, Storage Yard, Formwork Storage Yard, Site Office, Lifting Yard, Cement/Sand/Aggregate Storage Yard, Store Room, Labor Residence, Electrical Water and Utility Control room were considered when calculating moving distances.

STEP 2. Analyzing progress schedule and the number of laborers each floor
Progress schedule and the number of laborers should be exactly figured out since it is directly related with working distances.

STEP 3. Modeling of traffic line of laborers
Vertical moving path was mathematically analyzed to calculate quantitatively, the fig. 2 shows a specific case when a temporary restaurant located in 1F and after additional installation inF_k.

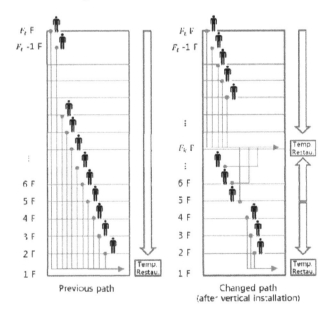

Fig. 2 Traffic line of laborers according to the position of restaurant

STEP 4. Formulate objective function for vertical distance(restaurant)

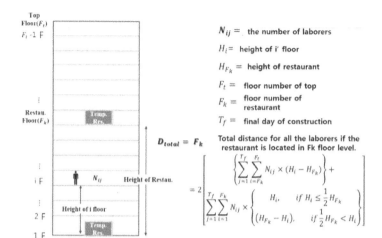

$N_{ij} = $ the number of laborers

$H_i = $ height of i' floor

$H_{F_k} = $ height of restaurant

$F_t = $ floor number of top

$F_k = $ floor number of restaurant

$T_f = $ final day of construction

$D_{total} = F_k$ Total distance for all the laborers if the restaurant is located in Fk floor level.

$$= 2 \begin{bmatrix} \left\{ \sum_{j=1}^{T_f} \sum_{i=F_k}^{F_t} N_{ij} \times (H_i - H_{F_k}) \right\} + \\ \sum_{j=1}^{T_f} \sum_{i=1}^{F_k} N_{ij} \times \left\{ \begin{array}{ll} H_i, & if\ H_i \leq \frac{1}{2}H_{F_k} \\ (H_{F_k} - H_i), & if\ \frac{1}{2}H_{F_k} < H_i \end{array} \right\} \end{bmatrix}$$

Fig. 3 Vertical distance for the traffic line of workers

In this paper, there are 10 TFs, and they are arranged in a floor level at the beginning of construction stage, and as the buildings getting higher, several or all of TFs are transferred to the inside of building. The optimization process involves the following steps: (1) identifying the time intervals or needed characteristics for the TFs as discussed earlier; (2) identifying facilities' objective function related with worker's working distance; (3) optimizing the location of the selected list of facilities in process or tasks.

3   Harmony Search Algorithm for the Site Layout Problem

There are 10 decision variables(the types of TFs), and each variables' range is 1~50F(assume). Objective function is working distances. The process of placing TFs inside of building uses the HSA which is one of famous heuristic algorithms for optimization problem. Researchers have reported the robustness of HSA and their ability to solve several engineering and construction management problems[6]. The procedure The procedure of HS is shown in below.

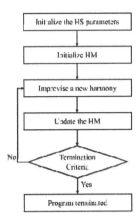

(1) Initialize harmony search algorithm parameters.
(2) Initialize harmony memory(HM).
(3) Generate a new harmony.
(4) Update harmony memory.
(5) Check for stopping criterion.

Fig. 4 Harmony search algorithm process

Step 1. Initialize Harmony Search Algorithm Parameters
Initializing to find solution of optimization problem. Size of decision variable, and their minimum, maximum value selection. Adjusting Harmony Memory Size(HMS), Harmony Memory Considering Rate(HMCR), Pitch Adjusting Rate(PAR), Iteration number. The HMCR value generally 0.7~0.9, and PAR is 0.2~0.5. The parameter value used in another researches are shown in below.

Table 1 HSA optimization parameter in other studies

Researcher	HMCR	PAR	HMS	NI
Mahdavi et al (2007)	0.95	0.35~0.45	4~7	3,000~300,000
Kang et al (2004)	0.85	0.3~0.45	10	50,000
Vasebi et al (2007)	0.8	0.5	6	30,000
Pan et al (2010)	0.9	0.3	5	50,000
Geem (2007)	0.95	0.05	30	30,000

Step 2. Initialize harmony memory(HM)
As shown in eq.(1). Harmony memory is in the form of the two-dimension matrix(HMS)$\times$ (N + 1), where each solution vector plus objective function value are kept inside.According to the random generation, the HM is initialized. Also, predetermined constraints conditions are check in this stage, and by giving penalty for them, bad solutions are deleted from the HM. At the end of this process, the HM contain optimal or near optimal solution vectors.

$$HM = \begin{bmatrix} x_1^1 & x_2^1 & \cdots & x_{N-1}^1 & x_N^1 & f(x^1) \\ x_1^2 & x_2^2 & \cdots & x_{N-1}^2 & x_N^2 & f(x^2) \\ \vdots & \cdots & \cdots & \cdots & \cdots & \vdots \\ x_1^{HMS} & x_2^{HMS} & \cdots & x_{N-1}^{HMS} & x_2^1 & f(x^1) \end{bmatrix} \qquad (1)$$

N: number of decision variable
HMS: harmony memory size

Step 3. Generate a new harmony

A new harmony memory vector $x' = (x_1', x_2', \cdots, x_N')$ is generated form the HM based on the HM consideration, random selection, pitch adjustment. For instance, the value of first decision variable (x_1') for the new vector can be chosen from any value in the HM range $(x_1^1 \sim x_1^{HMS})$. And the other decision variables also can be chosen with same rule. There is a possibility that the newly generated value can be chosen using HMCR rule, and that varies between 0 and 1 as follows:

$$x_i^{new} = \begin{cases} \{x_{i,1}, x_{i,2}, \cdots, x_{i,hms}\} \text{ with probability } HMCR \\ \{x_1, x_2, \cdots, x_N\} \text{ with probability } 1 - HMCR \end{cases} \qquad (2)$$

The HMCR sets the rate of choosing one value from the historic values stored in HM, and (1-HMCR) sets the rate of randomly choosing one value from the entire possible domain. For instance, a HMCR 0.9 indicates that the newly generated value is chosen from the HM with a 90% probability and from the entire domain with a 10% probability. Every component obtained from the memory consideration is examined to determine whether it should be pitch-adjusted.

$$x_i^{new} = \begin{cases} Yes \text{ with probability } HMCR \\ No \text{ with probability } 1 - HMCR \end{cases} \qquad (3)$$

The value of (1-PAR) sets the rate of doing nothing. If the pitch adjustment decision for x_i' is YES, x_i' is replaced as follows.

$$x_i' \leftarrow x_i' \pm rand(\ ) * bw \qquad (4)$$

Where
bw   is an arbitrary distance bandwidth
rand() is a random number between 0 and 1

Step 4: Update harmony memory

If newly generated harmony vector x^{new} is better than worst harmony x^{new} in HM(the evaluation is based on fitness function), then exclude x^{worst} from the HM. As a result, HM will be updated with better solutions as the iteration keep going.

Step 5: Check for stopping criterion
Repeat Step 3 and 4 until stopping criterion the criterion could be a certain time or number of iteration(NI). this paper used NI=1500

4 Numerical Experiments(Comparative Analysis Between GA)

4.1 Description of Case

The case site is 50th floor building, and 10 different facilities must be assigned to inside of the building. Numerical experiments are conducted to justify the proposed optimization model. Also, comparison between GA was conducted to show the efficiency of HSA. After applying HSA and GA at the same example case, the solution was quantitatively analyzed.

Table 2 Decision variables and its volume(assume)

Num	Facilities	Volume(m^2)
1	Temporary Restaurant	750
2	Storage Yard	800
3	Formwork Storage Yard	950
4	Site Office 1	1,000
5	Site Office 2	1,000
6	Lifting Yard	900
7	Cement, Sand, Aggregate Storage Yard	820
8	Store Room	1,300
9	Labor Residence	850
10	Electrical Water and other Utilities Control Room	980

Table 3 The height of each floor and their volumetric constraints(assume)

Floor Num	Height(m)	Volume(m^2)
1	6	1,500
2	12	1,300
3	18	1,200
⋮		
48	210	1,200
49	214	1,000
50	218	800

Table 4 The distance calculation formula of each facilities(assume)

Num	Facilities	Distance Calculation Formula
1	Temporary Restaurant	$\sum \sum_{i=1}^{F_k} N_i * (H_i - F_i) + \sum \sum_{i=f_k}^{50} N_i * (H_i \text{ or } H_k - H_i)$
2	Storage Yard	$\sum_{i=1}^{50} 2 * (i - F_i) * N_i$
3	Formwork Storage Yard	$(50\text{-}F_i)*N_{50}+(49\text{-}F_i)*N_{49}+(48\text{-}F_i)*N_{48}$
4	Site Office 1	$(50 - F_i) * N_t * 3.5 * 2 + F_i * N_n$
5	Site Office 2	$(50 - F_i) * N_t * 3.5 * 2 + F_i * N_n$
6	Lifting Yard	$\sum 2 * (F_i - F_{i-1})$
7	Cement, Sand, Aggregate Storage Yard	Better near N_{10}, N_2 farN_1
8	Store Room	$\sum 2 * (F_i)$
9	Labor Residence & Restroom	Distance between Restaurant and Labor Residence, Restroom
10	Electrical, Water and other Utilities Control Room	Summation of distance to another facilities position.

- *Constraints (Hard)*

1. Total installed volume of facilities in any floor must be less than the maximum space of the floor.

- *Constraints (Soft)*

1. Restaurant cannot be installed at the top floor.

2. Restrooms and Labor residence should be installed near one of offices.

3. Utility Control Systems should be installed lower than 25th

- *Optimization Parameter set*

Optimization Parameter	Value
HMCR	0.1, 0.5, 0.8
HMS	15, 50
PAR	0.1, 0.5, 0.8
Crossover probability	0.5, 0.8(Optimum)
Population Size	15, 50
Mutation probability	0.2(optimum)
Stopping criterion	500, 1000, 1500

4.2   Results

These tables are comparison between GA and HS.

Table 5 Optimum solution using HS

Facilities	X_1	X_2	X_3	X_4	X_5	X_6	X_7	X_8	X_9	X_{10}
Floor	22	25	11	1	14	11	18	2	10	9

Table 6 Optimum solution using GA

Facilities	X_1	X_2	X_3	X_4	X_5	X_6	X_7	X_8	X_9	X_{10}
Floor	23	26	25	1	49	27	50	2	24	3

In a 50th floor building, each facilities should be installed when starting construction of the i floor. For example, X_5 is Office, and to minimize the total distance of laborers, the office should be installed when 14th floor construction is finished (In HSA).

Table 7 Comparison between HMS and GA

HMS	HMCR	PAR	Iteration	Objective Function (HS)	Objective Function (GA)
15	0.8	0.8	500	53012 km	
			1000	52893 km	
			1500	52809 km	
	0.5	0.5	500	54152 km	
			1000	52800 km	
			1500	**52781 km**	
	0.1	0.1	500	55181 km	
			1000	54036 km	**53834 km** Minimum at (NI=1500, CR=0.5, M=0.2)
			1500	53383 km	
50	0.8	0.8	500	54387 km	**54110 km** Minimum at (NI=1500, CR=0.8, M=0.2)
			1000	53451 km	
			1500	54113 km	
	0.5	0.5	500	54099 km	
			1000	53451 km	
			1500	53069 km	
	0.1	0.1	500	53276 km	
			1000	52907 km	
			1500	**52675 km**	

In HMS=50, if the HMCR and PAR are higher, the objective function is increased

In HMS=15, if the HMCR and PAR are higher, the objective function is decreased

In the same HMS, HMCR, PAR condition, If the NI is increased, then the objective function continuously decreased. (Better result)

The best result, came from the GA, was **53,834km** and the best results came from HS were **52,781km(HMS=15, HMCR=0.5, PAR=0.5)** and **52,675 km(HMS=50, HMCR=0.1, PAR=0.1)** each. That means HS is better Algorithm than GA for this optimization problem within 1500 iteration.

5   Conclusion

The objective of this study is to provide a methodology for developing dynamic, vertical layout planning that are optimized over the duration of the project, while reflecting the actual changes on the site, in terms of object requirements and relationships between objects and their unique objective function which is made based on laborer's working traffic lines. The HSA was utilized to solve the problem. On the other hands, vertical layout can be conceptually viewed as a problem in which a multitude of objects, with different temporal and spatial dimensions, and different proximity relationships, compete over best locations in a given space.

A key feature of the model is that it considers the actual duration for which objects are required on the site in the process of optimization. This feature enables the reuse of the same space by different objects over the course of time. Furthermore, previous approaches are only focused on horizontal layout planning and they are not any meaning in the tall building projects since there is no space in horizontally. Another important aspect of the model is that it allows for a simultaneous search for the optimum location of all the objects that are required in different periods of the project. In other words, it allows all objects, regardless of the time and order in which they arrive on the site, to have an equal chance to compete over optimum locations for the specific time that they are required on the site.

5.1   Limitation and Further Study

The stopping criterion was set to be a 500, 1000, 1500 to investigate the efficiency of algorithm in a very short time(only 0.3sec searching time could be vulnerable to probability terms). Also, this model only considered 1day working distance, so it could not guarantee that is a global optimum, broader investigation is required.

Even though there are lots of TFs which should be considered when site layout planning, only 10 facilities are considered. Therefore, the decision variables should be more various. Also, parameter optimization was not conducted enough, more experiments and more realistic modeling is required.

Acknowledgement This research was supported by a grant(Code#09 R&D A01from High-Tech Urban Development Program funded by Ministry of Land, Infrastructure and Transport of Korean government.

References

1. Tommelein, I.D., Zouein, P.P.: Interactive Dynamic Layout Planning. J Conster. Eng. M. ASCE **119**, 266–287 (1993)
2. Ning, X., Lam, K.C., Lam, M.C.K.: Dynamic construction site layout planning using max-min ant system. Automation in Construction **19**, 55–65 (2010)
3. Geem, Z.W., Kim, J.H., Loganathan, G.: A new heuristic optimization algorithm: harmony search. Simulation **76**, 60–68 (2001)
4. Zhang, J., Liu, L., Coble, R.: Hybrid intelligence utilization for construction site layout. Automation in Construction **11**, 511–519 (2002)
5. Zhou, F., AbouRizk, S.M., Al-Battaineh, H.: Optimisation of construction site layout using a hybrid simulation-based system. Simulation Modelling Practice and Theory **17**, 348–363 (2009)
6. Zouein, P., Harmanani, H., Hajar, A.: Genetic algorithm for solving site layout problem with unequal-size and constrained facilities. J. Comput. Civil Eng. **16**, 143–151 (2002)
7. Xu, J., Li, Z.: Multi-objective dynamic construction site layout planning in fuzzy random environment. Automation in Construction **27**, 155–169 (2012)
8. Yeh, I.C.: Construction-site layout using annealed neural network. J. Comput. Civil Eng. 201–208 (1995)
9. Hegazy, T.: EvoSite: Evolution-Based Model for Site Layout Planning. J. Comput. Civil Eng. **13**, 198–206 (1999)
10. Zouein, P.P., Tommelein, I.D.: Dynamic Layout Planning Using a Hybrid Incremental Solution Method. J. Constr. Eng. M. ASCE **125**, 400–408 (1999)

Economic Optimization of Hydropower Storage Projects Using Alternative Thermal Powerplant Approach

Sina Raeisi, S. Jamshid Mousavi, Mahmoud Taleb Beidokhti,
Bentolhoda A. Rousta and Joong Hoon Kim

Abstract This paper presents a simulation-optimization model integrating particle swarm optimization (PSO) algorithm and sequential streamflow routing (SSR) method to maximize the net present value (NPV) of a hydropower storage development project. In the PSO-SSR model, the SSR method simulates the operation of reservoir and its powerplant on a monthly basis over long term for each set of controllable design and operational variables, which includes dam reservoir and powerplant capacities as well as reservoir rule curve parameters, being searched for by the PSO algorithm. To evaluate the project NPV for each set of the controllable variables, the "alternative thermal powerplant (ATP)" approach is employed to determine the benefit term of the project NPV. The PSO-SSR model has been used in the problem of optimal design and operation of Garsha hydropower development project in Iran. Results show that the model with a simple, hydropower standard operating policy results in an NPV comparable to another model optimizing operating policies.

Keywords Hydropower · Particle swarm optimization · Alternative thermal powerplant · Operating policy

S. Raeisi · S.J. Mousavi · B.A. Rousta(✉)
School of Civil and Environmental Engineering,
Amirkabir University of Technology (Tehran Polytechnic), Tehran, Iran
e-mail: raeisisina@yahoo.com, {jmosavi,roosta.hoda}@aut.ac.ir

M.T. Beidokhti
Iran Water and Power Resources Development Company (IWPC), Tehran, Iran
e-mail: beidokhty@gmail.com

J.H. Kim
School of Civil, Environmental and Architectural Engineering, Korea University,
Seoul 136-713, South Korea
e-mail: jaykim@korea.ac.kr

© Springer-Verlag Berlin Heidelberg 2016
J.H. Kim and Z.W. Geem (eds.), *Harmony Search Algorithm*,
Advances in Intelligent Systems and Computing 382,
DOI: 10.1007/978-3-662-47926-1_34

353

1   Introduction

System analysis approaches including simulation and optimization have been used
for decades in design and operation problems of hydropower systems. Optimization models as gradient-based and evolutionary algorithms have attracted lots of
attention as they have shown the ability to find promising solutions in such problems.

The optimization model of reservoir operation and hydropower problems taking
time reliability of water or energy supply into account could be a mixed integer
non-linear program (MINLP) that is difficult to solve by gradient-based optimization algorithms. In such problems, the random search of population-based evolutionary algorithms has shown some advantages. In this regard, genetic algorithms
[1], shuffled complex evolution [2], ant colony optimization [3] and honey-bee
mating optimization [4] have been used to optimize the operation of single and
multi-reservoir systems. Among the evolutionary algorithms, particle swarm optimization (PSO) has also attracted researchers' attention. Meraji et al. [5], Kumar
& Reddy [6,7] and Baltar & Fontane [8] have used PSO for reservoir systems
operation optimization. Mousavi & Shourian [9] have employed PSO in combination with the sequential streamflow routing (SSR) simulation model in the design
and operation optimization of Bakhtiari hydropower dam in Iran.

This study extends the Mousavi & Shourian's work [9] in terms of economic
evaluation of hydropower projects. Taking advantage of linking the PSO algorithm, as optimizer, to a hydropower reservoir simulation model based on the SSR
method, maximization of the net present value (NPV) of Garsha hydropower development project is considered. "Alternative thermal powerplant" (ATP) method
is then used in order to evaluate the benefit term of energy production in the
project's economic evaluation.

This paper is organized as follows: Model description and the methods used as
well as the case study are presented in section 2. The results of the developed optimization model and conclusions are then presented in sections 3 and 4.

2   Model Description

The amount of hydroelectricity production depends on the net water head on tur-
bines and the flow passing through them. In a hydropower design problem, reservoir's normal water level (N.W.L), the minimum operating level (M.O.L) and the
powerplant's production capacity (Pcap) are the main design unknowns that can
take any values in their predefined continuous interval. Hence, there are a lot of
combinations of these variables that each will result in a certain value of performance and economic efficiency. To deal with the problem of choosing the best
combination, we can take advantage of a systematic search or optimization algorithm. In this line, a reservoir operation simulation model is also needed by which
the variables of interest, which the system's performance or economic value depend on, are determined for any set of candidate design variables being searched

for by the optimization algorithm. We, therefore, integrate the PSO algorithm, a hydropower reservoir simulation model and the ATP economic evaluation technique to solve the problem of the optimal design of a hydropower storage development project.

2.1 PSO Algorithm

PSO proposed by Kennedy and Eberhart [10] is one of the meta-heuristic methods that is used to solve global optimization problems. By PSO, a problem is optimized by having a population of individuals (so-called particles), and moving these particles around in the search space. Each particle calculates the objective function of the optimization model based on its coordination in the search space. The particle chooses a direction using the current position and its best position in the previous iterations (pbest) and also the best position of the whole population or swarm (gbest). The particles move to new positions, and the objective function values of particles (solutions) are evaluated resulting in pbest and gbest in each iteration. This process is repeated until some stopping criteria are met.

In the PSO algorithm, each particle consists of three D-dimensioned vectors, which D is the dimension of the search space. For the ith particle, these vectors are:

i.   The current position of the particle, $x_i = (x_{i1}, x_{i2}, \ldots, x_{iD})$.

ii.   Velocity of the particle, $V_i = (V_{i1}, V_{i2}, \ldots, V_{iD})$.

iii. The best position that the particle has found so far (pbest),

$p_i = (p_{i1}, p_{i2}, \ldots, p_{iD})$

There is also another vector that shows the global best position identified in the entire population (gbest), $p_g = (p_{g1}, p_{g2}, \ldots, p_{gD})$.

Equation (1) updates the velocity and Equation (2) updates the position of each particle in the next iteration.

$$v_{id}^{n+1} = \chi\left(\omega v_{id}^n + c_1 r_1^n (p_{id}^n - x_{id}^n) + c_2 r_2^n (p_{gd}^n - x_{id}^n)\right) \tag{1}$$

$$x_{id}^{n+1} = x_{id}^n + v_{id}^{n+1} \tag{2}$$

where $d=1,2,\ldots,D$; $i=1,2,\ldots,N$; N is the swarm size; χ is the constriction coefficient; ω is the inertia weight; c_1 and c_2 are positive constant parameters called cognitive and social parameters, respectively; r_1 and r_2 are random numbers uniformly distributed in [0,1]; n is iteration number. The value of ω is changed in each iteration according to the Equation (3):

$$\omega = \omega_{max} - (\omega_{max} - \omega_{min}) \times \frac{it}{it\,max} \tag{3}$$

where $it\,max =$ the maximum number of iterations; $it =$ the current iteration number. Values of ω_{min} and ω_{max} are determined by trial and error.

Stretched PSO Algorithm

In order to escape from local optimum solutions in the PSO algorithm, Parsopoulos et al. [11] offered to equip PSO with "stretching" function. This means that as soon as a local maximum is found, the objective function's face will be transformed artificially by Equations (4) and (5). Assume that $\bar{x}$ is the position of local maximum which is found during a PSO generation, the two-stage transform functions are as follows:

$$G(x) = f(x) - \frac{\gamma_1}{2}\|\bar{x} - x\| \times (sgn(f(\bar{x}) - f(x)) + 1) \tag{4}$$

$$H(x) = G(x) - \frac{\gamma_2(sgn(f(\bar{x}) - f(x)) + 1)}{2tgh(\mu(G(\bar{x}) - G(x)))} \tag{5}$$

where γ_1, γ_2 and μ are arbitrary chosen positive constants. Equation (4) transform the objective function's face in a way that makes disappear all the local maxima whose objective function values are less than that of $\bar{x}$, and Equation (5) lowers the neighborhood of $\bar{x}$. It is worth mentioning that the latest equations don't affect the points with higher objective function values than that of $\bar{x}$, and the global optimum will be remained untouched.

2.2   SSR Method

Reservoir simulation model based on the SSR method is like a black box in which a given combination of hydropower system's design decision variables (N.W.L, M.O.L and Pcap) are its inputs, and the amount of generated energy (primary and secondary) and the reliability of meeting the energy yield are its outputs. The balance equation, energy production equation, equations related to geometry of the reservoir and upper and lower bounds on the variables are fulfilled in this simulation model. The iterative application of the SSR simulation model while changing Pcap in each iteration so that a certain level of reliability of meeting the energy is reached, is called reliability based simulation (RBS) model. Therefore, the difference between the SSR and RBS models is that Pcap in the RBS is associated with a predefined reliability level of the energy yield. More details on the RBS model may be found in Mousavi and Shourian [9].

2.3   Project's Economic Evaluation Using ATP Approach

ATP is an approach for estimating the hydropower plant's economic benefits. The concept of ATP is that if the hydropower plant was not built, a thermal powerplant would be constructed instead to supply an equivalent level of production in terms of both energy and power. So, the benefits of hydropower plant are equal to the costs of its alternative thermal powerplant. After calculating the ATP's costs, the cash-flow of the hydropower plant's benefits can be formed. On the other hand, the cash flow of the costs will be constructed using the cost estimation items regarding dam and power plant construction. Subsequently, the economic index of the project's net present value (NPV) is estimated. Figure 1 demonstrates the general steps of the ATP approach.

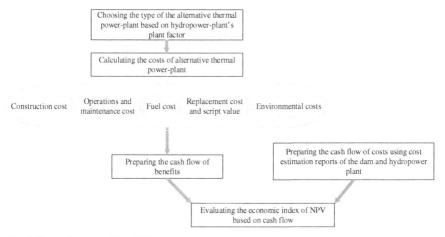

Fig. 1 General steps of the ATP approach

2.4 PSO-SSR Model

As it was mentioned before, in order to solve the optimization model, an integration of PSO with reservoir simulator and the ATP module is used. Figure 2 shows the connections between the three modules.

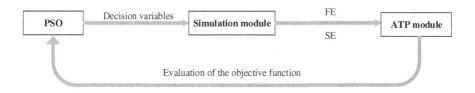

Fig. 2 Schematic flow diagram of the PSO-SSR model

One can see in the figure that first PSO generates random values for each decision variables in the search space for each particle. These values are used as inputs to the SSR or RBS simulation model. The outputs of the simulation module are firm and secondary energy values, i.e FE and SE, respectively. The ATP module uses these energy values to evaluate the project's NPV for each particle and pbest and gbest of the first generation will be defined. Afterwards, each particle moves through the search space using Equations (1) and (2) to obtain new values for decision variables, and the process is continued until the stopping criteria are met.

In this respect, two models were evaluated. The difference between these models is originated from whether the model optimizes the design variables or both design and operational variables. This means that the number of decision variables and their operating policy are different.

Table 1 Characteristics of the two models developed

Model	PSO variable	No. of PSO variables	Operating policy
1	N.W.L M.O.L	2	HSOP
2	N.W.L M.O.L Production capacity Monthly min and max of operating storage	27	parametric Seasonal HSOP

In model 1, by using the hydropower standard operating policy (HSOP) as the predefined operating policy, the design variables are optimized. In this model, N.W.L and M.O.L are considered as decision variables and the RBS model as the simulator. Therefore, by changing the Pcap iteratively in the RBS model, the reliability constraint on the energy yield is satisfied. Figure 3 shows the flow diagram of model 1 whose objective function is to maximize the NPV evaluated by the ATP approach.

In model 2, the PSO and SSR models interact to optimize design and operational variables simultaneously. The SSR model requires that Pcap has been defined in advance. Therefore, besides N.W.L and M.O.L, the production capacity of hydropower plant is considered as the PSO decision variable. Since the reliability constraint on energy yield is not met in the SSR model, the PSO objective function will take care of that constraint using a penalty approach as follows:

$$O.F = Max[\, NPV - (\,|Re\, l - T\, arg\, Re\, l|\,) \times P\,] \qquad (6)$$

where $O.F$ is the model 2's objective function, NPV is the net present value calculated by ATP module, Rel is the estimated reliability of meeting energy yield, $TargRel$ is the desired reliability level equal to 90%, and P is a penalty factor that is tuned by trial and error.

The operating policy of model 2 is parametric, seasonal HSOP whose structure is similar to HSOP, but it works on a monthly basis. Parametric seasonal HSOP determines an upper (S_{maxop}) and a lower (S_{minop}) bound of reservoir storage levels for each month of the year, whereas HSOP uses constant minimum and maximum storage levels for all the months. This bounds that play the role of rule curves are optimized in model 2. In other words, 24 operational variables (12 minimum and 12 maximum storage levels) are added to the PSO decision variables resulting in a 27-dimensional search space.

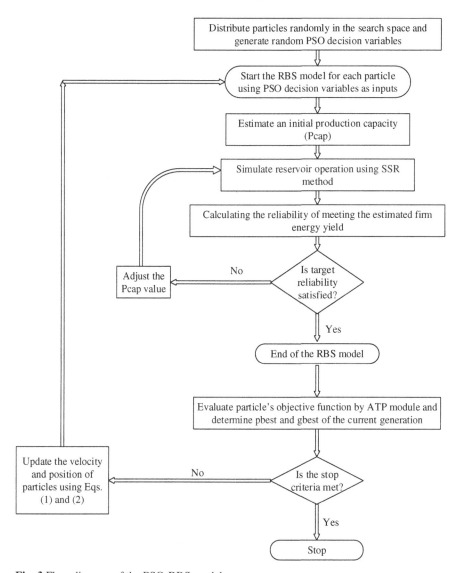

Fig. 3 Flow diagram of the PSO-RBS model

2.5   *Case Study*

The case study of this paper is the Garsha Dam, which is one of the water resources development projects on Seymareh River in Karkheh River Basin located in the west of Iran (Figure 4). Table 2 presents the upper and lower bounds of the decision variables of the project.

Table 2 Upper and lower bounds of decision variables [12]

Decision Variable	Minimum	Maximum
Normal Water Level (N.W.L)	1215	1244
Minimum operating level (M.O.L)	1170	N.W.L
Powerplant's production capacity	20	500

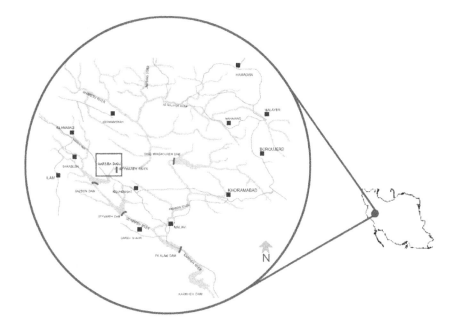

Fig. 4 Geographical location of the case study, Garsha Dam [12]

The PSO parameter values as selected for each model by trial and error are presented in Table 3.

Table 3 PSO parameters values

Model	Swarm size	c_1	c_2	w_{min}	w_{max}
1	15	2.1	1.8	0.4	0.9
2	50	2.2	2	0.4	0.9

3   Results

Table 4 presents optimal values of the objective function, design variables and annual firm and secondary energies obtained by the two models.

Table 4 Optimal values of the objective function, design variables and annual energy production

Item	Model 1	Model 2
NPV (10^6 Rials*)	3134182	3214733
N.W.L (masl**)	1244	1244
M.O.L (masl)	1199.1	1170
Pcap (MW)	245.05	248.67
Annual firm energy (GWh)	354.83	358.8
Annual secondary energy (GWh)	186.63	188.87

*Iran's monetary unit
**meters above sea level

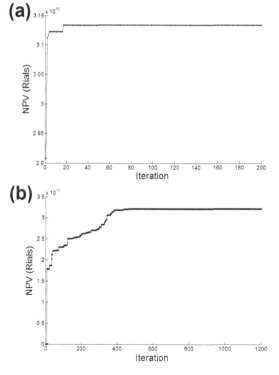

Fig. 5 Convergence of the best objective (fitness) function value (gbest) over PSO generations: (a) model 1, (b) model 2

One can see from the table that the results of the two models in terms of their objective function are comparable. Nevertheless, it is expectable that model 2 with parametric seasonal HSOP performs better than model 1 because it benefits from a larger degree of freedom in optimizing the controllable operational variables. The difference (2.57%) is, however, insignificant. This means that model 1 with a simple operating policy (HSOP) performs well enough. Figure 6 shows the variation of the gbest's objective function value against PSO iterations.

4   Conclusions

We presented an application of the PSO algorithm in the problem of economic optimization of design and operation of Garsha hydropower development project in Iran while taking the reliability of energy production into account. The "alternative thermal powerplant (ATP)" approach was employed to determine the benefit term of the project's net present value. The problem was solved by integrating PSO, the sequential streamflow routing (SSR) method simulating the hydropower systems' operation and the ATP approach. The PSO-SSR optimization-simulation model was used to optimize the Garsha systems characteristics in two cases of design and design-operation optimization of the system's components. Results indicated that the developed model is able to locate good solutions to the difficult to solve nonconvex, nonlinear optimization problem. Moreover, the simple, hydropower standard operating policy (HSOP) was found to be a promising policy that performs well enough compared to optimized seasonal parametric rule curves. However, the amount of secondary energy production can increases by further optimizing the operation policy because the ATP approach gives higher economic values to the firm energy than the secondary energy.

References

1. Oliveira, R., Loucks, D.P.: Operating Rules for Multireservoir Systems. Water Resour. Res. 33(4), 839–852 (1997)
2. Le Ngo, L., Madsen, H., Rosbjerg, D.: Simulation and Optimisation modeling approach for operation of the Hoa Binh reservoir. Vietnam. J. Hydro. 336, 269–281 (2007)
3. Jalali, M.R., Afshar, A., Marino, M.A.: Reservoir Operation by Ant Colony Optimization Algorithms. Iran J. Sci. Technol. Trans. B-Eng. 30(B1), 107–117 (2006)
4. Haddad, O.B., Afshar, A., Marino, M.A.: Honey-Bees Mating Optimization (HBMO) Algorithm: a New Heuristic Approach for Water Resources Optimization. Water Resour. Manag. 20(5), 661–680 (2006)
5. Meraji, S.H., Afshar, M.H., Afshar, A.: Reservoir operation by particle swarm optimization algorithm. In: Proceedings of the 7th International Conference of Civil Engineering (Icce7th), Tehran, Iran, vol. 9, pp. 8–10 (2005)
6. Kumar, D.N., Reddy, M.J.: Ant Colony Optimization for Multi-Purpose Reservoir Operation. Water Resour. Manag. 20(6), 879–898 (2006)

7. Kumar, D.N., Reddy, M.J.: Multipurpose Reservoir Operation Using Particle Swarm Optimization. J. Water Resour. Plan. Manage. ASCE **133**(3), 192–201 (2007)
8. Baltar, A.M., Fontane, D.G.: A multiobjective particle swarm optimization model for reservoir operations and planning. In: Joint International Conference on Computing and Decision Making in Civil and Building Engineering, Montréal, Canada (2006)
9. Mousavi, S.J., Shourian, M.: Capacity optimization of hydropower storage projects using particle swarm optimization algorithm. J. Hydroinform. **12**(3), 275–291 (2010)
10. Eberhart, R.C., Kennedy, J.: A new optimizer using particle swarm theory. In: Proceeding of the Sixth Symposium on Micro Machine and Human Science IEEE Service Centre, Piscataway, NJ, pp. 39–43 (1995)
11. Parsopoulos, K.E., Plagianakos, V.P., Magoulas, G.D., Vrahatis, M.N.: Stretching technique for obtaining global minimizers through particle swarm optimization. In: Proc. Particle Swarm Optimization Workshop, Indianapolis, IN, USA, pp. 22–29 (2001)
12. Iran Water and Power Resources Development Company: Garsha Economic Report, Tehran, Iran (2009)

Simulation Optimization for Optimal Sizing of Water Transfer Systems

Nasrin Rafiee Anzab, S. Jamshid Mousavi,
Bentolhoda A. Rousta and Joong Hoon Kim

Abstract Water transfer development projects (WTDPs) could be considered in arid and semi-arid areas in response to uneven distribution of available water resources over space. This paper presents a simulation-optimization model by linking Water Evaluation and Planning System (WEAP) to particle swarm optimization (PSO) algorithm for optimal design and operation of the Karoon-to- Zohreh Basin WTDP in Iran. PSO searches for optimal values of design and operation variables including capacities of water storage and transfer components as well as priority numbers of reservoirs target storage levels, respectively; And WAEP evaluates the system operation for any combinations of the design and operation variables. The results indicate that the water transfer project under consideration can supply water for the development of Dehdash and Choram Cropland (DCCL) in an undeveloped area located in Kohkiloyeh Province.

Keywords Water transfer systems · Simulation-optimization · WEAP · PSO

1 Introduction

Unevenly distribution of freshwater over space and time, along with rapid population growth and its consequent increase in per capita water consumption has led to

N.R. Anzab · S.J. Mousavi · B.A. Rousta(✉)
School of Civil and Environmental Engineering,
Amirkabir University of Technology, Tehran, Iran
e-mail: nasrin_rafiee66@yahoo.com, {jmosavi,roosta.hoda}@aut.ac.ir

J.H. Kim
School of Civil, Environmental and Architectural Engineering,
Korea University, Seoul, South Korea
e-mail: jaykim@korea.ac.kr

© Springer-Verlag Berlin Heidelberg 2016
J.H. Kim and Z.W. Geem (eds.), *Harmony Search Algorithm*,
Advances in Intelligent Systems and Computing 382,
DOI: 10.1007/978-3-662-47926-1_35

an inconsistency between water supplies and demands. Managing water resources requires planning, development, distribution, and optimal consumption of water resources. Such management would be recognized as a set of technical, institutional, and legal measures, the purpose of which is to balance the water supply and demand [1].

Iran is located in an arid and semi-arid area where water supply and demand is highly uneven over the space. Water availability is subject to considerable variations in different basins. While few basins are rich in water resources, the others suffer from significant water shortages. Hence, Water Transfer Development Projects (WTDPs) could be considered in order to alleviate spatial imbalance between water supplies and demands. A WTDP or an interbasin water transfer project is defined as transferring water from a distinct catchment or river reach to another one [2]. System analysis techniques including simulation and optimization models can be used to help investigate the technical aspects of water transfer projects ([3-7]). Some studies assess water transfer projects from social and environmental prospect ([8]), and some other studies incorporate both the socio-environmental aspects of water transfers and the technical ones ([9-11]). In this study, a mixed integer non-linear programming (MINLP) model is developed to determine the design parameters of the WTDP from Bashar Basin (one of Khersan River's tributaries flowing in Karoon Basin) to Dehdasht and Choram Cropland located in Zohreh Basin in Iran. Since it is not easy to solve the model by using a gradient-based optimization algorithm, we have made an attempt to solve it by a simulation-optimization technique through the linkage of the well-known Water Evaluation and Planning System (WEAP) water allocation simulation model and the PSO algorithm.

The remainder of this paper is organized as follows: A description of the study area is given in section 2. Section 3 describes the PSO-WEAP model and its application to the problem under study. The results and conclusions are then discussed in sections 4 and 5.

2   Study Area and Problem Definition

The target of this study is Dehdasht and Choram Cropland (DCCL), which is located in Kohgilouye and Boyerahmad Province, Iran, as one of the most potential land and soil resources. In spite of the fact that DCCL is located near the Maroon and Kheirabad Rivers, farmers have not been able to divert water from these surface resources to DCCL as the land is on a relatively high terrain. As a result, farming encounters water shortages and is mostly rain fed. From groundwater perspective, the land also has limited resources [12]. Therefore, in order to supply DCCL demands, a water transfer project has been proposed to transfer water from Bashar (one of Karoon's sub-basins) to Zohreh Basin encompassing the DCCL. The project is designed to pump water from Kabkian reservoir to Sepidar diversion dam (Fig. 1.) and flowing water by gravity through a tunnel to Shahbahram reservoir in Zohreh Basin (Fig. 1.). Kabkian and Shahbahram are designed to serve as regulating reservoirs. The water regulated by Shahbahram is intended to supply agricultural water demand of DCCL. In this system, the existing Kosar reservoir with, respectively, normal and minimum storage levels of 492.8 and 74.17 MCM provides water to municipalities along the coastline of the Persian Gulf and Kohgilouye and Boyerahmad, Khuzestan, Boushehr, Fars, and Hormozgan provinces as well as the Lishtar croplands of Gachsaran [12]. There are also other environmental, industrial and agricultural demands in the system which are supposed to be supplied by available surface water resources.

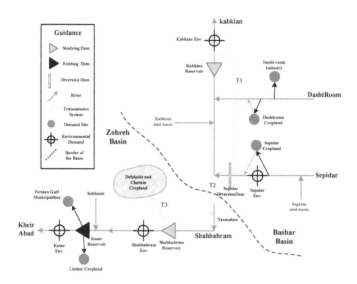

Fig. 1 Schematic representation of the system and the Bashar-to-Zoreh Basin water transfer project [12]

3   Model Description

The principal objective of this study is to develop a simulation-optimization model in order to optimize design and operation of Bashar-to-Zohreh WTDP. The motivation behind developing this model is the incapability of classical optimization methods for solving mixed integer nonlinear programs including binary variables controlling temporal reliability of water supply. Therefore, WEAP as the simulation model is linked with PSO as the optimization algorithm to construct such a simulation-based optimization tool.

3.1   WEAP Model

Developed by Stockholm Environment Institute (SEI) in 1988, WEAP is a physically based model that incorporates water supply projects and demand-side issues into a practical tool in order to assist water resources planners [13]. WEAP primarily operates based on the water balance accounting principle and can be applied to simulate either a single small sub-basin or a large scale complex basin as well as agricultural and municipal systems [13]. Even though WEAP solves several linear programs to determine the optimal allocations at a single time step, it is not capable of performing multiple time step optimization to determine the optimal decision variables. However, it can be linked to other process-based models using programming languages such as VB.net.

WEAP utilizes standard linear programs solved iteratively to calculate water allocation at each time step. The objective function of the LP is to maximize supplies to demand sites subject to supply preferences, mass balance and other constraints [13].

3.2   PSO Algorithm

First proposed by Kennedy and Eberhart [14], PSO is a stochastic evolutionary algorithm that adheres to the social behavior of bird flocks [15] to search through multi-dimensional decision spaces. Flexible operators, absence of gradients, and easily found solutions to mixed integer and combinatorial problems are some of outstanding characteristics of PSO. Providing the search space is D-dimensional, the i-th particle of the swarm is identified by the D-dimensional vector $x_i = (x_{i1}, x_{i2}, ..., x_{iD})$; the best former position of this particle is identified by $p_i = (p_{i1}, p_{i2}, ..., p_{iD})$; the particle's velocity change is identified by $V_i = (V_{i1}, V_{i2}, ..., V_{iD})$, and the swarm's best particle is denoted by g. Particles of the swarm will move according to the following equations:

$$v_{id}^{n+1} = \chi\left(\omega v_{id}^{n} + c_1 r_1^{n}(p_{id}^{n} - x_{id}^{n}) + c_2 r_2^{n}(p_{gd}^{n} - x_{id}^{n})\right) \tag{1}$$

$$x_{id}^{n+1} = x_{id}^{n} + v_{id}^{n+1} \tag{2}$$

where $d = 1, 2, ..., D$; $i = 1, 2, ..., N$; N =size of population; ω =weight of inertia; n = number of iterations; c_1, c_2 = two positive constants called cognitive and social coefficient; χ = constriction coefficient; r_1, r_2 = random values uniformly distributed in the range [0 1] [15]. In each iteration, ω is changed according to the equation (3):

$$\omega = \omega_{max} - (\omega_{max} - \omega_{min}) \times \frac{Iter}{Iter\,max} \tag{3}$$

where $Iter$ max = total number of iterations; $Iter$ = the current iteration number. Values of ω_{min} and ω_{max} are determined by trial and error. Like most evolutionary optimization techniques, PSO faces the problem of convergence to the local minima. Function Stretching ([15],[16]), a technique for escaping from the local minima, is used in this study to alleviate PSO's problem of local minima. This modified version is called SPSO. In the SPSO, as soon as a local minimum has been detected, a two-stage transformation will be performed on the objective function. In the first stage where G_i is produced, the original objective function (FC_i) is elevated using the equation (4). In the second stage, equation (5) is applied to stretch FC_i neighborhood upward.

$$G_i = FC_i + \gamma_1 \|i - \bar{i}\| \left(sign\,(FC_i - \overline{FC}_i) + 1 \right) \tag{4}$$

$$H_i = G_i + \gamma_2 \frac{sign\,(FC_i - \overline{FC}_i)}{\tanh\,(\mu\,(G_i - \overline{G}_i))} \tag{5}$$

where i = one of the local minima; FC_i = the objective function corresponding to i th particle; G_i = the first transformation function; H_i = the second transformation function; γ_1, γ_2, μ = positive constant values. The local minima located below i are not altered through aforesaid stages; therefore, the location of the global minimum remains unchanged.

3.3 PSO-WEAP Model

Given the WEAP's ability of linking with other programs, one can input the desired values of decision variables into the WEAP model in each iteration. Through coding the PSO algorithm in MATLAB environment and calling WEAP solver in the PSO algorithm, one can attempt to generate the values of variables of interest by the PSO algorithm and input them into the WEAP model. Once WEAP is performed, the objective function of the PSO-WEAP model is evaluated. The objective function is to minimize the design capacity of the proposed reservoirs and water transfer systems while maximizing the DCCL cultivable area and temporal reliability of DCCL supplies. Hence, given the components of the schematic plan illustrated in Fig. 1, the objective function (O.F) is formulated as follows:

$$\begin{aligned}
O.F = Min\,(cap_{kab} + cap_{sb} + T_1\max + T_2\max + (30000 - A_{\max})^2 \times \alpha \\
+ ((100 - reliability_{choram})^2 \times \beta)
\end{aligned} \tag{6}$$

where cap_{kab} = Kabkian reservoir storage capacity; cap_{sb} = Shahbahram reservoir storage capacity; $T_1\max$ = capacity of water transfer system from Kabkian reservoir to Sepidar diversion dam; $T_2\max$ = capacity of water transfer system from Bashar to Zohreh basin and $A_{\max}$ = maximum cultivable area of DCCL. α and β are coefficients for adjusting values of the two last terms to values of the other terms of the O.F and are determined by trial and error. $reliability_{choram}$ = reliability of DCCL supply. Temporal reliability is defined as the frequency of the periods during which DCCL is fully supplied when divided by the entire simulation periods. Temporal reliability is defined using binary variables as follows:

$$T3_t \geq Z_t \times D_t^{choram} \qquad \forall t = 1,2,\ldots,T \tag{7}$$

$$reliability_{choram} = \frac{\sum_{t=1}^{T} Z_t}{T} \tag{8}$$

where $T3_t$ = amount of water transferred to DCCL; D_t^{choram} = irrigation water demanded by DCCL; T = all periods of simulation (648 months) and Z_t =binary variable defined as:

$$Z_t = \begin{cases} 1 & \text{if DCCL is fully supplied in time period t} \\ 0 & \text{Otherwise} \end{cases} \tag{9}$$

Since we cannot call and run WEAP directly from MATLAB, , we used Excel as an interface between WEAP and MATLAB; that is, the generated-by-PSO values of variables are saved into the Excel; then, these values are called from Excel, and WEAP is executed; finally, the results are saved into the Excel and are called by PSO in MATLAB. This procedure is repeated up to the maximum number of iterations defined aiming at minimizing the objective function. The flow diagram of the PSO-WEAP model is presented in Fig. 2.

Table 1 Upper and lower bounds of decision variables

Decision Variable	Maximum	Minimum
Storage capacity of Kabkian reservoir (MCM)	201.89	14.7
Storage capacity of Shahbahram reservoir (MCM)	126.03	35
capacity of water transfer system from Kabkian reservoir to Sepidar diversion dam (cms)	9	0
capacity of water transfer system from Bashar to Zohreh Basin (cms)	18	0
DCCL's maximum cultivable area (ha)	30000	5000

Table 2 PSO parameters values

parameter	swarm	iterations	χ	ω_{min}	ω_{max}	c_1	c_2
value	20	200	1	0.4	0.9	1.5	1.5

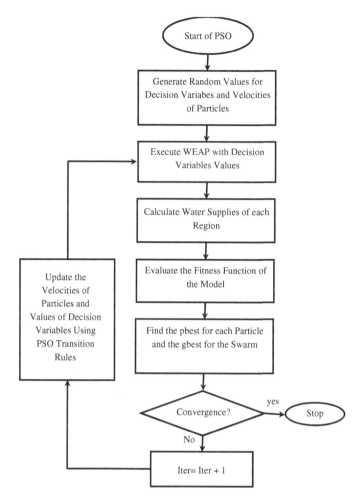

Fig. 2 The flow diagram of the PSO-WEAP model

The decision variables of the PSO algorithm are storage capacities of Kabkian and Shahbahram reservoirs, capacities of Kabkian-to-Sepidar and Bashar-to-Zohreh water transfer systems and the DCCL's maximum cultivable area. The upper and lower bounds considered for these variables are reported in Table 1 [12]. It is worth noting that in WEAP, the ordinal priorities of demands to be met were considered as environmental, municipal, industrial, and finally agricultural. Moreover, the simulation was done for a period of 54 years from 1935 to 1988. In the developed PSO-WEAP model, PSO feeds the decision variables into the inner linear programs of WEAP. Afterwards, the resulting water allocations are returned back from WEAP to the PSO where the objective function for each set of generated decision variables is evaluated. Using the PSO algorithm's evolutionary transition rules, this procedure ([1]) is continued between PSO and WEAP until the PSO objective function converges to a minimum value. The PSO parameter values are reported in Table 2.

In the first scenario, the PSO-WEAP model is applied based on its primary assumptions. The second scenario introduces other assumptions into the basic model where in addition to the capacity of storage elements, the priority numbers of the reservoirs target storage volumes are considered as operational decision variables. In other words, the priority numbers of the reservoirs target storage volumes which are fed by the PSO algorithm will be optimized by the PSO-WEAP simulation-optimization model. If the priority number of a reservoir target level becomes lower than that of the downstream demand, water will first be stored in the reservoir and the excess water will be released to the downstream. Conversely, when the number becomes higher than that of the downstream demand, water will first be released to meet the downstream demand after which the excess water will be stored in the reservoir.

4   Results and Discussions

Although reliability index is evaluated just for DCCL demand site, there are 648 integer variables in the model resulting in a relatively large scale MINLP model whose solution is difficult to obtain by classical optimization methods. Table 3

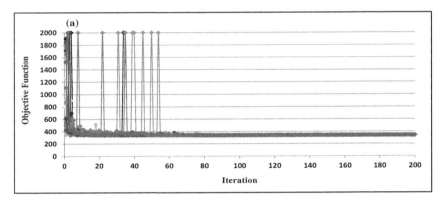

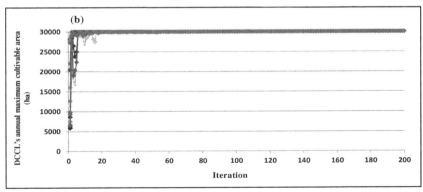

Fig. 3 Convergence trend of the PSO particles for (a) the objective function value and (b) the maximum cultivable area of DCCL

reports optimal values of decision variables of Bashar-to- Zohreh Basin WTDP obtained by the PSO-WEAP model for the first scenario. It is seen from Table 3 that the water transfer project under consideration can supply water for the development of 30,000 ha of the DCCL. Convergence curves of the particles' O.F value and the maximum cultivable area of DCCL over subsequent iterations is illustrated in Fig. 3.

Note that the O.F value goes up at some iterations in Fig. 3a which is due to function stretching; otherwise such fluctuations would not have been happened.

Table 3 PSO-WEAP model results, first scenario

Parameter	Basic scenario	Second scenario
Storage capacity of Kabkian (MCM)	57.3	53.93
Storage capacity of Shahbahram (MCM)	126.03	126.03
Capacity of Kabkian-to-Sepidar water transfer system (cms)	9	7.15
Capacity of Bashar-to-Zohreh water transfer system (cms)	9.17	7.9
DCCL's maximum cultivable area (ha)	30000	30000
Temporal reliability of DCCL demand (%)	73.92	73.46
Best objective function value of PSO-WEAP	337.54	335.92

Table 4 Volumetric and temporal reliability of water supplies to different demand sites

Demand site	Annual demand (MCM)	Volumetric reliability index (%)		Temporal reliability index (%)	
		Basic scenario	Second scenario	Basic scenario	Second scenario
Dashtroom croplands	6.95	98.2	98.2	96.76	96.75
Dashtroom industry	0.28	99.7	99.7	99.7	99.7
Sepidar lands	9.72	91.5	91.47	89.66	89.66
Dehdasht and Choram croplands (D_{DCCL})	variable	76	75.84	73.92	73.45
Lishtar lands	70.1	85.7	85.53	88.9	88.73
Persian Gulf municipals	210	95.27	95.2	90.12	89.96

In the second scenario, the increase of the number of decision variables has resulted in a larger number of function evaluations before the model convergence and therefore higher execution time of the PSO-WEAP model. It is, however, seen that the best O.F value obtained for the second scenario is almost the same as that for the basic model. This shows that optimizing the operational variables has not had a significant effect on the improvement of the model performance compared to capacity optimization of the project's storage and water transfer components. Table 4 presents the model results in terms of reliability of meeting different types of demands represented by both volumetric and temporal reliability indices.

5   Conclusions

This study was about formulating and solving an optimization model for optimally sizing the components of Bashar-to-Zohre water transfer system supplying water to Dehdasht and Choram Cropland (DCCL) area in Zohreh Basin located in Kohgilouye and Boyerahmad undeveloped Province, Iran. Considering the temporal reliability of meeting water demands, the formulation of the model was a mixed integer non-linear program, being difficult to solve by gradient-based optimization approaches. We, therefore, developed a simulation-optimization approach by linking the PSO algorithm to the well-known river basin water allocation model of WEAP. The PSO-WEAP model results indicated that the project can supply water to develop 30,000 ha of DCCL area for agricultural development. It is, however, of the utmost importance to consider socio-economic aspects of the proposed development plan, focusing on the target areas of DCCL area located in Kohgilouye and Boyerahmad undeveloped province as well as its negative effects on Karoon Basin.

Acknowledgement Mr. Jack Sieber, a senior scientist at SEI, is acknowledged for his technical help in employing WEAP.

References

1. Shourian, M., Mousavi, S., Tahershamsi, A.: Basin-wide water resources planning by integrating PSO algorithm and MODSIM. Water Resour. Manage. **22**(10), 1347–1366 (2008)
2. Gupta, J., van der Zaag, P.: Interbasin water transfers and integrated water resources management: Where engineering, science and politics interlock. Phys. Chem. Earth Parts A/B/C **33**(1), 28–40 (2008)
3. Wilchfort, O., Lund, J.R.: Shortage management modeling for urban water supply systems. J. Water Resour. Plan. Manage. **123**(4), 250–258 (1997)
4. Jain, S.K., Reddy, N., Chaube, U.: Analysis of a large inter-basin water transfer system in In-dia/Analyse d'un grand système de transfert d'eau inter-bassins en Inde. Hydrol. Sci. J. **50**(1), 125–137 (2005)
5. Mahjouri, N., Ardestani, M.: A game theoretic approach for interbasin water resources allocation considering the water quality issues. Environ. Monit. Assess. **167**(1–4), 527–544 (2010)

6. Zhang, C., Wang, G., Peng, Y., Tang, G., Liang, G.: A negotiation-based multi-objective, multi-party decision-making model for inter-basin water transfer scheme optimization. Water Resour. Manage. **26**(14), 4029–4038 (2012)
7. Wang, Y., Shi, H.S., Wang, J., Zhang, Y.: Research and application of water resources opti-mized distribution model in inter-basin water transfer project. In: Applied Mechanics and Materials. Trans. Tech. Publ., vol. 737, pp. 683–687 (2015)
8. Snaddon, C.D.: A global overview of inter-basin water transfer schemes, with an appraisal of their ecological, socio-economic and socio-political implications, and recommendations for their management. Water Research Commission (1999)
9. Feng, S., Li, L.X., Duan, Z.G., Zhang, J.L.: Assessing the impacts of South-to-North Water Transfer Project with decision support systems. Decis. Support Syst. **42**(4), 1989–2003 (2007)
10. Gohari, A., Eslamian, S., Mirchi, A., Abedi-Koupaei, J., Massah-Bavani, A., Madani, K.: Water transfer as a solution to water shortage: a fix that can backfire. J. Hydrol. **491**, 23–39 (2013)
11. Fang, X., Roe, T.L., Smith, R., Xin, X.: Water shortages, intersectoral water allocation and economic growth: the case of China. China Agr. Econ. Rev. **7**(1), 2–26 (2015)
12. Mahab Ghods Consulting Engineering Company: Water Master Plan of Kohgilouye and Boy-erahmad province- Case study: Dehdasht and Choram Cropland, Preliminary Water Resources Planning Studies. Technical Report, Tehran, Iran (2012)
13. Sieber, J., Purkey, D.: Water Evaluation And Planning System, User Guide. Stockholm Envi-ronment Institute, U.S. Center, Somerville, MA (2011)
14. Kennedy, J. Eberhart, R.: Particle swarm optimization. In: Proceedings of 2004 International Conference on Machine Learning and Cybernetics, pp. 1942–1948. IEEE Press (1995)
15. Parsopoulos, K.E., Plagianakos, V.P., Magoulas, G.D., Vrahatis, M.N.: Stretching technique for obtaining global minimizers through particle swarm optimization. In: Proceedings of the Particle Swarm Optimization Workshop, Indianapolis, USA (2001)
16. Kannan, S., Slochanal, S.M.R., Subbaraj, P., Padhy, N.P.: Application of particle swarm optimization technique and its variants to generation expansion planning problem. Electr. Pow. Syst. Res. **70**(3), 203–210 (2004)

Performance Evaluation of the Genetic Landscape Evolution (GLE) Model with Respect to Crossover Schemes

JongChun Kim and Kyungrock Paik

Abstract We investigate performance of the Genetic Landscape Evolution (GLE) model by changing number of crossover points, which controls spatial cohesiveness of topological information in generated offspring. Simulation results show that 1) GLE performance is insensitive to the number of crossover points, implying that the spatial cohesiveness does not significantly affect efficiency to find better solution sets; and 2) the method to generate randomness in GLE is a significant element for its performance.

Keywords Optimal channel network · Genetic landscape evolution · Genetic algorithm · 2-D crossover

1 Introduction

One of fundamental questions in geomorphology is whether natural river networks are organized in a certain optimal manner [1,2]. A particular hypothesis named Optimal Channel Network (OCN) states that landscape evolves toward a minimum of Total Energy Expenditure (TEE) which can be expressed as [1]:

$$\text{Minimize} \sum \eta Q_i^{0.5} L_i \tag{1}$$

where Q_i and L_i are flow discharge and the length of reach i, respectively, and η is a constant.

J. Kim · K. Paik(✉)
School of Civil, Environmental, and Architectural Engineering, Korea University,
145 Anam-ro, Seongbuk-gu, Seoul 136-713, South Korea
e-mail: arz6oiof@naver.com, paik@korea.ac.kr

© Springer-Verlag Berlin Heidelberg 2016
J.H. Kim and Z.W. Geem (eds.), *Harmony Search Algorithm,*
Advances in Intelligent Systems and Computing 382,
DOI: 10.1007/978-3-662-47926-1_36

To validate the minimum TEE hypothesis, Rodríguez-Iturbe et al. [1] demonstrated an optimized river network on 2-d grids. They obtained the optimal network configuration through an approach similar to the travelling salesman problem. Nevertheless, real river networks are features projected on 3-d landscapes. Dealing only with 2-d projected feature is limited because of the one missed dimension.

Paik [3] developed a powerful optimization algorithm named Genetic Landscape Evolution (GLE) to address this problem, which seeks an optimal 3-d topography for a given goal function. GLE is a unique tool which involves 2-d genetic algorithm components handling 3-d topography information varying over time dealing with 4-d landscape evolution problems. The emphasis by this far has been on the ability of 3-d optimization of GLE. GLE is born to test optimality hypotheses, i.e. dealing with metaphysical problems. In this regard, an important feature was whether we can see converging patterns toward optimal state. Optimality hypotheses such as OCN pose great emphasis on physical implications. Scientists are interested in whether landscape pursues optimal arrangement, but natural landscape does not necessarily have global optimum organization.

On the other side, from a perspective of optimization algorithm, it would be of interest to evaluate the optimization performance of GLE based on certain metric. Nevertheless, GLE has been only used for scientific interpretations and its detailed performance has never been thoroughly tested up to now. The main goal of this study is a quantitative evaluation of the performance of GLE. In particular, we focus on performance sensitivity with respect to algorithm parameters.

It is known that the number of crossover points is one of the parameters which are sensitive to the performance of Genetic Algorithm (GA) [e.g., 4]. However, their evaluation schemes are valid for limited 1-d crossover process. GLE deals with 4-d domains, it is difficult to find any comparable algorithms with GLE. Accordingly, it is of our interest to find a way to improve GLE performance, rather than comparing GLE with other algorithms. Here, we focus on the 2-d crossover module of GLE and seek rooms of performance improvements.

On the basis of this background, we implement sensitivity tests by varying the number of crossover points, which controls spatial cohesiveness of topological information in generated offspring. We organize the remainder of this paper as three sections. In the next section 2, we explain details of the 2-d crossover module used in GLE. In the section 3, performances of GLE depending on the number of crossover point are described. The summary and conclusions are given in the section 4.

2   2-d Crossover in GLE

In the conventional GA, crossover types are categorized as one point, two points, and uniform crossover [5]. It is known as one of critical parameters determining convergence performances in optimizations [e.g., 6]. However, to handle 3-d topography information recorded as 2-d matricies, it is required for GLE is to possess distinguished procedures compared to the conventional GA. The 2-d crossover with a number of crossover points is one of them. With n crossover points, two offspring are created by combinations of $(n+1)^2$ of alternate blocks from each parent (Fig. 1).

If n is zero, the parent's topography information (genes in GA) is delivered to the offspring without any combination. In other words, spatial cohesiveness is preserved to the next generation. If n is one, averagely 25% of spatial cohesiveness is maintained during the crossover procedure. When n reaches to the maximum value, i.e., domain size minus one, the crossover is equal to random mixing processes. In summary, n is closely connected to the degree of preservation for the spatial cohesiveness.

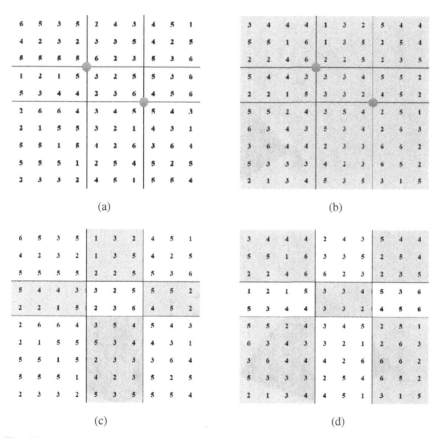

(a) (b)

(c) (d)

Fig. 1 Example showing how the 2-d crossover can be operated in GLE. (a) and (b) indicate a pair of parents where (c) and (d) are two offspring. Two crossover points divide the domain into 9 blocks.

3 Simulation and Results

We arbitrarily employ a Pyramid-shaped landscape consisting of 51 by 51 cells (Fig. 2a). Each cell has an area of 1 km^2. The other conditions are prepared as same as those used in the original GLE simulation [3], i.e., population size is 30 and elapsed condition is no improvement over 200 consecutive iterations. For more details, readers may refer to original paper [3].

Here, we make changes in n from 1 to 20. This range presents 0.23% to 25% of spatial cohesiveness for topograpy information. For better reliable comparisons, 10 independent simulations are conducted. Final outputs are calculated by averaging of   results from the 10 simulations.

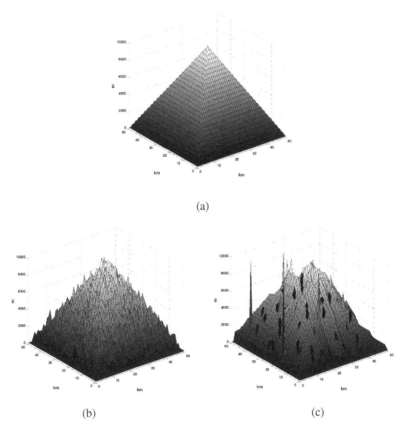

(a)

(b)   (c)

Fig. 2 (a) Initial condition of a Pyramid island domain used in this study and optimized landscapes following the minimum TEE criteria with conditions of (b) constant random seed and (c) variable random seeds.

GLE successfully performs to find the better landscape meeting minimum TEE criterion (Fig. 3). Paik [3] discussed that global optimum landscape should be symmetric (i.e., a same shape for all four triangular sides). The final pyramid does not show a perfect symmetric formation (Fig. 2b), and it means that it still does not reach to the global optimum.

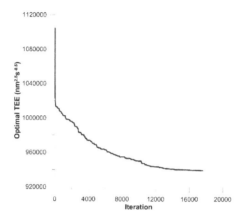

Fig. 3 Convergence to the minimum TEE criterion. Here, iterations are cumulated over 15 time steps.

3.1 Spatial Cohesiveness

As presented in Fig. 4a, optimal TEEs fluctuate with number of crossover points in which any significant trend is not observed. From the results, it seems that the

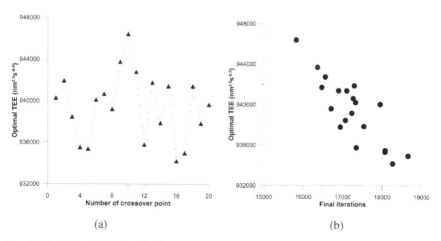

(a)   (b)

Fig. 4 (a) Optimal TEEs with the number of crossover points and (b) trade-off relation between the optimal TEEs and the number of iterations using a constant seed number

number of crossover points do not affect the performance of GLE. On the other hand, we find a trade-off trend between the iteration number and the optimal TEE (Fig. 4b). This relation means GLE performance may be same as a random search even GLE does not contain any random procedure.

3.2 Role of Randomness (Constant Random Seed vs. Variable Random Seeds)

In the previous section, we see the possibility that randomness in the modules of GLE may be an influential element on its performance efficiency. To test that, we repeat same simulations with various random seed numbers. Unlike the constant random seed number, we generate seed numbers varying over time. However, similar tendencies with the previous test, i.e., fluctuating TEE; and trade-off between cumulative iteration and optimal TEE, are observed (Fig. 5a-b).

One interesting point is that optimal TEEs with variable random seeds are smaller than those from the constant random seed (Fig. 4a and Fig. 5a). To support this finding, it is necessary to compare the two final landscapes from both cases. While the optimum pyramid from the constant seed number shows undulating surfaces (Fig. 2b), the pyramid from the variable seeds is optimized to smooth surfaces (Fig. 2c). The former brings diverse river networks resulting in a relatively large TEE compared to the other case.

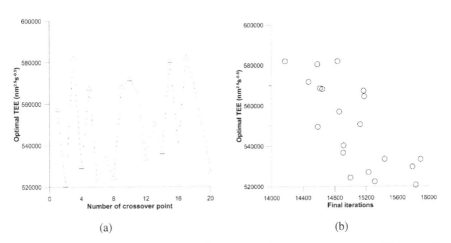

(a)   (b)

Fig. 5 (a) Optimal TEEs depending on the number of crossover points and (b) trade-off relation between the optimal TEEs and the number of the iterations using variable seed numbers.

4   Summary and Conclusions

In this study, we deal with GLE which is a unique and powerful optimization algorithm. Our interest is the algorithmic performance of GLE, especially focused on parameters used in the 2-d crossover module. We repeat simulations within the same conditions with GA by changing number of crossover points. Simulation results exhibit that 1) GLE's performances are insensitive to the crossover process; 2) the trade-off relationship between the number of iterations and the optimum TEEs implies that way to find the global solution in GLE is similar with random searches; and 3) the randomness in the GLE modules improves the performance.

Acknowledgement. This research was supported by the Basic Science Research Program through the National Research Foundation of Korea (NRF) funded by the Ministry of Science, ICT and Future Planning (grant number 2015R1A2A2A05001592).

References

1. Rodríguez-Iturbe, I., Ijjasz-Vasquez, E.J., Bras, R.L., Tarboton, D.G.: Power law distributions of discharge mass and energy in river basins. Water Resour. Res. **28**(4), 1089–1093 (1992)
2. Paik, K., Kumar, P.: Emergence of self-similar tree network organization. Complexity **13**(4), 30–37 (2008)
3. Paik, K.: Optimization approach for 4-D natural landscape evolution. IEEE Trans. Evol. Comput. **15**(5), 684–691 (2011)
4. De Jong, K.A., Spears, W.M.: A formal analysis of the role of multi-point crossover in genetic algorithms. Ann. Math. Artif. Intell. **5**(1), 1–26 (1992)
5. Goldberg, D.E.: Genetic algorithms in search, optimization, and machine learning. Addion Wesley (1989)
6. Rand, W., Riolo, R., Holland, J.H.: The effect of crossover on the behavior of the GA in dynamic environments: a case study using the shaky ladder hyperplane-defined functions. In: Proceedings of the 8th annual conference on Genetic and evolutionary computation, pp. 1289–1296 (2006)

Optimal Design of Permeable Pavement Using Harmony Search Algorithm with SWMM

Young-wook Jung, Shin-in Han and Deokjun Jo

Abstract The permeable pavement is one of representative Low Impact Development (LID) facilities which were used to reduce flooding and recover the water cycle in urban environments. Since the unit cost of porous pavement is greater than that of non-porous pavement, the designs of permeable pavement need to consider reduction effect of rainwater runoff and cost of facilities. These are determined by the size and location of facilities. In this study, the optimal design of permeable pavement, considering the size and location of that, was simulated in a developed optimization model using the Harmony Search (HS)algorithm connected to the Storm Water Management Model (SWMM) to calculate urban Rainfall-Runoff.

Keywords Harmony search algorithm · LID · Permeable pavement · SWMM

1 Introduction

Today, the continuously increasing amount of impermeable increases the risk of inundation by reducing the length of time of inundation concentration and increasing the amount of runoff. To reduce the risk of inundation, many underground reservoirs have been constructed and studies on the economics and effectiveness of underground reservoirs have been conducted [1,2]. However, the construction of underground reservoirs involves many constraints such as increasing the cost, traffic congestion during the construction and civil complaints, is still in difficult

Y.-w. Jung · S.-i. Han
R&D Team, Seoyeong Engineering Co., Ltd, Seongnamsi, Gyeonggi-do,
Seoul 463-825, South Korea
e-mail: {ywjung815,sihan}@seoyeong.co.kr

D. Jo(✉)
Civil Engineering Department, Dongseo University, Busan 617-716, South Korea
e-mail: water21c@gdsu.dongseo.ac.kr

© Springer-Verlag Berlin Heidelberg 2016
J.H. Kim and Z.W. Geem (eds.), *Harmony Search Algorithm,*
Advances in Intelligent Systems and Computing 382,
DOI: 10.1007/978-3-662-47926-1_37

385

situation. Therefore, a distributed basin management technology, the Low Impact Development (LID), has recently been proposed as an alternative to traditional centralized rainwater management facilities [3,4]. LID is a development technology to minimizing the impact on water circulating system of nature. The reduction of impermeable area resulting from comprehensive land using plan and application of technology (including water cycling function, such as reserve, infiltration, filtration and evaporation), can cause increasing infiltration and reducing surface flow of rainwater. The aims of LID to improve the water cycling system of nature and reduce the pollution can be achieved simultaneously through these measures [5].The permeable pavement is representative of one of the LID technology. However, using the permeable pavement is limited as it involves a higher cost than traditional pavement, and for safety reasons. Therefore, the designs of permeable pavement need to consider the effect of the reduction of rainwater runoff and cost of facilities based on the size and location of facilities

The main goal of this paper is to develop an optimal design model of permeable pavement using the Harmony Search(HS)algorithm connected to the Storm Water Management Mode (SWMM) to calculate the urban Rainfall-Runoff. This model determines the locations, the type of LID and the size of permeable pavement that could minimize runoff. Also, the effectiveness of the model is verified through its application to specific catchment basin.

2   Optimal Design Model of Permeable Pavement

In this study, the optimal design model of permeable pavement consists of the HS algorithm and SWMM. HS algorithm [6,7] is used to reduce repeated processes to search for the optimal solution. SWMM developed by the United Stated Environmental Protection Agency (EPA) is a dynamic rainfall-runoff-subsurface runoff simulation model used for simulation of the surface and subsurface hydrology quantity from primarily urban and suburban areas [8]. Figure 1 shows the optimal design process of permeable pavement.

The HS algorithm selects locations where the permeable pavement should be installed and determines the pavement type and size to meet each condition of selected location. SWMM performs a rainfall-runoff analysis using the parameters determined by the HS algorithm and send the runoff result back to the HS algorithm. Then, the HS algorithm generates the more optimal solution from the runoff result. From these repeated process, the location for the installation of the permeable pavement, pavement type and size to satisfy the minimum runoff can be determined.

Figure 2 shows the process of the HS algorithm. An initial randomly generated population of harmony vectors is stored in a Harmony Memory (HM). A new harmony candidate is then generated from all of solutions in the HM by adopting a memory consideration rule, a pitch adjustment rule, and a random re-initialization. Finally, the HM is updated by comparing the new candidate vectors value and worst harmony vector in HM. The worst harmony vector is replaced by the new candidate vector if it is better than the worst harmony vector in the HM. The above process is repeated until a certain termination criterion is met[9].

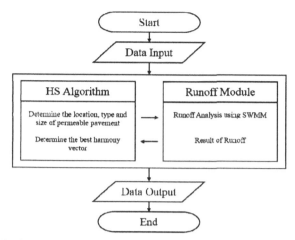

Fig. 1 Optimal Design Process for Permeable Pavement HS Algorithm

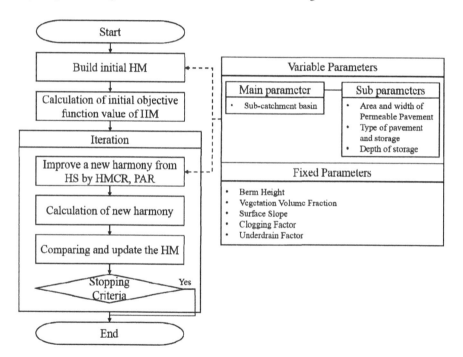

Fig. 2 HS Algorithm

In this study, the harmony vectors consist of location (C) to install the permeable pavement, the pavement area (A), pavement width (W), pavement (P), storage (S) type, and depth of storage (D). The optimal solution, the runoff, is the maximum reduction runoff ($Max\ Q$) calculated by SWMM. When the government plans to install the permeable pavement, the construction cost limit is given.

The construction cost is thus regarded as a constraint. Based on the standard unit of estimate for construction, the construction cost calculated for each harmony vectors is set that it cannot excess the specific cost ($Cost_{max}$).The objective function and cost constraint of this study are as follows(1,2).

$$Max\ Q = Q_o - Q_i[C, A, W, P, S, D] \qquad (1)$$

$$Cost_{max} = 100,000 \qquad (2)$$

In this harmony search algorithm, the harmony memory size (HMS) is 20, the harmony memory considering ratio (HMCR) is 0.95, and the pitch adjusting ratio (PAR) is 0.7. The maximum iteration to search optimal solution is set at 5000.

2.1 Runoff Analysis Module

The location, type and size of permeable pavement determined by the HS algorithm are used to input data for runoff analysis module. The role of runoff module is to link to SWMM 5.0 DLL and to calculate the rainfall-runoff result. The HS algorithm determines a goodness of fit from this runoff result and determines the optimal harmony vector.

3 Application and Optimization Results

To search the best harmony vector to result in the least amount of runoff, a variety permeable pavement type was used. A new town under construction near Seoul was selected to application basin and the best harmony vector including locations, type and size of permeable pavement was determined according the number of permeable pavement.

3.1 Permeable Pavements

To search the optimal pavement type to reduce runoff, a variety of permeable pavements was used. The permeable pavement was divided according to the pavement layer and storage layer, and the HS algorithm combined these for optimal solution. A porous concrete (PC), a porous asphalt (PA), a porous block (PB), and a porous polymer concrete (PP), which are mostly used in permeable pavement, were applied to the pavement layer. In the storage layer, three storage types were used: general gravel storage (SG), reservoir gravel storage (SR), and plastic reservoir (SP). The reservoir gravel storage has a higher permeability and porosity than general gravel storage, and the plastic reservoir has 50% porosity as shown in Table 1.

Table 1 Properties of Pavement Material and Storage

Layer	Type	Permeability (mm/sec)	Porosity (%)	Thickness (mm)
Pavement Layer	Porous Concrete(PC)	0.31	12	250
	Porous Asphalt(PA)	1	12	175
	PorousBlock(PB)	1	12	110
	Porous PolymerConcrete (PP)	30	20	100
Storage Layer	General GravelStorage (SG)	0.1	8	-
	Reservoir GravelStorage (SR)	1	12	-
	Plastic Reservoir(SP)	100	50	-

3.2 Catchment Basin for Application

A new town under construction near Seoul was selected to apply the optimal design model for permeable pavement. The total area of basin is 11.38ha, but only the road and parking lot were selected for installation of the permeable pavement, and were divided into 33 sub-catchment basins with a total area of 2.44ha (Fig. 3).

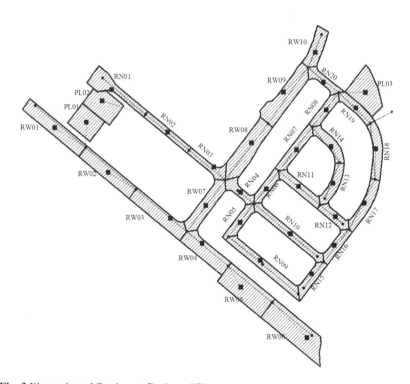

Fig. 3 Illustration of Catchment Basin and Pipe

The 33 sub-catchment basins were classified as a main street, a narrow street, and a parking lot according to their role. Some pavement and storage were limited for application to narrow streets or parking lots according to safety issues as shown in Table 2.

The probability of rainfall intensity of 5 year frequency and 60 minutes duration was selected as the input rainfall intensity.

Table 2 Classification of Sub-Catchment Basin

Role	Sub-catchment Basin	Applicable Pavement	Applicable Storage	Max. Depth of Permeable Pavement (mm)
Main Street	RW01-10	PC, PA, PB, PP	SG	500
Narrow Street	RN01-20	PB, PP	SG, SR	800
Parking Lots	PL01-03	PB, PP	SG, SR, SP	3,000

3.3 Construction Cost

The standard process and unit construction cost for each type of pavement and storage were calculated based on the construction standard production unit system [10] for construction. The material cost is the average of commercial products. The labor and machine expenses were calculated from the price information of Korea (Table 3). The total construction cost is the price added the profit and general management expenses. In this study, the construction cost is a constraint such that it cannot excess USD$100,000.

Table 3 Unit Cost of Pavement and Storage Construction

Type	Process	Unit Cost (USD)			Unit
		Material	Labor	Machine	
SG	Excavation	-	48.327	13.434	m^3
	Laying Subbase	17	2.341	7.612	m^3
SR	Excavation	-	48.327	13.434	m^3
	Laying Subbase	24	2.341	7.612	m^3
SP	Excavation	-	48.327	13.434	m^3
	Install Reservoir	450	3.249	3.503	m^3
PC	Aggregate Base	24	2.602	8.457	m^3
	Porous Concrete	130	0.929	2.328	m^3
PA	Aggregate Base	24	2.602	8,457	m^3
	Porous Asphalt	15	0.992	3.624	m^2
PB	Porous Block	25	1.892	0.364	m^2
PP	Porous Polymer	150	9.432	14.137	m^2

3.4   Result of Application

To review the application of optimal design of permeable pavement, the optimal design was performed with the condition that the permeable pavement should be installed in two sub-catchments. The result showed that the best sizes and locations of permeable pavement are 687m^2 in RW05 and 330m^2 in RW06, respectively. Figure 4 shows a vertical section of the design of permeable pavement from the result of optimal design. Figure 5 shows the reduction runoff of the generated harmony vectors to improve HM, the average of improved HM, and the best solution. While the reduction runoff of generated harmony vector is distributed, the average of improved HM and solution stably approached 8.14mm after 2,000 iterations.

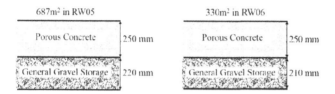

Fig. 4 The Vertical section of design of permeable pavement

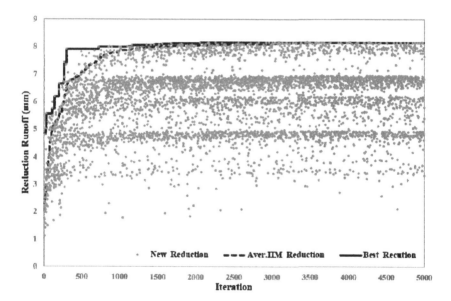

Fig. 5 Reduction Runoff Generated Harmony Vector (Average HM, Best Solution)

The reductions of runoff and construction cost were compared according to the amount of installed permeable pavement (Table 4). The reduction of runoff increased from 4.46mm to 16.73mm when the number of the installed facilities is 1 to 6. Generally, as the number of the sites where the permeable pavement is installed increases at the same total area of permeable pavement, the total construction cost increases due to the additional facilities and increasing amount of excavation. However, the result of analysis shows that multiple permeable pavements are more effective to reduce runoff than using a large area of permeable pavement, as shown figure 6.This demonstrates the basic concept of LID, whereby distributed water management is more efficient than centralized water management.

Table 4 Reduced Runoff and Construction Cost according to Number of Facilities

Num[1]	Sub-catch-ment	Area (m²)	Width (m)	TP[2] (Thickness (mm))	TS[3] (Thick-ness(mm))	Reduc-tion Runoff (mm)	Approx-imate Cost (USD)
1	RW06	824	0.8	PC (250)	SG (210)	4.464	93,505
2	RW05	687	2.2	PC (250)	SG (220)	8.144	97,465
	RW06	330	1.6	PC (250)	SG (210)		
3	RW05	216	1.9	PC (250)	SG (240)	11.135	97,837
	RW06	367	3.5	PA (175)	SG (200)		
	RW08	368	2.1	PA (175)	SG (240)		
4	RW05	234	3.6	PA (175)	SG (220)	13.656	93,139
	RW06	174	0.3	PA (175)	SG (220)		
	RW08	116	1.9	PC (250)	SG (230)		
	RW09	265	0.9	PA (175)	SG (210)		
5	RW01	150	2.7	PA (175)	SG (300)	15.896	83,759
	RW04	86	2.0	PC (250)	SG (200)		
	RW05	344	3.9	PC (250)	SG (200)		
	RW06	169	3.2	PA (175)	SG (220)		
	RW09	105	2.5	PA (175)	SG (210)		
6	RW01	169	1.3	PA(175)	SG(220)	16.735	85,261
	RW03	161	1.4	PA(175)	SG(300)		
	RW05	120	2.9	PA(175)	SG(260)		
	RW06	160	3.3	PA(175)	SG(220)		
	RN01	45	0.7	PA(175)	SR(500)		
	RN09	22	0.4	PB(100)	SR(510)		

[1]The number of installed permeable pavements.
[2]Type of pavement layer.
[3]Type of storage layer.

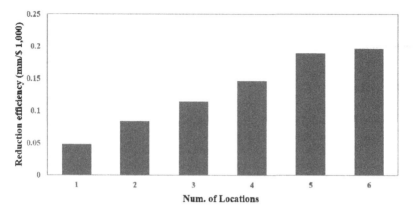

Fig. 6 Reduction Efficiency According to the Number of Installed Permeable Pavements

However, the optimal design for the installation of a large number of permeable pavements needs the long time to build the initial harmony search memory as shown in table 5, because the range of construction cost calculated from randomly generated population is significantly greater than constraint condition. Therefore, this optimal design needs to be improved in order to reduce range of population to operate in many installed locations.

Table 5 The Number of Repeat to Build Initial Harmony Search Memory

The Num. of Installed Permeable Pavements	1EA	2EA	3EA	4EA	5EA	6EA
The Num. of Repeat	24	49	99	319	1,608	9,169

4 Conclusions

In this study, the optimal model for permeable pavement using the HS algorithm connected to SWMM 5.0 DLL was developed and applied to a new town to determine the location, type, and size of permeable pavement in an urban environment.

From the analysis of comparing the reduction of runoff and construction cost according to the number of installed permeable pavement, as the number of installed permeable pavement increased, the amount of reduction runoff increased under the same conditions construction cost. This demonstrates the basic concept of LID, whereby distributed water management is more efficient than centralized water management.

However, the installation of a large number of pavement locations increases the time of building the initial HS memory. The Building the initial harmony search memory thus need to be improved. Nevertheless, the optimal model for permeable pavement developed in this study was useful for LID design.

Acknowledgement This research was supported by a grant (12 Technology Innovation C04) from the Advanced Water Management Research Program funded by Ministry of Land, Infrastructure and Transport of the Korean government.

References

1. Chung, J.H., Han, K.Y., Kim, K.S.: Optimization of detention facilities by using multi-objective genetic algorithms. J KWRA **41**, 1211–1218 (2008)
2. Ryu, S.H., Lee, J.H.: Determination of optimal location and size of storage in the urban sub-surface using genetic algorithm. J KOSHAM **12**, 285–290 (2012)
3. Shin, D.S., Park, J.B., Kang, D.K., Jo, D.J.: An analysis of runoff mitigation effect using SWMM-LID model for frequently inundated basin. J KOSHAM **13**, 303–309 (2013)
4. Ministry of Environment: Low Impact Development (LID) technique element for building healthy water circulation system. South Korea (2013)
5. Prince George's County: Low-Impact Development Design strategies An Integrated Design Approach. Prince George's County, Maryland (1999)
6. Geem, Z.W., Kim, J.H., Loganathan, G.V.: A new heuristic optimization algorithm: harmony search. Simulation **76**, 60–68 (2001)
7. Lee, K.S., Geem, Z.W.: A new structural optimization method based on the harmony search algorithm. Computer & Structures **82**, 781–798 (2004)
8. Rossman, L.A.: Storm Water Management Model User's Manual Version 5.0, United States Environmental Protection Agency, USA (2010)
9. Banerjee, A., Mukherjee, V., Ghoshal, S.P.: An opposition-based harmony search algorithm for engineering optimization problems. Ain Shams Engineering Journal **5**, 85–101 (2014)
10. Journal of Construction and Transportation: Construction standard production unit system. South Korea (2015)

Development of Mathematical Model Using Group Contribution Method to Predict Exposure Limit Values in Air for Safeguarding Health

Mohanad El-Harbawi and Phung Thi Kieu Trang

Abstract Occupational Exposure Limits (OELs) are representing the amount of a workplace health hazard that most workers can be exposed to without harming their health. In this work, a new Quantitative Structure Property Relationships (QSPR) model to estimate occupational exposure limits values has been developed. The model was developed based on a set of 100 exposure limit values, which were published by the American Conference of Governmental Industrial Hygienists (ACGIH). MATLAB software was employed to develop the model based on a combination between Multiple Linear Regression (MLR) and polynomial models. The results showed that the model is able to predict the exposure limits with high accuracy, $R^2 = 0.9998$. The model can be considered scientifically useful and convenient alternative to experimental assessments.

Keywords OELs · Group contribution method · QSPR · MATLAB

1 Introduction

Harmful substances can be defined as any substances in the air which can cause health problems. These harmful substances can be inhaled and cause harm to human

M. El-Harbawi(✉)
Department of Chemical Engineering, College of Engineering, King Saud University, Riyadh 11421, Kingdom of Saudi Arabia
e-mail: melharbawi@ksu.edu.sa

P.T.K. Trang
Chemical Engineering Department, Universiti Teknologi PETRONAS, Bandar Seri Iskandar 31750, Tronoh, Perak, Malaysia
e-mail: phungkieutrang@gmail.com

© Springer-Verlag Berlin Heidelberg 2016
J.H. Kim and Z.W. Geem (eds.), *Harmony Search Algorithm,*
Advances in Intelligent Systems and Computing 382,
DOI: 10.1007/978-3-662-47926-1_38

beings at workplace. Therefore, the occupational health major goal is to prevent health impairment from expose to harmful substances in the workplace [1].

There are several different agencies and organizations worldwide have established and regulated exposure limit of chemical substances at workplace. The most widely used limits are; Threshold Limit Values (TLVs), which were issued in the USA by the American Conference of Governmental Industrial Hygienists (ACGIH), Permissible Exposure Limits (PELs) or usually called a Time-Weighted Average (TWA) concentrations, which were established in the USA by the Occupational Safety and Health Administration (OSHA), Recommended Exposure Limits (RELs), which were recommended by the United States National Institute for Occupational Safety and Health (NIOSH), and maximum workplace concentration values (MAK), which were established by Federal Republic of Germany (DFG). These exposure limit values were obtained from industrial experience, experimental animal studies, and from epidemiologic surveys [2]. Summary of these data can be found in the work of Yaws [3]. However, there are many substances for which safety and health organizations do not provide workplace exposure limits. Therefore, there is a need to provide a faster and an alternative approach to predict these limits. Group contribution methods have been widely used for the estimation and prediction of the physical and chemical properties of pure substances [4]. Several structural group contribution models have been developed in the past to predict different properties, include Flash Point (FP) [5-9], Auto-Ignition Temperature (AIT) [8, 10-14], Lower and Upper Flammability Limits (LFL & UFL) [15-19]. However, up-to-date, predictions of the exposure limits using theoretical methods are poorly appeared in the literature. To the best of our knowledge, there were only two models which were developed in the past to estimate the exposure limit values. Whaley et al. [20] developed equations based on linear regressions method to estimate TLV/WEEL (Threshold Limit Value/Workplace Environmental Exposure Levels) values.

Debia and Krishnan [21] developed QSPR model for computing the Occupational Exposure Limits (OELs).

Today, there are about 2000 published exposure limits values for chemicals at workplace [20]. However, there are about 100,000 chemicals are presently handled and hundreds of new chemicals are added each year to be used in the industry. Thus, evaluating exposure limits for all existing chemicals can exceed the capacity of the toxicology profession worldwide [20,22]. There is, therefore, a need to provide a faster and more cost-effective approach to estimate the exposure limit values.

The aim of the present work is to develop a novel mathematical model using a group contribution method that is capable of predicting the exposure limit values at a workplace.

2 Methodology

2.1 Dataset Preparation

The data set used for developing the proposed model was taken from Yaws' handbook of thermodynamic and physical properties of chemical compounds [3].

In this handbook, Yaws collected exposure limit values for four different organisations, which are: TLVs (of ACGIH), PELs or TWA (of OSHA), REL (of NIOSH), and MAK (of DFG). In this work, we have selected the TLVs set to develop our model. This dataset contains 100 data for organic compounds encompass various families: hydrocarbons, halogenated compounds, ethers, aldehydes, alcohols, esters, ketones, amides, nitriles, acids, amines, nitro compounds, heterocyclic compounds.

2.2   Model Development

Quantitative structure activity (or property) relationship (QSAR/QSPR) is the process by which chemical structure is quantitatively correlated with a well-defined process, such as biological activity or chemical reactivity [23].

Modelling methods used in the development of structure–property relationships are two types: correlative and pattern recognition. The most common correlative method is regression analysis. The model that will be developed in this work are based on a combination between Multiple Linear Regression (MLR) method and Polynomial method. We have used 45 functional groups as molecular descriptors, which were defined according to the Valderrama and Robles [24] group contribution method. The group contribution method expresses the exposure limit values as a function of a sum of contributions of all the functional groups constituting the molecules.

Generally, in QSAR, to develop any model, the data set consist of a response variable, which can be denoted as Y, while the predictor variables will be denoted as X_1, X_2, ..., X_p, where p represents the total number of predictor variables. The true relationship between Y and X_1, X_2, ..., X_p is approximated by a regression model [25] expressed as:

$$Y = f(X_1, X_2, \dots X_p) + \varepsilon \tag{1}$$

where ε is defined as a normal random error expressing the discrepancy in the approximation. The linear form of equation (1) can be expressed as:

$$Y = \beta_o + \beta_1 X_1 + \beta_2 X_2 + \dots + \beta_p X_p + \varepsilon \tag{2}$$

where $\beta_0, \beta_1, \dots \beta_p$, are defined as regression coefficients, *i.e.* constants to be estimated from the data. Equation (2) was modified and integrated with a polynomial model [equation (3)] and the interaction between the two models $\left(\alpha X_i\right)$ can be described by equation (4):

$$Y = \gamma_0 + \gamma_1 \omega^1 + \gamma_2 \omega^2 + \dots\dots + \gamma_m \omega^m + \varepsilon \tag{3}$$

$$Y = \alpha_0 + \sum_{i=1}^{n} \alpha_i X_i + \sum_{i=1}^{m} \gamma_i \varpi^i + \sum_{i=n+1}^{k} \delta_i \varpi X_i \tag{4}$$

where, $\alpha_i = \beta_i$ for $i \leq n$; $\gamma_i = \beta_{i+n}$ for $i \leq m$; $\delta_i = \beta_{i+n+m}$ for $i \leq k$; $n + m + k = p$

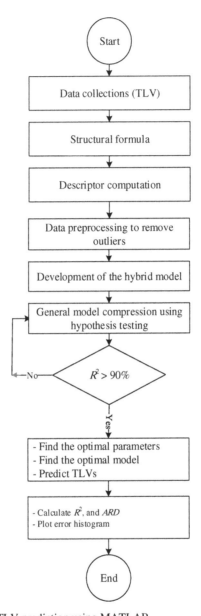

Fig. 1 Algorithm for the TLV prediction using MATLAB

The MATLAB software was used to develop the code and estimate the exposure limit values by means of the algorithm illustrated in Figure 1 and the following steps:

- Pre-processing the data to remove outliers or misplaced data.
- Combine MLR and polynomial models to obtain the best fitting model.

- Optimize the most accurate QSAR model (based on the best coefficients).
- Use the developed model to estimate TLVs.
- Validate the results obtained using the developed model with others obtained from experimental work and published literatures.
- Check the accuracy of the model (R^2).
- Obtain error histogram.

The accuracy of the develop model will be checked using the average relative deviation (ARD) (Eq. 5):

$$ARD = \frac{100}{N} \sum_{i=1}^{N} \left(\frac{\Phi_{Cal} - \Phi_{Exp}}{\Phi_{Exp}} \right)_i \tag{5}$$

where,

N is the number of the substances,

Φ_{Cal} is the predicted values, and

Φ_{Exp} is the experiment values

3   Results and Discussion

MATLAB code utilized the experimental data and computed the best coefficients of the proposed model ($\alpha_i, \gamma_i, \delta_i$). Then MATLAB tested the accuracy of the proposed model by comparing the predicted TLV with the other values obtained from Yaws [3]. As it can be seen from Figure 2 and Figure 3, a good fit is achieved with squared correlation coefficient, $R^2=99.98\%$ and the standard deviation error = 7.32. In addition, it can be concluded from Figure 3 that out of 100 components, there are 96 components possess 0% errors between the TLV predicted and experimental values. The errors between the predicted results and others TLV values could be due to the different methods adopted in measuring the TLVs as well as the conditions of the experiments.

Table 1 Comparison between several models and their correlation coefficients

Method	No. of substances used	$R^2 (\%)$	Reference
Linear regression	598 (TLV)	0.513-0.898	[20]
Linear univariate regression	68 (WEEL) 88 (TLV)	0.71	[21]
MLR+ polynomial	100 (TLV)	0.9998	This work

The accuracy of the proposed model was compared with other two models, which are models of Whaley et al. [20] and Debia and Krishnan [21]. As it can be seen from Table 1, the proposed model is able to predict TLV with higher accuracy than the other two models.

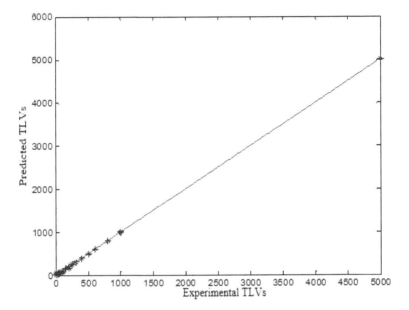

Fig. 2 Comparison between predicted and experimental TLV values

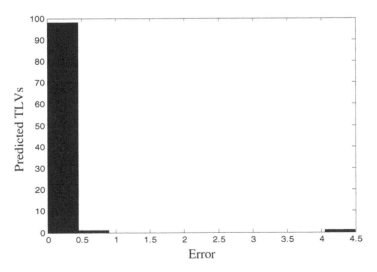

Fig. 3 Results error histogram

4 Conclusion

In this paper, a new QSPR model for the prediction of exposure limit values based on ACGIH data has been developed. The model has been developed based on the combination of MLR and polynomial models and the code was written using MATLAB software. A data set comprising of 100 experimental TLV values was used for developing and verifying the model. The results show that the proposed model was able to predict TLV values with very high accuracy, $R^2 = 99.98\%$. Thus, the model can be considered reliable and can be used in the absence of experimental measurements for determination of exposure limit values.

References

1. Ong, C.N.: Reference values and action levels of biological monitoring in occupational exposure. Toxicol. Lett. **108**, 127–135 (1999)
2. Weisburger, E.K.: History and background of the Threshold Limit Value Committee of the American Conference of Governmental Industrial Hygienists. Chem. Heath. Saf. **8**, 10–12 (2001)
3. Yaws, C.L.: Yaws' Handbook of Thermodynamic and Physical Properties of Chemical Compounds. Knovel (2003)
4. Corrêa, R.O., Telles, A.S., Ourique, J.E.: A graph-structural method for prediction of polymer properties. Br. J. Chem. Eng. **21**, 621–628 (2004)
5. Suzuki, T., Ohtaguchi, K., Koide, K.: A Method for Estimating Flash Points of Organic Compounds from Molecular Structures. J. Chem. Eng. Jap. **24**, 258–261 (1991)
6. Katritzky, A.R., Petrukhin, R., Jain, R., Karelson, M.: QSPR Analysis of Flash Points. J. Chem. Inf. Comput. Sci. **41**, 1521–1530 (2001)
7. Wang, K., Du, Z., Wang, J.: A new method for predicting the flashpoints of organic compounds from the information of molecular component. Bull. Sci. Technol. **18**, 235–239 (2002)
8. Albahri, T.A.: Flammability characteristics of pure hydrocarbons. Chem. Eng. Sci. **58**, 3629–3641 (2003)
9. Albahri, T.A.: MNLR and ANN structural group contribution methods for predicting the flash point temperature of pure compounds in the transportation fuels range. Process Saf. Environ. Prot. (in press)
10. Brooke, E.M., Jurs, P.C.: Prediction of Autoignition Temperatures of Organic Compounds from Molecular Structure. J. Chem. Inf. Comput. Sci. **37**, 538–547 (1997)
11. Pan, Y., Jiang, J., Wang, R., Cao, H., Cui, Y.: A novel QSPR model for prediction of lower flammability limits of organic compounds based on support vector machine. J. Hazard. Mater. **168**, 962–969 (2009)
12. Chen, C.C., Liaw, H.J., Kuo, Y.Y.: Prediction of autoignition temperatures of organic compounds by the structural group contribution approach. J. Hazard. Mater. **162**, 746–762 (2009)
13. Jingjie, S., Liping, C., Wanghua, C.: Prediction on the Auto-ignition Temperature Using Substructural Molecular Fragments. Pro. Eng. **84**, 879–886 (2014)
14. Gharagheizi, F.: An accurate model for prediction of autoignition temperature of pure compounds. J. Hazard. Mater. **189**, 211–221 (2011)

15. Gharagheizi, F.: Prediction of upper flammability limit percent of pure compounds from their molecular structures. J. Hazard. Mater. **167**, 507–510 (2009)
16. Gharagheizi, F.: A QSPR model for estimation of lower flammability limit temperature of pure compounds based on molecular structure. J. Hazard. Mater. **169**, 217–220 (2009)
17. Gharagheizi, F.: A new group contribution-based model for estimation of lower flammability limit of pure compounds. J. Hazard. Mater. **170**, 595–604 (2009)
18. Lazzús, J.A.: Neural network/particle swarm method to predict flammability limits in air of organic compounds. Thermochim. Acta. **512**, 150–156 (2011)
19. Rowley, J.R., Rowley, R.L., Wilding, W.V.: Estimation of the lower flammability limit of organic compounds as a function of temperature. J. Hazard. Mater. **18**(1), 551–557 (2011)
20. Whaley, D.A., Attfield, M.D., Bedillion, E.J., Walter, K.M., Quilong, Y.: Regression Method to Estimate Provisional TLV/WEEL-equivalents for Non-carcinogens. Ann. Occup. Hyg. **44**, 361–374 (2000)
21. Debia, M., Krishnan, K.: Quantitative property–property relationships for computing Occupational Exposure Limits and Vapour Hazard Ratios of organic solvents. SAR QSAR Environ. Res. **21**, 583–601 (2010)
22. United States Environmental Protection Agency: EPA Toxicology Handbook. Government Institutes, Inc., Rockville (1986)
23. Kleandrova, V.V., Speck-Planche, A.: Regulatory issues in management of chemicals in OECD member countries. Fron. in Biosc. **E5**, 375–398 (2013)
24. Valderrama, J.O., Robles, P.A.: Critical properties, normal boiling temperatures, and acentric factors of fifty ionic liquids. Ind. Eng. Chem. Res. **46**, 1338–1344 (2007)
25. Chatterjee, S., Hadi, A.S.: Regression Analysis by Example, 4th edn. John Wiley, New York (2006)

Retracted Chapter: Optimization of Water Distribution Networks with Differential Evolution (DE)

Ramin Mansouri, Hasan Torabi and Hosein Morshedzadeh

Abstract Nowadays, due to increasing population and water shortage and competition for its consumption, especially in the agricultural that is the largest consumer of water, proper and suitable utilization and optimal use of water resources is essential. One of the important parameters in agriculture field is water distribution network. In this research, differential evolution algorithm (DE) was used to optimize Ismail Abad water supply network. This network that is pressurized network and includes 19 pipes and nodes 18. Optimization of the network has been evaluated by developing an optimization model based on DE algorithm in MATLAB and the dynamic connection with EPANET software for network hydraulic calculation. The developing model was run for the scale factor (F), the crossover constant (Cr), initial population (N) and the number of generations (G) and was identified best adeptness for DE algorithm is 0.6, 0.5, 100 and 200 for F and Cr, N and G, respectively. The optimal solution was compared with the classical empirical method and results showed that Implementation cost of the network by DE algorithm 10.66% lower than the classical empirical method.

Keywords Differential evolution algorithm · Optimization · Distribution systems · Crossover constant · Scale factor

R. Mansouri(✉) · H. Torabi
Water Engineering Department, Lorestan University, Khoramabad, Iran
e-mail: ramin_mansouri@yahoo.com

H. Morshedzadeh
Economics and Management Department, Tehran University, Tehran, Iran

© Springer-Verlag Berlin Heidelberg 2016
J.H. Kim and Z.W. Geem (eds.), *Harmony Search Algorithm*,
Advances in Intelligent Systems and Computing 382,
DOI: 10.1007/978-3-662-47926-1_39

1   Introduction

Nowadays, due to increasing population and water shortage and competition for its consumption, proper and suitable utilization and optimal use of water resources is essential. Distribution networks are an essential part of all water supply systems. A water distribution network is a system containing pipes, reservoirs, pumps, and valves of different types, which are connected to each other to provide water to consumers.

The water distribution system is one of the major requirements in urban and regional economic development. For any agency dealing with the design of the water distribution network, an economic design will be an objective. Attempts should be made to reduce the cost and energy consumption of the distribution system through optimization in analysis and design. A water distribution network that includes booster pumps mounted in the pipes, pressure reducing valves, and check-valves can be analyzed by several common methods such as Hardy-Cross, linear theory, and Newton-Raphson (Stephenson, 1984).

Traditionally, pipe diameters are chosen according to the average economical velocities (Hardy-Cross method) (Cross, 1936). This procedure is cumbersome, uneconomical, and requires trials, seldom leading to an economical and technical optimum.

In the case of the design of a pipe network the optimization problem can be stated as follows: minimize the cost of the network components subject to the satisfactory performance of the water distribution system (mainly, the satisfaction of the allowable pressures).

Numerous optimization techniques are used in water distribution systems. These include the deterministic optimization techniques such as linear programming (for separable objective functions and linear constraints), and non-linear programming (when the objective function and the constraints are not all in the linear form), and the stochastic optimization techniques such as genetic algorithms, simulated annealing, Deferential Algorithm, Particle Swarm Optimization and etc.

Numerous works were reported in the literature for optimal design and some of them considered certain reliability aspects too. In optimization models, continuous diameters (Pitchai 1966; Jacoby 1968; Varma et al. 1997) and split pipes (Alperovits & Shamir 1977; Quindry et al. 1979; Goulter et al. 1986; Fujiwara et al. 1987; Kessler & Shamir 1989; Bhave & Sonak 1992) were more prominently used.

Mays and Tung (1992) recommended strongly the use of the linear programming (LP) technique in designing the pipe networks due to the capability of the LP in handling more decision variables than other optimization techniques. Dandy and Hassanli (1996) developed a nonlinear model for optimum design and operation of multiple subunit drip irrigation systems on flat terrains.

Application of the genetic algorithm (Hassanli and Dandy. 1996; Savic & Walters 1997; Vairavamoorthy & Ali 2000, 2005), the modified genetic algorithm (Montesinos et al. 1999; Neelakantan & Suribabu 2005; Kadu et al. 2008), the simulated annealing algorithm (Cunha & Sousa 1999), the shuffled leapfrog algorithm (Eusuff & Lansey 2003), ant colony optimization (Maier et al. 2003; Zecchin et al. 2007; Ostfeld & Tubaltzev 2008), novel cellular automata (Keedwell & Khu 2006) and the particle swarm algorithm (Suribabu & Neelakantan 2006a,b) for optimal design of water distribution systems are some of them.

Samani and Mottaghi (2006) used the integer linear programming for obtaining the optimum pipe sizes and reservoir elevations in pipe networks. Kale, Sing and Mahar (2008) presented a linear programming (LP) model for optimization a design of a pressurized irrigation system subunit. Cisty and Bajtek (2009) proposed a new hybrid GA-LP approach for determining the least-cost design of a water distribution system.

Dercas and Valiantzas (2011) presented two explicit optimum design methods for simple irrigation delivery systems. In the first method, a simple equation calculates explicitly the critical values of discharges corresponding to the available pipe diameters. The second method calculates the optimum economic diameter for every pipeline of the network.

Mansouri et al. (2014) by using differential evolution algorithm (DE), CU equation (water distribution uniformity coefficient in zb sprinkler irrigation) was optimized and the best optimized coefficients obtained.

Shahinezhad et al. (2011) presented a mixed integer linear programming (MILP) model for optimization of pressurized branched irrigation networks. Detailed analysis of the results is reported and compared with those generated based on trial-and-error method. The proposed method results in a reduction of 12.5% in costs.

In this paper, DE algorithm is developed to obtain the optimum pipe size and inlet pressure head that produce the least cost design of Shahinezhad et al. (2011) networks. In this study, the hydraulic analysis of the network is based on continuity at nodes and Hazen-Williams formula for head loss calculations by using link between Epanet and Matlab Software. The results of this investigation compared with absolute optimization is obtained by mixed integer linear programming (MILP) model that is presented by Shahinezhad et al. (2011).

2   Material and Methods

2.1 Case Study

The Ismail Abad irrigation network is located in 7 kilometers North West of Noorabad city in Lorestan province. Land area of this project is 1000 ha. Fig. 1 depicts the schematic network of Ismael Abad. This network consists of 18 pipes and 19 nodes are. In Table 1, the hydraulic details and arrangement of pipes for water distribution networks Ismael Abad is presented.

This project consists of two kinds of steel pipe that is used. Polyethylene pipe material is used for pipe sizes equal or less than 500 mm and GRP for greater sizes. Pipe specifications are given in Table 2.

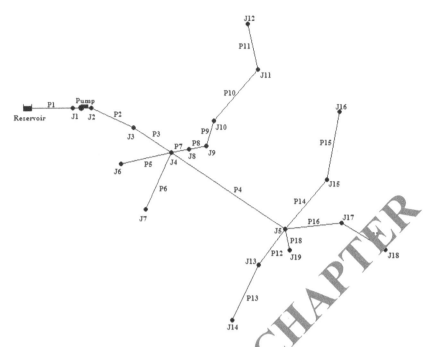

Fig. 1 Ismael Abad water distribution network

Table 1 Main and sub main pipe line data of Ismail Abad Network

Pipe	Pipe No.	Length (m)	Discharge (L/s)	Beginning Elevation (m)	End Elevation (m)
Res.-J1	P1	-	-	-	-
J2-J3	P2	558	856.56	1791	1816.54
J3-J4	P3	558	856.56	1816.54	1842.08
J4-J5	P4	450	429.8	1842.08	1847.57
J4-J6	P5	955	52.9	1842.08	1838.71
J4-J7	P6	1100	128.94	1842.08	1856.52
J4-J8	P7	200	244.92	1842.08	1847.05
J8-J9	P8	201	190.34	1847.05	1846.32
J9-J10	P9	390	128.94	1846.32	1841.18
J10-J11	P10	806	58.33	1841.18	1811.32
J11-J12	P11	575	21.49	1811.32	1810.94
J5-J13	P12	550	165.8	1847.57	1853.21
J13-J14	P13	700	132	1853.21	1861.89
J5-J15	P14	670	98.24	1847.57	1821.48
J15-J16	P15	840	33.77	1821.48	1814.43
J5-J17	P16	720	119.73	1847.57	1826.47
J17-J18	P17	660	49.12	1826.47	1847.95
J5-J19	P18	110	46.05	1847.57	1847.57

Table 2 Pipe specifications data of Ismail Abad Network

No.	Material	Internal diameter (mm)	Outer diameter (mm)	Cost ($/m)
1	PE80	93.8	110	5.895
2	PE80	106.6	125	7.895
3	PE80	119.4	140	9.495
4	PE80	136.4	160	12.375
5	PE80	153.4	180	15.705
6	PE80	170.6	200	19.305
7	PE80	191.8	225	24.525
8	PE80	213.2	250	30.150
9	PE80	238.8	280	7.800
10	PE80	268.6	319	7.700
11	PE80	302.8	355	60.525
12	PE80	341.2	400	76.725
13	PE80	383.8	50	97.200
14	PE80	426.4	500	108.820
15	GRP	600.0	600	111.323
16	GRP	700.0	700	137.997
17	GRP	800.0	800	170.633
18	GRP	900.0	900	204.289

2.2 Water Distribution Network Constraints

2.2.1 Pressure Constraint

Minimum Allowable pressure head required for each node is considered to be 50 m.

2.2.2 Velocity Constraint

In order to prevent sediment deposition in low flow velocities and avoid water hammer at high velocities, minimum and maximum allowable flow velocities in pipes are considered to be 0.7 m/s and 2m/s, respectively.

2.3 Differential Evolution Algorithm (DE)

Differential Evolution (DE) algorithm is a branch of evolutionary programming developed by Rainer Storn and Kenneth Price (1995) for optimization problems over continuous domains. In DE, each variable's value is represented by a real number. The advantages of DE are its simple structure, ease of use, speed and robustness. DE is one of the best genetic type algorithms for solving problems with the real valued variables. Differential Evolution is a design tool of great utility that is immediately accessible for practical applications. DE has been used in several science and engineering applications to discover effective solutions to nearly intractable problems without appealing to expert knowledge or complex design algorithms. Differential Evolution uses mutation as a search mechanism and selection to direct the search toward the prospective regions in the feasible region. Genetic Algorithms generate a sequence of populations by using selection mechanisms. Genetic Algorithms use crossover and mutation as search mechanisms. The principal difference between Genetic Algorithms and Differential Evolution is that Genetic Algorithms rely on crossover, a mechanism of probabilistic and useful exchange of information among solutions to locate better solutions, while evolutionary strategies use mutation as the primary search mechanism.

Differential Evolution (DE) is a parallel direct search method which utilizes NP D-dimensional parameter vectors.

$$x_{i,G}, \quad i=1,2,....,NP \tag{1}$$

As a population for each generation G. NP does not change during the minimization process. The initial vector population is chosen randomly and should cover the entire parameter space. As a rule, we will assume a uniform probability distribution for all random decisions unless otherwise stated. In case a preliminary solution is available, the initial population might be generated by adding normally distributed random deviations to the nominal solution $x_{nom,0}$. DE generates new parameter vectors by adding the weighted difference between two population vectors to a third vector. Let this operation be called mutation. The mutated vector's parameters are then mixed with the parameters of another predetermined vector, the target vector, to yield the so-called trial vector. Parameter mixing is often referred to as "crossover" in the ES-community and will be explained later in more detail. If the trial vector yields a lower cost function value than the target vector, the trial vector replaces the target vector in the following generation. This last operation is called selection. Each population vector has to serve once as the target vector so that NP competitions take place in one generation. More specifically DE's basic strategy can be described as follows:

2.3.1 Mutation

For each target vector $x_{i,G}, \quad i=1,2,....,NP$, a mutant vector is generated according:

$$V_{i,G+1} = x_{r1,G} + F \times (x_{r2,G} - x_{r3,G}) \tag{2}$$

With random indexes $r_1, r_2, r_3 \in \{1, 2 \ldots NP\}$ integer, mutually different and F > 0. The randomly chosen integers r1, r2 and r3 are also chosen to be different from the running index i, so that NP must be greater or equal to four to allow for this condition. F is a real and constant factor$\in [0, 2]$ which controls the amplification of the differential variation ($x_{r2,G}$ - $x_{r3,G}$). Fig.2 shows a two-dimensional example that illustrates the different vectors which play a part in the generation of $V_{i,G+1}$.

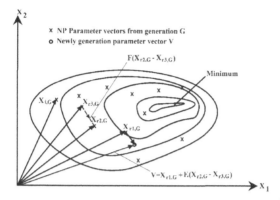

Fig. 2 An example of a two-dimensional cost function showing its contour lines and the process for generating $V_{i,G+1}$

2.3.2   Crossover

In order to increase the diversity of the perturbed parameter vectors, crossover is introduced. To this end, the trial vector:

$$u_{i,G+1} = (u_{1i,G+1}, u_{2i,G+1} \cdots u_{Di,G+1})$$

(3)

Is formed, where:

$$u_{ji,G+1} = \begin{cases} V_{ji,G+1} & \text{if } \text{randb}(j) \leq CR \text{ or } j = \text{ranbr}(i) \\ x_{ji,G} & \text{otherwise} \end{cases}$$

$$j = 1,2 \cdots, D.$$

(4)

In Eq. (5), randb(j) is the jth evaluation of a uniform random number generator with outcome $\in [0; 1]$. CR is the crossover constant $\in [0; 1]$ which has to be determined by the user. ranbr(i) is a randomly chosen index $\in 1, 2, \ldots, D$ which ensures that $u_{i,G+1}$ gets at least one parameter from $V_{i,G+1}$.

2.3.3   Selection

To decide whether or not it should become a member of generation G+1, the trial vector $u_{i,G+1}$ is compared to the target vector $x_{i,G}$ using the greedy criterion. If vector $u_{i,G+1}$ yields a smaller cost function value than $x_{i,G}$, then $x_{i,G+1}$ is set to $u_{i,G+1}$; otherwise, the old value $x_{i,G}$ is retained.

$$x_{ji,G+1} = \begin{cases} u_{ji,G+1} & \text{if } f(u_{i,G+1}) \leq f(x_{i,G}) \\ x_{ji,G} & \text{if } \text{otherwise} \end{cases}$$

(5)

Finally, this process continues to reach new generations to the number of NP. Then the same process is repeated to reach termination condition.

Fig. 3 schematically overview of differential evolution algorithm for numerical model, the entire above process is specified numerically in this Fig. 3.

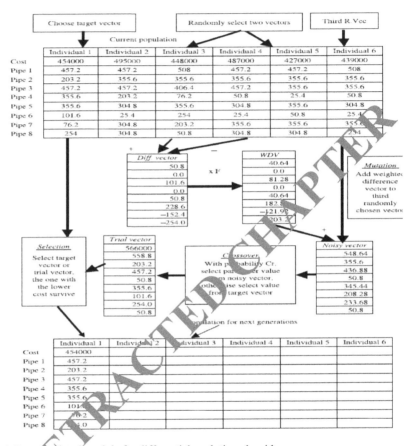

Fig. 3 Computational module for differential evolution algorithm

According the literature review in the differential evolution algorithm (Suribabu, 2010) and other evolutionary algorithms, to find the best conditions for optimizing water distribution network, at first considering an initial population of 100 member (N=100) and generation of 500 (G=500) to find the coefficients of F and CR, 18 different combinations of these factors was examined. It should be mentioned, at study each of the condition in this algorithm, three runs were conducted and the optimal run was chosen for that.

In general, in this study, in total 120 runs with different conditions of the algorithm was implemented, in order to derive the optimal of water distribution networks by using differential evolution algorithm.

3   Results and Discussion

3.1   F and CR Factor

In the first step, to obtain the best conditions for algorithm that provide the most optimum and do not face local optimum problem, 18 combinations of different modes for the coefficients F and CR were examined. The results are shown in Table 3.

Table 3 Study F and CR

No. Combination	F	Cr	Optimal Cost ($)
1	F=0.1	Cr =0.1	115427393
2		Cr =0.3	832628
3	F=0.5	Cr =0.4	768561
4		Cr =0.5	758917
5		Cr =0.6	740000
6		Cr =0.3	738039
7	F=0.6	Cr =0.4	737931
8		Cr =0.5	737920
9		Cr =0.6	737992
10		Cr =0.3	737924
11	F=0.7	Cr =0.4	737988
12		Cr =0.5	740588
13		Cr =0.6	758028
14		Cr =0.3	786416
15	F=0.8	Cr =0.4	824850
16		Cr =0.5	832628
17		Cr =0.6	833455
18	F=1	Cr =1	55293902

The Results show that median values for the coefficients of F and Cr provide the optimum situation and cause DE algorithm not to be trapped in local optimum. The most optimal answer for coefficients are 0.6 and 0.5 for F and Cr coefficients, respectively. These values matched with the results of Suribabu (2010).

Scale factor (F) can increase the accuracy of the search. The smaller coefficient, the shorter steps needs to be taken for an accurate research. But the problem is that the algorithm may be trapped in local optimum and it cannot be withdrawn. On the other hand, the higher value of F, the more area will be searched, but the best optimum situation may not be obtained.

3.2   Population and Generation

After finding the best combination of coefficients values F and CR, algorithms for solving the independent populations were examined. For this purpose, the population of 4, 25, 50, 100, 500 and 1000 members were studied in two generations (G=50 and 100). Fig. 3 shows these results.

Based on the DE algorithm, the initial population is very important to select the initial three members, when the population gets more, the selection of four initial members has more variety, which causes the algorithm to reach convergence.

According to Fig.   4, it is clear that by increasing population, the optimal cost will be lower. In addition it is proved that the increasing population will extend the domain of the search; and more members are used for optimization.

Finally, the best combination of coefficients and population were used to examine the effect of generations' number, so ten generations (30, 40, 50, 100, 200, 300, 500, 1000, 2000, and 3000) were studied. The results are shown in Table 4.

Table 4 indicates that the generation number 200 is suitable for optimizing water distribution networks.

This results show that DE algorithm for optimizing water distribution networks in the generation of 200 gives acceptable results.

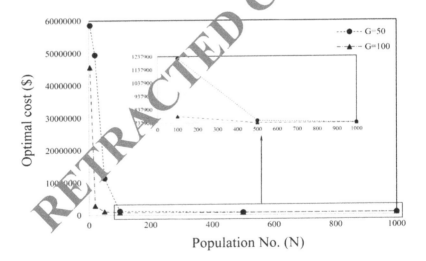

Fig. 4 Optimization cost in different populations

Table 4 The effect of generation on optimization cost

No. Generation	Optimal cost ($)	Runtime (s)
30	55293902	629
40	2065347	780
50	1222823	1005
100	786416	1950
200	737920	4024
300	737931	6164
500	737920	9324
1000	737924	20163
2000	737920	39826
3000	737920	53911

The increase in time per the number of population has almost a linear trend, which indicates the effect of population in the runtime algorithm. Hence specifying suitable population to obtain an optimal results is very important.

The runtime algorithm for 100 members of population and 50 generations is 935s and 100 generation is 1950s. According to the numbers, the running time of the algorithm to reach new member in each generation takes an average of 0.19s (Fig. 5).

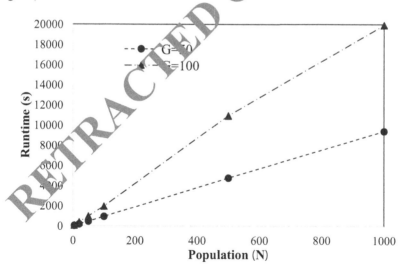

Fig. 5 Runtime in different population

Results of Fig. 6 indicates a fairly linear relationship between runtime and number of generations.

In general it can be said that the population and number of generations to run the algorithm, in order to optimize water distribution network is 100 and 200, respectively that requires nearly an hour to reach the optimal answer.

So it can be revealed that one of the advantages of this algorithm is the high speed runtime. Another advantage is rapid convergence of the algorithm, that takes 16 minutes (G =50 and N =100) to reach convergence.

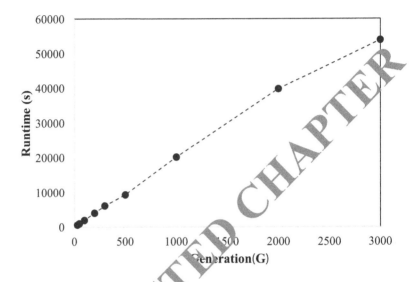

Fig. 6 Runtime in different generation

3.3 *Differential Evolution Algorithm Optimization*

The network has been optimized with conditions Cr=0.5, F=0.6, 100 members of population and 200 generations in the differential evolution algorithm. The algorithm makes relationship between Epanet and MATLAB software to optimize the water distribution network. The combination of optimum pipe diameter is shown in Table

Table 5 Optimum pipe diameter in Differential Evolution Algorithm

Pipe	No. Pipe	Optimum Diameter (inch)	Internal Optimum Diameter (mm)	Outer Optimum Diameter (mm)
Res.-J1	P1	10.575	268.6	315
J2-J3	P2	31.496	800	800
J3-J4	P3	31.496	800	800
J4-J5	P4	23.622	600	600
J4-J6	P5	7.551	191.8	225
J4-J7	P6	11.921	302.8	355
J4-J8	P7	16.787	426.4	500
J8-J9	P8	15.110	383.8	450
J9-J10	P9	11.921	302.8	355
J10-J11	P10	8.394	213.2	250
J11-J12	P11	4.701	119.2	140
J5-J13	P12	15.110	383.8	450
J13-J14	P13	11.921	302.8	355
J5-J15	P14	10.575	268.6	315
J15-J16	P15	6.039	153.4	180
J5-J17	P16	11.921	302.8	355
J17-J18	P17	7.551	191.8	225
J5-J19	P18	7.551	191.8	225
Optimal cost ($)			737920	
Runtime			1:07:00	

This combination of optimal diameter is the best diameter to have the optimal costs. According to these network diameters, hydraulic conditions in Tables 6 and 7 are for pipes and nodes.

Due to the hydraulic conditions in the pipes, it can be seen from Table 6, each pipe is in standard conditions and velocity in each pipe is in permitted range. Table 7 shows pressure in each node in permitted range. So it can be said in this optimized network the constraint of pressure and velocity is considered.

Table 6 Hydraulic conditions optimal diameters in pipes

Pipe	No. Pipe	Optimum Diameter (mm)	Discharge (L/s)	Velocity (m/s)	Losses in 1000 (m)
Res.-J1	P1	315	-	-	-
J2-J3	P2	800	856.56	1.70	0.72
J3-J4	P3	800	856.56	1.70	0.72
J4-J5	P4	600	429.8	1.52	0.81
J4-J6	P5	225	52.9	1.83	4.64
J4-J7	P6	355	128.94	1.79	2.45
J4-J8	P7	500	244.92	1.72	1.62
J8-J9	P8	450	190.34	1.65	1.69
J9-J10	P9	355	128.94	1.79	2.6
J10-J11	P10	250	58.33	1.63	3.12
J11-J12	P11	140	21.49	1.92	8.80
J5-J13	P12	450	165.8	1.43	1.31
J13-J14	P13	355	132	1.83	2.73
J5-J15	P14	315	98.24	1.73	2.84
J15-J16	P15	180	33.77	1.33	6.00
J5-J17	P16	355	119.73	1.66	2.28
J17-J18	P17	225	49.12	1.70	4.04
J5-J19	P18	225	46.05	1.59	3.59

Table 7 Hydraulic conditions optimal diameters in nodes

No. Node	Discharge (L/s)	Hydraulic Elevation (m)	Pressure (m-water)
Res.	-856.69	1789.00	0.00
J1	0.00	1788.71	-1.29
J2	0.00	1926.02	134.69
J3	0.00	1924.71	109.44
J4	0.00	1923.39	84.98
J5	0.00	1919.57	73.19
J6	52.90	1908.85	73.73
J7	128.94	1914.55	60.20
J8	54.58	1922.32	80.93
J9	61.40	1921.21	77.91
J10	70.61	1917.86	79.37
J11	36.84	1909.08	100.03
J12	21.49	1892.48	84.97
J13	33.80	1917.20	65.51
J14	132.01	1910.93	50.48
J15	64.57	1913.33	91.90
J16	33.77	1896.79	83.39
J17	70.61	1914.18	88.77
J18	49.12	1905.42	59.08
J19	46.05	1918.27	70.25

3.4 Comparison of Differential Evolution Algorithm Optimization and Classical Methods

In this paper the network is optimized by differential evolution algorithm (DE) and the results are compared with classic method. Table 8 shows the results of optimizing from differential evolution algorithm classic method.

In this study, we compared the algorithm (DE) with this method, Therefore, According to great potential of DE, the algorithm can be used in the loop, branch and complex network.

In table 9 optimal cost obtained by each method can be seen.

According to Table 9, it can be said that algorithm presents very good results for optimizing water distribution network. So that the optimal solution was compared with the classical empirical method and results showed that Implementation cost of the network by DE algorithm 10.66% lower than the classical empirical method.

Table 8 Optimum diameter from DE algorithm and classic method

Pipe	No. Pipe	Optimum Diameter (mm) DE Algorithm	Optimum Diameter (mm) Classic Method
Res.-J1	P1	-	-
J2-J3	P2	800	900
J3-J4	P3	800	900
J4-J5	P4	630	700
J4-J6	P5	225	250
J4-J7	P6	355	355
J4-J8	P7	500	500
J8-J9	P8	450	500
J9-J10	P9	355	400
J10-J11	P10	250	250
J11-J12	P11	140	160
J5-J13	P12	450	400
J13-J14	P13	355	315
J5-J15	P14	315	315
J5-J16	P15	180	200
J5-J17	P16	355	400
J17-J18	P17	225	250
J5-J19	P18	225	160

Table 9 Inlet pressure head and network cost by DE algorithm, Classic method and MILP Method

Methods	Inlet Pressure Head (m)	Cost ($)
DE Algorithm	134.69	737920
Classic Method	140	825935

4   Conclusions

In this study, to optimize water distribution network by DE algorithm, the best scale and probability coefficients (F and Cr) are 0.6 and 0.5, respectively. About the initial population and the number of generations investigation revealed that the initial population of 100 members and generations 200 are the best, in terms of time and efficiency.

Conclusions show DE algorithm runtime is less than the other method that provides absolute optimum. While optimization of differential evolution algorithm (737,920 \$) is %10.66 more than the classic method.

About major networks with many pipes, using differential evolution algorithm is recommended compared with other evolutionary algorithms, because of high-speed runtime and convergence to reach the optimum.

References

1. Alperovits, E., Shamir, U.: Design of optimal water distribution systems. Water Resour. Res. **13**(6), 885–900 (1977)
2. Babu, B.V., Angira, R.: Optimization of water pumping system using differential evolution strategies. In: Proceedings of the Second International Conference on Computational Intelligence, Robotics, and Autonomous Systems, pp. 25–30 (2003)
3. Bhave, P.R., Sonak, V.V.: A critical study of the linear programming gradient method for optimal designof water supply networks. Water Resour. Res. **28**(6), 1577–1584 (1992)
4. Cross, H.: Analysis of flow in networks of conduits or conductors, Bulletin no. 286, Univ. of Illinois Engrg. Experiment Station, III (1936)
5. Cunha, M., Sousa, J.: Water distribution network design optimization: simulated annealing approach. J. Water Resour. Plann. Manage. **125**(4), 215–221 (1999)
6. Dandy, G.C., Simpson, A.R., Murphy, L.J.: An improved genetic algorithm for pipe network optimization. Water Resour. Res. **32**(2), 449–458 (1996)
7. Eusuff, M.M., Lansey, K.E.: Optimization of water distribution network design using the shuffled frog leaping algorithm. J. Water Resour. Plann. Manage. **129**(3), 210–225 (2003)
8. Fujiwara, O., Khang, D.B.: A two-phase decomposition method for optimal design of looped water distribution networks. Water Resour. Res. **27**(5), 985–986 (1990)
9. Fujiwara, O., Jenchaimahakoon, B., Edirisinghe, N.C.P.: A modified linear programming gradient method for optimal design of looped water distribution networks. Water Resour. Res. **23**(6), 977–982 (1987)
10. Goulter, I.C., Lussier, B.M., Morgan, D.R.: Implications of head loss path choice in the optimization of water distribution networks. Water Resour. Res. **22**(5), 819–822 (1986)
11. Jacoby, S.L.S.: Design of optimal hydraulic networks. J. Hydraul. Div. **94**(3), 641–661 (1968)
12. Janga Reddy, M., Nagesh Kumar, D.: Multi-objective differential evolution with application to reservoir system optimization. J. Comput. Civil Eng. **21**(2), 136–146 (2007)
13. Kadu, M.S., Rajesh, G., Bhave, P.R.: Optimal design of water networks using a modified genetic algorithm with reduction in search space. J. Water Resour. Plann. Manage. **134**(2), 147–160 (2008)

14. Keedwell, E., Khu, S.-T.: Novel cellular automata approach to optimal water distribution network design. J. Comput. Civil Eng. **20**(1), 49–56 (2006)
15. Kessler, A., Shamir, U.: Analysis of linear programming gradient method for optimal design of water supply networks. Water Resour. Res. **25**(7), 1469–1480 (1989)
16. Maier, H.R., Simpson, A.R., Zecchin, A.C., Foong, W.K., Phang, K.Y., Seah, H.Y., Tan, C.L.: Ant colony optimization for design of water distribution systems. J. Water Resour. Plann. Manage. **129**(3), 200–209 (2003)
17. Mansouri, R., Torabi, H., Mirshahi, D.: Differential Evolution Algorithm (DE) to Estimate the Coefficients of Uniformity of Water Distribution in Sprinkler Irrigation. Scientific Journal of Pure and Applied Sciences **5**(2) (2014)
18. Montesinos, P., Guzman, A.G., Ayuso, J.L.: Water distribution network optimization using a modified genetic algorithm. Water Resour. Res. **35**(11), 3467–3473 (1999)
19. Neelakantan, T.R., Suribabu, C.R.: Optimal design of water distribution networks by a modified genetic algorithm. J. Civil Environ. Eng. **1**(1), 20–34 (2005)
20. Ostfeld, A., Tubaltzev, A.: Ant colony optimization for least cost design and operation of pumping and operation of pumping water distribution systems. Water Resour. Plann. Manage. **134**(2), 107–118 (2008)
21. Pitchai, R.: Model for Designing Water Distribution Pipe Networks. PhD Thesis, Harvard University, Cambridge (1966)
22. Quindry, G., Brill, E.D., Lienman, J.: Water Distribution System Design Criteria. Department of Civil Engineering, University of Illinois at Urbana-Champaign
23. Savic, D.A., Walters, G.A.: 1997 Genetic algorithms for least cost design of water distribution networks. J. Water Resour. Plann. Manage. **123**(2), 67–77 (1979)
24. ShahineZhad, B.: Optimal design of water distribution networks using mixed integer linear programming. Ph.D. thesis, Chamran University, Ahvaz, Iran (2011)
25. Stephenson, D.: Pipe flow analysis. Elsevier Science Publishers B.V. (1984)
26. Storn, R., Price, K.: Differential evolution - a simple and efficient heuristic for global optimization over continuous spaces. Journal of Global Optimization **11**, 341–359 (1997)
27. Storn, R., Price, K.: Differential Evolution-A Simple and Efficient Adaptive Scheme for Global Optimization over Continuous Spaces. Technical report, International Computer Science Institute, Berkeley (1995)
28. Suribabu, C.R.: Differential evolution algorithm for optimal design of water distribution networks. Journal of Hydroinformatics **12**(1), 66–82 (2010)
29. Suribabu, C.R., Neelakantan, T.R.: Design of water distribution networks using particle swarm optimization. J. Urban Water **3**(2), 111–120 (2006)
30. Suribabu, C.R., Neelakantan, T.R.: Particle swarm optimization compared to other heuristic search techniques for pipe sizing. J. Environ. Informatics **8**(1), 1–9 (2006)
31. Vairavamoorthy, K., Ali, M.: Pipe index vector: a method to improve genetic-algorithm-based pipe optimization. J. Hydraul. Eng. **131**(12), 1117–1125 (2005)
32. Vairavamoorthy, K., Ali, M.: Optimal design of water distribution systems using genetic algorithms. Comput. Aided Civil Infrastruc. Eng. **15**(2), 374–382 (2000)
33. Varma, K.V., Narasimhan, S., Bhallamudi, S.M.: Optimal design of water distribution systems using an NLP method. J. Environ. Eng. **123**(4), 381–388 (1997)
34. Vasan, A., Raju, K.: Application of differential evolution for irrigation planning: an Indian case study. Water Res. Manage. **21**(8), 1393–1407 (2007)
35. Zecchin, A.C., Maier, H.C., Simpson, A.R., Leonard, M., Nixon, J.B.: Ant colony optimization applied to water distribution system design: comparative study of five algorithms. J. Water Resour. Plann. Manage. **133**(1), 87–92 (2007)

Part VII
Multi-objectives Variants of HSA

Artificial Satellite Heat Pipe Design Using Harmony Search

Zong Woo Geem

Abstract The design of an artificial satellite requires an optimization of multiple objectives with respect to performance, reliability, and weight. In order to consider these objectives simultaneously, multi-objective optimization technique can be considered. In this chapter, a multi-objective method considering both thermal conductance and heat pipe mass is explained for the design of a satellite heat pipe. This method has two steps: at first, each single objective function is optimized; then multi-objective function, which is the sum of individual error between current function value and optimal value in terms of single objective, is minimized. Here, the multi-objective function, representing thermal conductance and heat pipe mass, has five design parameters such as 1) length of conduction fin, 2) cutting length of adhesive attached area, 3) thickness of fin, 4) adhesive thickness, and 5) operation temperature of the heat pipe. Study results showed that the approach using harmony search found better solution than traditional calculus-based algorithm, BFGS.

Keywords Artificial satellite heat pipe · Harmony search · Optimization

1 Introduction

Traditionally researchers have utilized calculus-based algorithms that give gradient information in order to find the right direction to the optimal solution of their own optimization problems. However, some researchers have recently turned their interests to nature-inspired meta-heuristic algorithms including genetic algorithm [1], simulated annealing [2], tabu search [3], ant colony optimization [4], and particle swarm optimization [5].

Z.W. Geem(✉)
Department of Energy IT, Gachon University, Seongnam-si,
Gyeonggi-do 461-701, South Korea
e-mail: geem@gachon.ac.kr

© Springer-Verlag Berlin Heidelberg 2016
J.H. Kim and Z.W. Geem (eds.), *Harmony Search Algorithm*,
Advances in Intelligent Systems and Computing 382,
DOI: 10.1007/978-3-662-47926-1_40

423

These meta-heuristic algorithms have advantages over calculus-based algorithms because they do not require complex calculus-derivative, massive enumeration, and sensitive initial solutions. Also, meta-heuristic algorithms can easily handle discrete decision variables as well as continuous decision variables.

Recently, another meta-heuristic algorithm, named harmony search (HS), has been emerging. The HS algorithm mimics Jazz improvisation and it contains a novel stochastic derivative [6] based on the experiences of musicians in Jazz improvisation process. Instead of the calculus information of an objective function, the stochastic derivative of HS gives a probability to be selected for each value of a decision variable. For example, if the decision variable   has three candidate values {3, 4, 5}, the partial stochastic derivative of the objective function with respect to   at each discrete value gives the selection probability for each value like 30% for 3; 50% for 4; and 20% for 5. While cumulative probability becomes unity (100%), the probability to be selected for each value is updated iteration by iteration based on musician's experience. Desirably the value, which is included in the optimal solution vector, has higher chance to be chosen as the iterations progress.

With this stochastic derivative information, the HS algorithm has been applied to various science and engineering optimization problems that include [7-9]:

- Real-world applications: music composition, sudoku puzzle, time-tabling, tour planning, logistics
- Computer science problems: web page clustering, text summarization, Internet routing, visual tracking, robotics
- Electrical engineering problems: energy system dispatch, photo-electronic detection, power system design, multi-level inverter optimization, cell phone network
- Civil engineering problems: structural design, water network design, dam scheduling, flood model calibration, groundwater management, soil stability analysis, ecological conservation, vehicle routing
- Mechanical engineering problems: heat exchanger design, satellite heat pipe design, offshore structure mooring
- Bio & medical applications: RNA structure prediction, hearing aids, medical physics

The goal of this chapter is to review the HS optimization in space engineering, especially, the optimal design of a satellite heat pipe.

2   Structure of Harmony Search Algorithm

As mentioned above, the HS algorithm was originally inspired by the improvisation process of Jazz musicians. Figure 1 [8] shows the analogy between improvisation and optimization: each musician corresponds to each decision variable; musical instrument's pitch range corresponds to decision variable's value range;

musical harmony at certain time corresponds to solution vector at certain iteration; and audience's aesthetics corresponds to objective function. Just like musical harmony is improved practice by practice, solution vector is improved iteration by iteration.

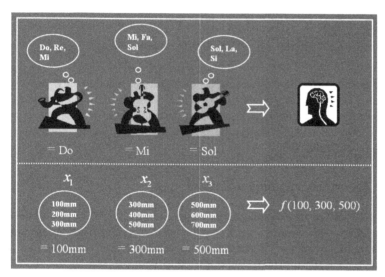

Fig. 1 Analogy between Improvisation and Optimization

This section introduces each step of the HS algorithm in detail, including 1) problem formulation, 2) algorithm parameter setting, 3) random tuning for memory initialization, 4) harmony improvisation (random selection, memory consideration, and pitch adjustment), 5) memory update, 6) performing termination, and 7) cadenza [8].

2.1 Problem Formulation

The HS algorithm was developed for solving optimization problems. Thus, in order to apply HS, problems should be formulated in the optimization environment, having objective function and constraints:

$$\text{Optimize (minimize or maximize)} \quad f(\mathbf{x}) \tag{1}$$

Subject to

$$h_i(\mathbf{x}) = 0; \quad i = 1,\ldots, p \; ; \tag{2}$$

$$g_i(\mathbf{x}) \geq 0; \quad i = 1,\ldots, q \; . \tag{3}$$

$$x_i \in \mathbf{X}_i = \{x_i(1),\ldots,x_i(k),\ldots,x_i(K_i)\} \quad \text{or} \quad x_i^L \leq x_i \leq x_i^U \tag{4}$$

The HS algorithm searches entire solution area in order to find globally optimal solution vector $\mathbf{x} = (x_1, \ldots, x_n)$, which optimizes (minimizes or maximizes) the objective function as in Equation 1. If the problem has equality and/or inequality conditions, these can be considered as constraints in Equations 2 and 3. If the decision variable has discrete values, the set of candidate values for the variable becomes $x_i \in \mathbf{X}_i = \{x_i(1), \ldots, x_i(k), \ldots, x_i(K_i)\}$; and if the decision variable has continuous values, the set of candidate values for the variable becomes $x_i^L \leq x_i \leq x_i^U$.

The HS algorithm basically considers the objective function only. However, if a solution vector generated violates any of the constraints, 1) the algorithm abandons the vector or 2) considers it by adding certain amount of penalty to the objective function value. Also, HS can be applied to multi-objective problems by conjugating with Pareto set.

2.2 Algorithm Parameter Setting

Once the problem formulation is ready, algorithm parameters should be set with certain values. HS contains algorithm parameters including harmony memory size (HMS), harmony memory considering rate (HMCR), pitch adjusting rate (PAR), maximum improvisation (MI), and fret width (FW).

HMS is the number of solution vectors simultaneously handled in the algorithm; HMCR is the rate ($0 \leq \text{HMCR} \leq 1$) where HS picks one value randomly from musician's memory. Thus, (1-HMCR) is the rate where HS picks one value randomly from total value range; PAR ($0 \leq \text{RAR} \leq 1$) is the rate where HS tweaks the value which was originally picked from memory. Thus, (1-PAR) is the rate where HS keeps the original value obtained from memory; MI is the number of iterations. HS improvises one harmony (= vector) each iteration; and FW is arbitrary length only for continuous variable, which was also known as bandwidth (BW). For more information of the term, a fret is the metallic ridge on the neck of a string instrument (such as guitar), which divides the neck into fixed segments, and each fret represents one semitone. In the context of the HS algorithm, frets mean arbitrary points which divide the total value range into fixed segments, and fret width (FW) is the length between two neighboring frets. Uniform FW is normally used in HS.

Originally, fixed parameter values were used. However, some researchers have proposed changeable parameter values. Mahdavi et al. [10] suggested that PAR increase linearly and FW decrease exponentially with iterations:

$$PAR(I) = PAR_{min} + (PAR_{max} - PAR_{min}) \times \frac{I-1}{MI-1} \tag{5}$$

$$FW(I) = FW_{max} \exp\left[\ln\left(\frac{FW_{min}}{FW_{max}}\right) \frac{I-1}{MI-1} \right] \tag{6}$$

Das et al. [11] suggested that FW be the standard deviation of the current population when HMCR is close to 1.

$$FW(I) = \sigma(\mathbf{x}_i) = \sqrt{\mathrm{var}(\mathbf{x}_i)} \tag{7}$$

Geem [12] tabulated fixed parameter values, such as number of variables, HMS, HMCR, PAR, and MI, after surveying various literatures. FW normally ranges from 1% to 10% of total value range.

Furthermore, some researchers have proposed adaptive parameter theories that enable HS to automatically have best parameter values at each iteration [13-17].

2.3 Random Tuning for Memory Initialization

After problem is formulated and the parameter values are set properly, random tuning process is performed.

In an orchestra concert, after oboe plays the note A (usually 440Hz), other instruments randomly play any pitches out of playable ranges. Likewise, the HS algorithm initially improvises many random harmonies. The number of random harmonies should be at least HMS. However, the number can be more than HMS, such as twice or three times as many as HMS [18]. Then, top-HMS harmonies are selected as starting vectors.

Musician's harmony memory (HM) can be considered as a matrix:

$$\mathbf{HM} = \begin{bmatrix} x_1^1 & x_2^1 & \cdots & x_n^1 & f(\mathbf{x}^1) \\ x_1^2 & x_2^2 & \cdots & x_n^2 & f(\mathbf{x}^2) \\ \vdots & \cdots & \cdots & \cdots & \vdots \\ x_1^{HMS} & x_2^{HMS} & \cdots & x_n^{HMS} & f(\mathbf{x}^{HMS}) \end{bmatrix} \tag{8}$$

Previously, the objective function values were sorted ($f(\mathbf{x}^1) \le f(\mathbf{x}^2) \le \ldots \le f(\mathbf{x}^{HMS})$) in HM, but current structure does not require it any more.

2.4 Harmony Improvization

In Jazz improvisation, a musician plays a note by randomly selecting it from total playable range (see Figure 2), or from musician's memory (see Figure 3), or by tweaking the note obtained from musician's memory (see Figure 4). Likewise, the HS algorithm improvises a value by choosing it from total value range or from HM, or tweaking the value which was originally chosen from HM.

Fig. 2 Total Playable Range of a Music Instrument

Fig. 3 Set of Good Notes in Musician's Memory

Fig. 4 Tweaking the Note chosen from Musician's Memory

Random Selection: When HS has to determine the value x_i^{New} for the new harmony $\mathbf{x}^{New} = (x_1^{New}, \ldots, x_n^{New})$, it randomly picks any value from total value range ($\{x_i(1), \ldots, x_i(K_i)\}$ or $x_i^L \le x_i \le x_i^U$) with probability of (1-HMCR). Random selection is also used for previous memory initialization.

Memory Consideration: When HS has to determine the value x_i^{New}, it randomly picks any value x_i^j from HM = $\{x_i^1, \ldots, x_i^{HMS}\}$ with probability of HMCR. The index j can be calculated using uniform distribution $U(0,1)$:

$$j \leftarrow \text{int}(U(0,1) \cdot HMS) + 1 \tag{9}$$

However, we may use different distributions. For example, if we use $[U(0,1)]^2$, HS chooses lower j more. If the objective function values are sorted by j, HS will behave similar to particle swarm algorithm, which prefers the best solution vector.

Pitch Adjustment: After the value x_i^{New} is randomly picked from HM in the above memory consideration process, it can be further adjusted into neighbouring values by adding certain amount to the value, with probability of PAR. For discrete variable, if $x_i(k) = x_i^{New}$, the pitch-adjusted value becomes $x_i(k+m)$ where $m \in \{-1, 1\}$ normally; and for continuous variable, the pitch-adjusted value becomes $x_i^{New} + \Delta$ where $\Delta = U(-1, 1) \cdot FW(i)$ normally.

The above-mentioned three basic operations (random selection, memory consideration and pitch adjustment) can be expressed as follows:

$$
x_i^{New} \leftarrow \begin{cases}
\begin{cases}
x_i \in \{x_i(1), ..., x_i(k), ..., x_i(K_i)\} \\
x_i \in [x_i^{Lower}, x_i^{Upper}]
\end{cases} & \text{w.p.} \quad (1 - HMCR) \\
x_i \in \mathbf{HM} = \{x_i^1, x_i^2, ..., x_i^{HMS}\} & \text{w.p.} \quad HMCR \cdot (1 - PAR) \quad (10) \\
\begin{cases}
x_i(k+m) \text{ if } x_i(k) \in \mathbf{HM} \\
x_i + \Delta \text{ if } x_i \in \mathbf{HM}
\end{cases} & \text{w.p.} \quad HMCR \cdot PAR
\end{cases}
$$

Especially for discrete variables, the HS algorithm has the following stochastic partial derivative which consists of three terms such as random selection, memory consideration and pitch adjustment [6]:

$$
\frac{\partial f}{\partial x_i} = \frac{1}{K_i}(1 - HMCR) + \frac{n(x_i(k))}{HMS} HMCR(1 - PAR) + \frac{n(x_i(k-m))}{HMS} HMCR\, PAR \quad (11)
$$

Also, the HS algorithm can consider the relationship among decision variables using ensemble consideration just as there exists stronger relationship among specific musicians. The value x_i^{New} can be determined based on x_j^{New} if the two has the strongest relationship [19]:

$$
x_i^{New} \leftarrow fn(x_j^{New}) \quad \text{where} \quad \max_{i \neq j}\left\{[Corr(\mathbf{x}_i, \mathbf{x}_j)]^2\right\} \quad (12)
$$

If the newly improvised harmony $\mathbf{x}^{New}$ violates any constraint, HS abandons it or still keeps it by adding penalty to the objective function value just like musicians sometimes still accept rule-violated harmony (e.g., parallel fifth violation).

2.5 Memory Update

If the new harmony $\mathbf{x}^{New}$ is better, in terms of objective function value, than the worst harmony in HM, the new harmony is included in HM and the worst harmony is excluded from HM:

$$
\mathbf{x}^{New} \in \mathbf{HM} \quad \wedge \quad \mathbf{x}^{Worst} \notin \mathbf{HM} \quad (13)
$$

However, for the diversity of harmonies in HM, other harmonies (in terms of least-similarity) can be considered. Also, the number of identical harmonies in HM can be limited in order to prevent premature HM.

2.6   Performing Termination

If HS satisfies termination criteria (for example, reaching MI), the computation is terminated. Otherwise, HS improvises another new harmony again.

2.7   Cadenza

Cadenza is a musical passage occurring at the end of a movement. In the context of the HS algorithm, cadenza can be referred to a process occurring at the end of the HS computing. In this process, HS returns the best harmony ever found and stored in HM.

3   Satellite Heat Pipe Design Example

For the satellite cooling system, heat pipes are generally used, and many researchers have applied optimization techniques to the heat pipe design [20-22]. The heat pipe model explained in this chapter is as shown in Figure 5 [22, 23], which has two objectives (thermal conductance and total mass) and five design parameters: 1) length of flange fin (L_f), 2) cutting length of adhesive attached area (L_c), 3) thickness of fin (t_f), 4) Adhesive thickness (t_b), and 5) operation temperature (T_{op}).

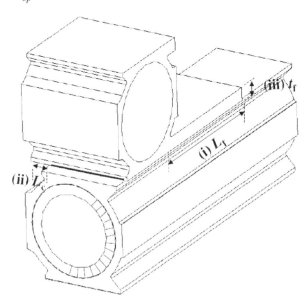

Fig. 5 Schematic of Satellite Heat Pipe

The parameter ranges are:

- L_f : 10.0mm ~ 25.4mm
- L_c : 1.5mm ~ 2.5mm
- t_f : 1.0mm ~ 1.7mm
- t_b : 0.12mm ~ 0.22mm
- T_{op} : -20.0°C ~ 60.0°C

The objective function of the thermal conductance across the thermal joint of the heat pipe, which is to be maximized, is as follows:

$$
\begin{aligned}
G = f(L_f, L_c, t_f, t_b, T_{op}) = {} & 0.3745378 - 0.9352909\,\text{tb} \\
& + 1.01612\,\text{tb}^2 + 0.02324128\,L_c - 0.007209993\,L_c^2 \\
& + 0.001838379\,L_f - 0.00005379707\,L_f^2 + 0.02447391\,t_f \\
& + 0.002304583\,t_f^2 - 0.0006483411\,T_{op} - 0.0000009232971\,T_{op}^2 \\
& - 0.02259702\,t_b\,L_c - 0.004735652\,t_b\,L_c^2 + 0.1102442\,t_b^2\,L_c \\
& - 0.009702533\,t_b^2 L_c^2 + 0.005382211\,t_b L_f - 0.00009540484\,t_b L_f^2 \\
& + 0.00515048\,t_b^2\,L_f - 0.0001232524\,t_b^2\,L_f^2 + 0.2972589\,t_b\,t_f \\
& - 0.1052935\,t_b\,t_f^2 - 0.5422262\,t_b^2\,t_f - 0.1829687\,t_b^2\,t_f^2
\end{aligned}
\tag{14}
$$

The objective function of the total mass, which is to be minimized, is as follows:

$$
\begin{aligned}
M = f(L_f, L_c, t_f, t_b) = {} & (1313.877 - 75.5\,L_c + 11.0\,L_c^2 \\
& + 1.402597\,L_f - 1.278314E\text{-}15\,L_f^2 + 62.38776\,t_f \\
& - 6.122449\,t_f^2 - 380.8\,t_b + 1120\,t_b^2) \times 21
\end{aligned}
\tag{15}
$$

The multi-objective function considering both thermal conductance and total mass, which is to be minimized, can be as follows:

$$
Z = f(L_f, L_c, t_f, t_b, T_{op}) = \left| \frac{G - G^*}{G^*} \right| + \left| \frac{M - M^*}{M^*} \right|
\tag{16}
$$

As seen in the above equation, the multi-objective function is expressed as the sum of the absolute values of relative errors between optimal value and current function value, where the optimal value for each objective function can be before-hand calculated using single objective functions for thermal conductance and total mass, respectively [23].

For the single objective optimization, HS found the maximal thermal conductance of 0.3945 (W/K) or total mass of 25.763 (kg) while BFGS, one of best mathematical techniques, found 0.3808 (W/K) or 25.868 (kg).

For the multiple objective optimization, HS found the maximal thermal conductance of 0.3810 (W/K) and total mass of 26.704 (kg) while BFGS found 0.3750 (W/K) and 26.854 (kg). The multi-objective solution of HS is Pareto optimal one because the conductance (0.3810) of HS is higher than that (0.3750) of BFGS and the mass (26.704) of HS is lower than that (26.854) of BFGS, that is, HS solution is better than BFGS one without disadvantaging at least one objective.

4 Conclusions

This chapter briefly explained the structure of the HS algorithm and also reviewed the HS application to the satellite heat pipe design which has two objectives of maximal thermal conductance and minimal total mass. The HS found better results than one of powerful calculus-based optimization technique, BFGS.

The HS algorithm is still growing. We hope other researchers to suggest new ideas to make better shape of the algorithm structure. Also, we hope to see various HS applications in aerospace engineering field in the future.

References

1. Goldberg, D.E.: Genetic Algorithms in Search Optimization and Machine Learning. Addison Wesley, MA (1989)
2. Kirkpatrick, S., Gelatt, C., Vecchi, M.: Optimization by Simulated Annealing. Science **220**, 671–680 (1983)
3. Glover, F.: Heuristic for Integer Programming using Surrogate Constraints. Decision Sci. **8**, 156–166 (1977)
4. Dorigo, M., Maniezzo, V., Colorni, A.: The Ant System: Optimization by a Colony of Cooperating Agents. IEEE Transactions on Systems, Man, and Cybernetics-Part B **26**, 29–41 (1996)
5. Kennedy, J., Eberhart, R.C.: Particle swarm optimization. In: Proceedings of the IEEE International Joint Conference on Neural Networks, pp. 1942–1948 (1995)
6. Geem, Z.W.: Novel derivative of harmony search algorithm for discrete design variables. Appl. Math. Comput. **199**, 223–230 (2008)
7. Geem, Z.W. (ed.): Music-Inspired Harmony Search Algorithm. SCI, vol. 191. Springer, Heidelberg (2009)
8. Geem, Z.W. (ed.): Recent Advances in Harmony Search Algorithm. SCI, vol. 270. Springer, Heidelberg (2009)
9. Geem, Z.W. (ed.): Harmony Search Algo. for Structural Design Optimization. SCI, vol. 239. Springer, Heidelberg (2009)
10. Mahdavi, M., Fesanghary, M., Damangir, E.: An improved harmony search algorithm for solving optimization problems. Appl. Math. Comput. **188**, 1567–1579 (2007)

11. Das, S., Mukhopadhyay, A., Roy, A., Abraham, A., Panigrahi, B.K.: Exploratory Power of the Harmony Search Algorithm: Analysis and Improvements for Global Numerical Optimization. IEEE Transactions on Systems, Man, and Cybernetics, Part B: Cybernetics **41**, 89–106 (2011)

12. Geem, Z.W.: Optimal cost design of water distribution networks using harmony search. Eng. Optimiz. **38**, 259–280 (2006)

13. Wang, C.M., Huang, Y.F.: Self-adaptive harmony search algorithm for optimization. Expert Syst. Appl. **37**, 2826–2837 (2010)

14. Hasancebi, O., Erdal, F., Saka, M.P.: Adaptive Harmony Search Method for Structural Optimization. J. Struct. Eng.-ASCE **136**, 419–431 (2010)

15. Geem, Z.W., Sim, K.B.: Parameter-Setting-Free Harmony Search Algorithm. Appl. Math. Comput. **217**, 3881–3889 (2010)

16. Geem, Z.W., Cho, Y.-H.: Optimal Design of Water Distribution Networks Using Parameter-Setting-Free Harmony Search for Two Major Parameters. J. Water Res. Pl-ASCE **137**, 377–380 (2011)

17. Geem, Z.W.: Parameter Estimation of the Nonlinear Muskingum Model Using Parameter-Setting-Free Harmony Search. Journal of Hydrologic Engineering – ASCE **16**, 684–688 (2011)

18. Degertekin, S.: Optimum design of steel frames using harmony search algorithm. Struct. Multidiscp. O. **36**, 393–401 (2008)

19. Geem, Z.W.: Improved harmony search from ensemble of music players. Lecture Notes in Artificial Intelligence **4251**, 86–93 (2006)

20. Rajesh, V.G., Ravindran, K.P.: Optimum Heat Pipe Design: A Nonlinear Programming Approach. International Communications in Heat and Mass Transfer **24**, 371–380 (1997)

21. de Sousa, F.L., Vlassov, V., Ramos, F.M.: Generalized Extremal Optimization: An Application in Heat Pipe Design. Appl. Math. Model. **28**, 911–931 (2004)

22. Huband, S., Barone, L., While, L., Hingston, P.: A scalable multi-objective test problem toolkit. In: Coello Coello, C.A., Hernández Aguirre, A., Zitzler, E. (eds.) EMO 2005. LNCS, vol. 3410, pp. 280–295. Springer, Heidelberg (2005)

23. Geem, Z.W., Hwangbo, H.: Application of harmony search to multi-objective optimization for satellite heat pipe design. In: Proceedings of US-Korea Conference on Science, Technology, & Entrepreneurship (UKC 2006), CD-ROM (2006)

A Pareto-Based Discrete Harmony Search Algorithm for Bi-objective Reentrant Hybrid Flowshop Scheduling Problem

Jingnan Shen, Ling Wang, Jin Deng and Xiaolong Zheng

Abstract In this paper, a Pareto-based discrete harmony search (P-DHS) algorithm is proposed to solve the reentrant hybrid flowshop scheduling problem (RHFSP) with the makespan and the total tardiness criteria. For each job, the operation set of each pass is regarded as a sub-job. To adopt the harmony search algorithm to solve the RHFSP, each harmony vector is represented by a discrete sub-job sequence, which determines the priority to allocate all the operations. To handle the discrete representation, a novel improvisation scheme is designed. During the search process, the explored non-dominated solutions are stored in the harmony memory with a dynamic size. The influence of the parameter setting is investigated, and numerical tests are carried out based on some benchmarking instances. The comparisons to some existing algorithms in terms of several performance metrics demonstrate the effectiveness of the P-DHS algorithm.

Keywords Harmony search · Improvisation scheme · Reentrant hybrid flowshop scheduling · Bi-objective · Pareto dominance

1 Introduction

The hybrid flowshop scheduling problem (HFSP) is of significant applications in the flexible manufacturing environment, which has been extensively studied by many researchers during the past few decades [1-3]. Classical HFSP assumes that each job visits each stage only once. However, due to some technical reasons or quality consideration, some jobs have to visit or reenter some stages several times.

J. Shen(✉) · L. Wang · J. Deng · X. Zheng
Tsinghua National Laboratory for Information Science and Technology (TNList),
Department of Automation, Tsinghua University, Beijing 100084,
People's Republic of China
e-mail: {chenjn12,dengj13,zhengxl11}@mails.tsinghua.edu.cn, wangling@tsinghua.edu.cn

© Springer-Verlag Berlin Heidelberg 2016
J.H. Kim and Z.W. Geem (eds.), *Harmony Search Algorithm*,
Advances in Intelligent Systems and Computing 382,
DOI: 10.1007/978-3-662-47926-1_41

435

Such a characteristic leads to an extension of the HFSP called the reentrant HFSP (RHFSP), which can be found in many real-world production systems, such as semiconductor wafer fabrication, printed circuit board fabrication, and thin film transistor-liquid crystal display (TFT-LCD) panel manufacturing [4]. The RHFSP is more complex than the classical HFSP, which has been proved to be NP-hard [5].

Recently, the RHFSP has attracted increasing attention due to its theoretical and practical importance. For a two-stage RHFSP with the makespan criterion, Choi et al. [6] presented some heuristics and a branch & bound algorithm. Then, Hekmatfar et al. [7] developed a hybrid genetic algorithm (GA). For a three-stage RHFSP, Bertel and Billaut [8] designed some heuristics to minimize the weighted number of tardy jobs. Besides, the multi-stage RHFSP has been studied in much literature [4, 9-14]. Choi et al. [9] presented two list-scheduling algorithms for the RHFSP with a set of orders to minimize the total tardiness criterion. Kim and Lee [10] designed two heuristics to minimize the makespan by considering the total tardiness as a constraint with a certain allowance. To minimize the makespan and the total tardiness simultaneously, Cho et al. [11] developed some local-search based Pareto GAs, and Ying et al. [4] proposed an iterated Pareto greedy (IPG) algorithm with better performances. To minimize the total workload, Pearn et al. [12] proposed three network algorithms to solve an IC final testing scheduling problem with reentry, which is a generalization of the RHFSP. Dugardin et al. [13] developed a Lorenz-dominance based multi-objective GA for RHFSP with stochastic characteristics to minimize the mean cycle time and maximize the utilization rate of the bottleneck facility. Choi et al. [14] proposed a decision tree based real-time scheduling mechanism for the dynamic RHFSP arising from the TFT-LCD panel manufacturing.

With the development of computational intelligence, nature-inspired meta-heuristics have been powerful tools to solve complex optimization problems in a variety of fields [15, 16]. The harmony search (HS) algorithm [17,18] is a meta-heuristic inspired by the behaviors of the musical improvisation. In the HS algorithm, it treats each solution as a harmony. By mimicking a musician seeking for the pleasing harmony, the HS algorithm searches the best solution based on memory consideration, pitch adjustment and random selection. During recent years, the HS algorithm has been applied to many optimization problems [19], such as structural design [20], power dispatch [21,22], feature selection [23], and production scheduling [24], etc. Inspired by the successful applications of the HS algorithm, a Pareto-based discrete HS (P-DHS) algorithm is proposed in this paper for the bi-objective RHFSP with the makespan and the total tardiness criteria. In the P-DHS, each harmony vector is represented by a sub-job sequence, where a sub-job denotes the operation set of each pass of a job. To handle the discrete representation, a novel improvisation scheme is designed. The harmony memory is used to update the Pareto solutions explored during the search process. The influence of the parameter setting is investigated, and numerical comparisons based on some benchmarking instances are presented to show the effectiveness of the proposed algorithm.

The remainder of the paper is organized as follows: In Section 2, the RHFSP with the makespan and the total tardiness criteria is described. In Section 3, the P-DHS for the RHFSP is presented in details. In Section 4, the influence of parameter setting on the performances of the P-DHS is investigated, and the comparative results are provided. Finally, the paper is ended with some conclusions and future work in Section 5.

2 Problem Statement

The RHFSP can be described as follows. There are n jobs to be processed at s sequential stages. Each stage contains at least one identical parallel machine, and at least one stage contains more than one machine. At each stage, each job can be processed on any one of the machines. Due to the reentrant characteristic, some jobs should pass certain stages more than once. Suppose all jobs and machines are available at time zero. Each machine can process only one job at a time, and each job can be processed on only one machine at a time. All operations of each job are known in advance, and each operation should be completed without interruption once it is started. The capacity of buffers between successive stages is infinite.

For the RHFSP, it needs to determine the start time and the processing machine for each operation at every stage to optimize certain scheduling objectives. In this paper, the makespan (C_{max}) and the total tardiness (TDD) criteria are considered simultaneously, as shown in Eqs. (1) and (2), respectively, where C_j denotes the completion time of job j and d_j denotes its due date.

$$f_1 = C_{max} = \max C_j \tag{1}$$

$$f_2 = TTD = \sum_{j=0}^{n} \max(0, C_j - d_j) \tag{2}$$

According to the concept of Pareto dominance, a solution X_1 dominates solution X_2 if $\forall i \in \{1,2\}, f_i(X_1) \le f_i(X_2)$ and $\exists i \in \{1,2\}, f_i(X_1) < f_i(X_2)$. A solution is Pareto optimal or called the non-dominated solution, if it cannot be dominated by any other solution. A collection of all the Pareto optimal solutions is called a Pareto set, which forms the Pareto front in objective space. Thus, for the above bi-objective RHFSP, it aims to obtain the Pareto optimal solutions or the Pareto front.

3 P-DHS for Bi-objective RHFSP

In this section, the Pareto-based discrete harmony search (P-DHS) algorithm is presented to solve the RHFSP. First, the basic HS algorithm is introduced briefly with its main steps. Then, the encoding and decoding schemes, the initialization and update rules for the HM, and the improvisation scheme of the P-DHS are introduced in detail.

3.1 Basic HS Algorithm

In the HS algorithm [17,18], it treats each solution as a harmony and imitates the behaviors of musical improvisation to seek for good solutions (fantastic harmonies). The main steps of the basic HS algorithm are as follows.

Step 1: Initialize a harmony memory (HM).

Step 2: Improvise a new harmony based on memory consideration, pitch adjustment and random selection. For the memory consideration, the value of a decision variable is randomly chosen from the HM. The pitch adjustment is performed after the memory consideration with a certain probability to make a slight change to the value of the current decision variable. As for the random selection, the value of a decision variable is randomly generated between the possible ranges.

Step 3: Update the HM. If the new harmony is better than the worst harmony in the HM, replace the worst harmony by the new one.

Step 4: Repeat Step 2 and Step 3 until the stopping criterion is met.

3.2 Encoding and Decoding Schemes

In the RHFSP, each job should pass all the stages one or more times. For each job, the operation set of each pass is regarded as a sub-job. Thus, each job can be treated as a set of sub-jobs with precedence constraints. In the P-DHS, each harmony denotes a solution of the RHFSP, which is represented by a sub-job sequence. In the sub-job sequence, each sub-job is denoted by the number of the job to which it belongs. The k-th occurrence of a job number refers to the k-th sub-job according to the processing sequence of the job. Considering a simple RHFSP with three jobs (each job should pass all the stages twice), the representation of a feasible solution is illustrated in Fig. 1.

Sub-job sequence | 1 | 1 | 2 | 3 | 2 | 3 |

Fig. 1 Example of the Encoding Scheme

To obtain feasible schedules, it needs to arrange machines for all the operations and then determine the processing sequence on all the machines. In the P-DHS, the sub-job sequence determines the order for allocating all the sub-jobs. When allocating a certain sub-job, all its operations are assigned to the machines that can process the sub-job with the earliest completion time.

3.3 Initialization and Update Rules for the HM

In the P-DHS, the HM is initialized with a random generated harmony. At each generation, the HM will be updated by a new harmony. To handle the bi-objective optimization problem, two rules based on the concept of Pareto dominance are used to update the HM: (1) if a new harmony is not dominated by any a harmony in the HM, then add the new one to the HM; (2) if a harmony in the HM is dominated by the new harmony, then remove the harmony from the HM. Thus, in the P-DHS the HM is used to store all the non-dominated solutions explored during the search process, and its size changes dynamically.

3.4 Improvisation Scheme

The HS algorithm adopts three rules to improvise a new harmony, including memory consideration, pitch adjustment and random selection. In the P-DHS, the pitch adjustment is modified to handle the sub-job based representation. Besides, a repair procedure is embedded to ensure the feasibility of a new solution. The Pseudo-code of the improvisation scheme is shown in Fig. 2.

```
Initialize job number set I= {1, 2, ..., n};
Random select a harmony X from the HM;
For k = 1 : length(X)
    If rand(0, 1) >= HMCR //Random selection
        Randomly select a job number j from I;
        new_X(k) = j;
    Else //Memory consideration
        If rand(0, 1) >= PAR
            k₁ = k;
        Else //Pitch adjustment
            k₁ = k - 1 or  k₁ = k + 1;
        End If
        While (X(k₁) ∉ I) //Repair
            k₁++;
            If  k₁ == length(X) +1
                k₁ = 1
            End If
        End While
        new_X(k) = X(k₁);
    End If
    If all sub-jobs of job new_X(k) are allocated
    //Update the job number set
        Delete new_X(k) from I;
    End If
End For
```

Fig. 2 Pseudo-code of the Improvisation Scheme

4   Computational Experiments

The proposed P-DHS algorithm was implemented in C++ and run on a personal computer with a 3.30 GHz processor. The benchmarking instances generated by Cho et al. [11] are used for numerical tests. Job number, stage number, machine number, reentrance times, and processing times are random integers from the uniform distributions U[10, 20], U[5, 10], U[1, 2], U[1, 2] and U[1, 10], respectively. The due dates are random numbers from the uniform distribution U[αP, βP], where P denotes the weighted sum of two approximations of makespan (see [11] for details) and the parameter pair (α, β) controls the tightness of due dates. Six pairs denoted as Case 1 ~ Case 6 are considered, including (0, 0.33), (0.33, 0.67), (0.67, 1), (0, 0.5), (0.5, 1) and (0, 1). For each case, there are 20 different instances, which results in a total of 120 instances.

4.1   Performance Metrics

Three performance metrics [4,11] for the bi-objective optimization are used for comparisons, including the convergence metric (γ), the diversity metric (Δ), and the dominance metric (Ω).

The convergence metric measures the convergence of the obtained solutions to the known optimal Pareto front. First, the minimum Euclidean distance from each obtained non-dominated solution to the known optimal Pareto solutions is calculated. Then, it calculates the average of the distances of all the obtained solutions as γ. Clearly, the smaller γ is, the better the convergence quality is.

The diversity metric shows the spread of the obtained non-dominated solutions, which is calculated as follows:

$$\Delta = \frac{d_f + d_l + \sum_{i=1}^{N-1}\left|d_i - \overline{d}\right|}{d_f + d_l + (N-1)\overline{d}} \tag{3}$$

where d_f and d_l are the Euclidean distances between the boundary solutions of the obtained Pareto solutions with the extreme solutions of the known optimal Pareto front; N denotes the number of the obtained Pareto solutions; d_i is the Euclidean distance between the i-th and the $(i+1)$-th consecutive solutions in the obtained Pareto set; and $\overline{d}$ denotes the average distance of all the solutions. Clearly, the value of Δ will decrease to zero, if all the solutions are uniformly located on the front. Thus, the smaller the value is, the better the diversity is.

The dominance metric is defined as the percentage of the Pareto solutions obtained by a certain algorithm in the known Pareto set. Let H_{A_k} denote the set of Pareto solutions obtained by a given algorithm A_k, the dominance metric Ω_{A_k} is calculated as follows:

$$\Omega_{A_k} = \frac{|P(\bigcup_{A_i} H_{A_i}) \setminus P(\bigcup_{A_i \neq A_k} H_{A_i})|}{|P(\bigcup_{A_i} H_{A_i})|} \tag{4}$$

where $|S|$ denotes the number of the Pareto solutions in a set S. Clearly, the value of Ω_{A_k} will increase to 1, if all the Pareto solutions are obtained by algorithm A_k. Thus, the larger the value is, the better the dominance of the obtained solutions by the corresponding algorithm is.

4.2   Parameter Setting

The P-DHS contains two key parameters: the harmony memory consideration rate *HMCR* and the pitch adjustment rate *PAR*. To investigate the influence of parameter setting on the performances of the P-DHS in solving the bi-objective RHFSP, the two-way analysis of variance (ANOVA) [25] is implemented by using the instance *Sproblem-01-01*.

Four levels are considered for each parameter, as listed in Table 1. For each parameter combination, we run the P-DHS with 200000 evaluations and record the obtained Pareto solutions. Then, it calculates the dominance metric for each parameter combination as the response variable (*RV*) of this parameter combination. Clearly, the larger the *RV* value is, the better the combination is. To yield more reliable information, we carry out the above steps 20 independent times, and present the ANOVA results in Table 2.

Table 1 Combinations of Parameter Values

Parameters	Factor Level			
	1	2	3	4
HMCR	0.6	0.7	0.8	0.9
PAR	0.1	0.2	0.3	0.4

Table 2 ANOVA Results

Source	Degree of Freedom	Sum of Squares	Mean Square	*F*-ratio	*P*-value
HMCR	3	0.43	0.14	7.26	0.00
PAR	3	0.53	0.18	8.91	0.00
Interaction	9	0.10	0.01	0.55	0.84
Error	304	5.97	0.02		
Total	319	7.02			

From Table 2, it can be seen that the *P*-values for *HMCR* and *PAR* are 0.00<0.05, that is, both parameters significantly affect the performances of the P-DHS at the 95% confidence level. Clearly, a large value of *HMCR* or a small value of *PAR* will lead to premature convergence, while a small value of *HMCR* or a large value of *PAR* will slow the convergence rate. As for the interaction effect of the two parameters, the *P*-value is 0.84>0.05, that is, there is no significant interaction effect between *HMCR* and *PAR* at the 95% confidence level. Then, we illustrate the level trends of the parameters in Fig. 3. According to

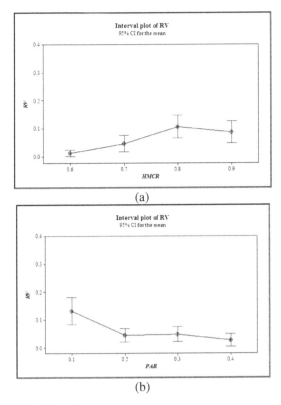

(a)

(b)

Fig. 3 Level Trend of the Parameters

Table 2 and Fig. 3, we set the parameter combination for the P-DHS algorithms as follows: *HMCR*=0.8, *PAR*=0.1.

4.3 Comparative Results

To evaluate the performances of the proposed P-DHS, we compare it with the IPG [4] and the MOHS [21] algorithms. The IPG was proposed by Ying et al, which has been shown as the state-of-the-art for the bi-objective RHFSP. To adopt the MOHS to the problem, the largest position value rule [24] is employed to transform a harmony vector to a sub-job sequence. Besides, the MOHS uses the same decoding scheme as the P-DHS. The parameters of MOHS are set as suggested in [21]: $HMS = 20$, $HMCR = 0.85$, $PAR_{min} = 0.2$, $PAR_{max} = 2$, $BW_{min} = 0.45$ and $BW_{max} = 0.9$. The three algorithms are run 20 independent times for each benchmarking instance. The IPG was run on a PC with a 2.40 GHz processor [4], while we run the P-DHS and the MOHS on a PC with a 3.30 GHz processor. After recording the Pareto sets, it calculates the performance metrics as section 4.1. The average results of the three performance metrics on the benchmarking instances grouped by the problem type (Case 1 ~ Case 6) are listed in Table 3.

Table 3 Comparative Results of P-DHS, MOHS and IPG

	P-DHS			MOHS			IPG		
	γ	Δ	Ω	γ	Δ	Ω	γ	Δ	Ω
Case 1	**11.709**	**0.763**	**0.660**	49.873	0.793	0.021	13.531	0.767	0.272
Case 2	**10.057**	**0.743**	**0.804**	43.203	0.792	0.021	20.149	0.783	0.138
Case 3	**4.730**	**0.692**	**0.755**	19.139	0.795	0.017	10.047	0.739	0.144
Case 4	**12.062**	**0.755**	**0.692**	54.382	0.790	0.013	16.088	0.764	0.276
Case 5	**7.166**	**0.637**	**0.635**	37.650	0.755	0.028	12.055	0.689	0.169
Case 6	**15.445**	0.828	**0.667**	72.579	0.836	0.007	18.086	**0.810**	0.326
Ave.	**10.195**	**0.736**	**0.702**	46.138	0.794	0.018	14.993	0.759	0.221

From Table 3, it can be seen that the P-DHS performs better than the MOHS and the IPG in considering all the performance metrics. Besides, Fig. 4 illustrates the Pareto fronts obtained by the P-DHS, the MOHS and the IPG in solving the instance *Sproblem-02-01*, which clearly illustrates the effectiveness of the P-DHS. As for the computational efficiency, the average CPU times used by the P-DHS, the MOHS and the IPG are 2.2s, 2.0s and 3.3s, respectively. Considering the different processors, it can be concluded that the computational costs of these algorithms are at the same level.

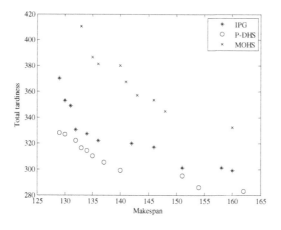

Fig. 4 Pareto Fronts of *Sproblem-02-01* Explored by the P-DHS, the MOHS and the IPG

Thus, it can be concluded that the proposed P-DHS is more effective than the MOHS and the IPG in solving the bi-objective RHFSP with similar computational costs.

5   Conclusions

In this paper, the reentrant hybrid flowshop scheduling problem (RHFSP) with the makespan and the total tardiness criteria was studied, and a Pareto-based discrete harmony search (P-DHS) algorithm was presented to solve the problem. With the sub-job sequence based representation, specific improvisation scheme and harmony memory updating, the P-DHS well solved the problem and achieved better statistical performance metrics than the existing algorithms. Future work could focus on developing the adaptive HS algorithm for the problem as well as generalizing the harmony search algorithm for the stochastic or dynamic RHFSP.

Acknowledgment The authors would like to thank Professor Cho HM and Professor Ying KC for providing us the benchmarking instances. This research is partially supported by the National Key Basic Research and Development Program of China (No. 2013CB329503), the National Science Foundation of China (No. 61174189), the Doctoral Program Foundation of Institutions of Higher Education of China (20130002110057).

References

1. Linn, R., Zhang, W.: Hybrid flow shop scheduling: a survey. Comput. Ind. Eng. **37**(1), 57–61 (1999)
2. Ruiz, R., Vázquez-Rodríguez, J.A.: The hybrid flow shop scheduling problem. Eur. J. Oper. Res. **205**(1), 1–18 (2010)
3. Ribas, I., Leisten, R., Framiñan, J.M.: Review and classification of hybrid flow shop scheduling problems from a production system and a solutions procedure perspective. Comput. Oper. Res. **37**(8), 1439–1454 (2010)
4. Ying, K.C., Lin, S.W., Wan, S.Y.: Bi-objective reentrant hybrid flowshop scheduling: an iterated Pareto greedy algorithm. Int. J. Prod. Res. **52**(19), 5735–5747 (2014)
5. Wang, M.Y., Sethi, S.P., Van De Velde, S.L.: Minimizing makespan in a class of reentrant shops. Oper. Res. **45**(5), 702–712 (1997)
6. Choi, H.S., Kim, H.W., Lee, D.H., Yoon, J., Yun, C.Y., Chae, K.B.: Scheduling algorithms for two-stage reentrant hybrid flow shops: minimizing makespan under the maximum allowable due dates. Int. J. Adv. Manuf. Technol. **42**(9–10), 963–973 (2009)
7. Hekmatfar, M., Ghomi, S.F., Karimi, B.: Two stage reentrant hybrid flow shop with setup times and the criterion of minimizing makespan. Appl. Soft Comput. **11**(8), 4530–4539 (2011)
8. Bertel, S., Billaut, J.C.: A genetic algorithm for an industrial multiprocessor flow shop scheduling problem with recirculation. Eur. J. Oper. Res. **159**(3), 651–662 (2004)
9. Choi, S.W., Kim, Y.D., Lee, G.C.: Minimizing total tardiness of orders with reentrant lots in a hybrid flowshop. Int. J. Prod. Res. **43**(11), 2149–2167 (2005)
10. Kim, H.W., Lee, D.H.: Heuristic algorithms for re-entrant hybrid flow shop scheduling with unrelated parallel machines. Proc. Inst. Mech. Eng. B. **223**(4), 433–442 (2009)
11. Cho, H.M., Bae, S.J., Kim, J., Jeong, I.J.: Bi-objective scheduling for reentrant hybrid flow shop using Pareto genetic algorithm. Comput. Ind. Eng. **61**(3), 529–541 (2011)

12. Pearn, W.L., Chung, S.H., Chen, A.Y., Yang, M.H.: A case study on the multistage IC final testing scheduling problem with reentry. Int. J. Prod. Econ. **88**(3), 257–267 (2004)
13. Dugardin, F., Yalaoui, F., Amodeo, L.: New multi-objective method to solve reentrant hybrid flow shop scheduling problem. Eur. J. Oper. Res. **203**(1), 22–31 (2010)
14. Choi, H.S., Kim, J.S., Lee, D.H.: Real-time scheduling for reentrant hybrid flow shops: a decision tree based mechanism and its application to a TFT-LCD line. Exp. Syst. Appl. **38**(4), 3514–3521 (2011)
15. Parpinelli, R.S., Lopes, H.S.: New inspirations in swarm intelligence: a survey. Int. J. Bio-Inspired Comput. **3**(1), 1–16 (2011)
16. Boussaïd, I., Lepagnot, J., Siarry, P.: A survey on optimization metaheuristics. Inform. Sciences. **237**, 82–117 (2013)
17. Geem, Z.W., Kim, J.H., Loganathan, G.V.: A new heuristic optimization algorithm: harmony search. Simul. **76**(2), 60–68 (2001)
18. Lee, K.S., Geem, Z.W.: A new meta-heuristic algorithm for continuous engineering optimization: harmony search theory and practice. Comput. Meth. Appl. Mech. Eng. **194**(36), 3902–3933 (2005)
19. Manjarres, D., Landa-Torres, I., Gil-Lopez, S., Del Ser, J., Bilbao, M.N., Salcedo-Sanz, S., Geem, Z.W.: A survey on applications of the harmony search algorithm. Eng. Appl. Artif. Intell. **26**(8), 1818–1831 (2013)
20. Degertekin, S.O.: Improved harmony search algorithms for sizing optimization of truss structures. Comput. Struct. **92**, 229–241 (2012)
21. Sivasubramani, S., Swarup, K.S.: Multi-objective harmony search algorithm for optimal power flow problem. Int. J. Electr. Power Energ. Syst. **33**(3), 745–752 (2011)
22. Khazali, A.H., Kalantar, M.: Optimal reactive power dispatch based on harmony search algorithm. Int. J. Electr. Power Energ. Syst. **33**(3), 684–692 (2011)
23. Diao, R., Shen, Q.: Feature selection with harmony search. IEEE Tran. Syst. Man. Cybernet. B. **42**(6), 1509–1523 (2012)
24. Wang, L., Pan, Q.K., Tasgetiren, M.F.: A hybrid harmony search algorithm for the blocking permutation flow shop scheduling problem. Comput. Ind. Eng. **61**(1), 76–83 (2011)
25. Casella, G., Berger, R.L.: Statistical Inference, vol. 2. Duxbury, Pacific Grove (2002)

A Multi-objective Optimisation Approach to Optimising Water Allocation in Urban Water Systems

S. Jamshid Mousavi, Kourosh Behzadian, Joong Hoon Kim and Zoran Kapelan

Abstract Lack of available surface water resources and increasing depletion of groundwater are the major challenges of urban water systems (UWSs) in semi-arid regions. This paper presents a long term multi-objective optimisation model to identify optimal water allocation in UWS. The objectives are to maximize reliability of water supply and minimize total operational costs while restricting the annual groundwater withdrawal. Pumping water from a dam reservoir and water recycling schemes are two alternatives for supplying water to increasing water demands over the planning horizon. The developed approach is demonstrated through its application to the UWS of Kerman City in Iran. The Pareto-optimal solutions are obtained as a trade-off between reliability of water supply and operational costs in the Kerman UWS. The results show that addition of recycled water to the water resources can provide the least cost-effective and most efficient way for meeting future water demands in different states of groundwater abstractions.

Keywords Urban water systems · Water allocation and multi-objective optimisation

S.J. Mousavi(✉)
School of Civil and Environmental Engineering, Amirkabir University of Technology, Tehran, Iran
e-mail: jmosavi@aut.ac.ir

K. Behzadian · Z. Kapelan
Centre for Water Systems, College of Engineering, Mathematics and Physical Sciences, University of Exeter, Exeter, UK
e-mail: {k.behzadian-moghadam,z.kapelan}@exeter.ac.uk

J.H. Kim
School of Civil, Environmental and Architectural Engineering, Korea University, Seoul 136-713, South Korea
e-mail: jaykim@korea.ac.kr

© Springer-Verlag Berlin Heidelberg 2016
J.H. Kim and Z.W. Geem (eds.), *Harmony Search Algorithm*,
Advances in Intelligent Systems and Computing 382,
DOI: 10.1007/978-3-662-47926-1_42

447

1   Introduction

Given the limited available fresh water resources, future planning of urban water systems (UWSs) is a challenging task for many water authorities when water demands increase owning principally to population growth [1]. This is especially a major challenge in arid and semi-arid regions where climate change intensifies water scarcity though severe droughts [2]. Climate change and subsequently water scarcity would not only prohibit further expansion of water resources but also impose some further constrains on the withdrawal of raw water from water resources (e.g. groundwater). In these circumstances, optimal development and operation of potential, available water resources would be inevitably a mandatory option for future planning. This is gradually becoming to be top of the agenda for many water companies that are in the high risk of water scarcity for critically rethinking a new infrastructure for development of urban water resources in order to handle imminent water deficits in the future. In addition to traditional water resources such as surface water (e.g. river and lake) and groundwater, water reuse as a non-traditional water resource has been receiving more attention in the recent decades as a viable option, which can be introduced as an alternative and supplementary water resource in the future planning of UWSs [3]. The advanced technologies of various water recycling schemes in the recent decades are converting these schemes into valuable commodities that can be highly competitive to the conventional water resources. Due to some concerns about public health and environmental issues, the application of water reuse has been primarily limited to non-potable uses including either the domestic use such as toilet flushing or outside uses such as agricultural and landscape irrigation and industrial uses [4].

Despite the existence of advanced technologies in water recycling, there remain issues relating to the extent of the performance improvement in conjunctive use of this water resource and other traditional sources [5]. Generally, previous researches have principally addressed the potential potable water reduction, environmental and economic criteria once water recycling is added to an UWS [5-7]. In one of the cases, Rozos et al. (2010) assessed the problem of optimal sizing of water recycling schemes in which the objective was to evaluate the trade-off between capital cost of installing water reuse schemes and potable water reduction in a water supply system [7]. However, as outlined above, construction of alternative water resources as a back-up plan is necessary, especially in vulnerable areas to climate change, which are heavily relied on available water resources. One of the recently built instances is the Sydney desalination plant that is kept on standby and will be in operation once the main dam storage level reaches 70% [8]. Given the availability of water recycling sources, the main challenge remains how to allocate water demand profiles from multiple water resources including traditional and non-traditional ones for long-term planning of UWSs. The impact assessment of recycling schemes within the context of water allocation in UWSs has yet to be included.

This paper presents a multi-objective optimisation model for identifying optimal water allocation strategies for long-term operation (e.g. 20-40 years) of Kerman UWS, Iran. Next section describes the methodology followed by illustrating the case study, and then results and discussion are presented. Key findings are finally summarised and recommendations are made.

2 Methodology

Optimised water allocation strategies derived from both conventional and non-conventional water resources are developed in this study for long-term operation of an UWS. The water allocation strategy is required once there exist multiple types of water resources each with different specifications. The types of water resources analysed here are groundwater and surface water for conventional and water recycling schemes for non-conventional type. This also demands an appropriate modelling approach for the simulation of water allocation from each source in different time steps and finally evaluate the performance of this operation for a long-term planning horizon. The simulation and performance assessment in Kerman UWS is undertaken by using the WaterMet^2 model. Further details of both simulation and optimisation models are described in the following.

2.1 Simulation Model

The simulation of the UWS operation is handled by using WaterMet^2 model, which is a demand-oriented, conceptual model for mass balance-based simulation of water and other flows in the main components of UWSs [2, 9]. The modelled UWS comprises three main subsystems of clean water supply, stormwater and wastewater collection. The main components modelled in these subsystems are various conveyance components (water supply conduits, distribution mains, trunk mains, wastewater collection) and storage components (water resources, water treatment works (WTWs) and service reservoirs, subcatchments, waste water treatment works (WWTWs)). Any arbitrary number of each type of these components can be defined in WaterMet^2 (see Fig. 2). WaterMet^2 also quantifies the principal water-related flows and other metabolism-based fluxes in an UWS such as delivered demand, energy, costs and so on for each component and each time step over the entire analysed period. A number of other indicators can be derived from these basic flows and fluxes for the analysis of the UWS performance. Additional details of the WaterMet^2 model and its application in UWSs can be found in [2, 9, 10]. In this study, reliability of water supply and total costs are two indicators for performance measures in UWSs, which will be used as objective functions in the next section.

In the water supply subsystem module of WaterMet^2, simulation of water allocation for multiple demand using multiple water sources follows a two-step demand-oriented approach [10]. The first step calculates water demands of subcatchments and then aggregate them up to the most upstream components (i.e. water resources) during each time step. The calculated water demands are added by the leakages from conveyance elements. For instance, the volume of water demand for water resource i and time step t (RD_{it}) is calculated in Eq. (1) by adding the leakage percentage pertaining to conduit SC_{ij} (CL_{ij}) to the water demand of that conduit:

$$RD_{it} = \sum_{j=1}^{m} CF_{ij} \times WD_{jt} \left(1 + CL_{ij}/100\right) \qquad (1)$$

where WD_{jt}=water demand of WTW j; CF_{ij}= pre-specified fraction (coefficient) of water demand of WTW j supplied from resource i by conduit SC_{ij}; m=number of WTWs. By identifying the total water demands from each water resource, the second step starts off by abstracting water and distribution among downstream elements sequentially in which capacity control of storage components (i.e. both minimum and maximum) are the only governing equations. The released/abstracted water is finally distributed among subcatchments. In this study, WaterMet^2 uses monthly time steps for simulation of Kerman UWS performance for a period of 30 years defined as the planning horizon.

Two common water reuse mechanisms, i.e. rainwater harvesting (RWH) and grey water recycling (GWR), can also be simulated and incorporated into the UWS in WaterMet^2. In this study, the many small RWH and GWR units located across the UWS are represented by using a single RWH and GWR scheme located in each subcatchment modelled. The water recycled from the two water reuse schemes analysed here can provide the collected rainwater and treated grey water only for toilet flushing, irrigation and industrial usages in their own subcatchments. In addition, RWH tank can collect rainwater from roof, road and pavement surfaces, whereas GWR tank can collect greywater from consumption of some household appliances and fittings (i.e. hand basin, shower, dish washer and washing machine), industrial and commercial users. Other main parameters for design of RHW and GWR schemes are determined by using the developed optimisation model outlined below.

2.2 Optimisation Model

The problem of optimised allocation from multiple water resources to multiple water demands (i.e. subcatchments) is formulated as a multi-objective optimisation model [11]. The NSGA-II multi-objective, evolutionary algorithm [12] is used here to solve the optimisation model. The assessment criteria considered in the optimisation model as objective functions are (1) maximizing the reliability of water supply and (2) minimizing the total operational cost. The reliability of water supply is expressed here as the ratio of the total water delivered to customers (S_i) to the total water demand (D_i) [13]:

$$\text{Max Reliability} = \sum_{t=1}^{ntimestep} S_t \bigg/ \sum_{t=1}^{ntimestep} D_t \times 100 \qquad (2)$$

The second objective comprises the annual average of total operational expenditure of the UWS components including water-recycling schemes over the planning horizon, discounted to the present value with a specified discount rate. Note that none of the capital costs related to any components in the UWS is included in this cost. The operational costs include any fixed (e.g. labour and maintenance) and variable (e.g. energy) costs incurred in different UWS components. More specifically, given fixed annual costs in time step t as An_t (e.g. labour and O&M) and variable costs per spent unit volume of water, including energy in time step t as En_t (e.g. pumps, treatment and distribution) and chemicals in time step t as Ch_t used over the entire time horizon, the cost objective function can be written as:

$$\text{Min Cost} = \left(\sum_{t=1}^{ntimestep} (An_t + En_t + Ch_t) \middle/ ntimestep \right) \times 12 \qquad (3)$$

The decision variables (genes) of each solution (chromosome) consist of two parts (Fig. 1): (1) water demand coefficients as noted in Eq. 1 for water allocation from various water resources; and (2) size of water recycling tanks and the start year of operation for water recycling options in each subcatchment. Note that the start year of operation is an integer value ranging between 1 and 30 (end year of the planning horizon), whereas other variables are real values in which demand coefficients range between 0 and 1. Further details of assumptions are described in the next section.

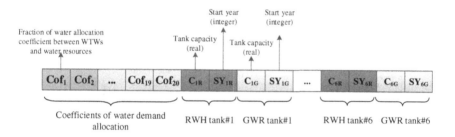

Fig. 1 Chromosome representation in the developed optimisation model

3   Case Study

The above methodology is demonstrated through its application to the real-world UWS of Kerman City located in the south-eastern part of Iran as a reference city. While the city is suffering from decreasing water resources due to overexploitation of groundwater resources, it is likely to face challenges in the future due to population growth and increasing urbanization. The city of Kerman with a population of ~640,000 in 2011 and a total area of 140 Km^2 is in an arid region. The main water resources of the Kerman UWS is currently groundwater (i.e. four sets of grouped wells as schematically shown in Fig. 2) [14, 15].

Increasing rate of population growth and numerous droughts in recent years have resulted in the sever depletion of the aquifer levels due to excessive water withdrawals. In addition, to deal with increasing water demand, a dam reservoir as a new water source will be constructed and added to the system. The new dam is 150 Km away from the city and the stored water needs to be pumped around 1000 meters causing a lot of energy consumed for pumping. Alternatively, non-traditional water resources (i.e. RWH and GWR schemes) can also be built to balance out the ever-increasing water demand over a long-term planning horizon. The electricity required for the operation of RWH and GWR schemes is estimated to be 0.54 and 1.84 kWh/m^3, respectively [2, 9]. As a result, multiple water sources including existing and potential ones are available to supply different water demand profiles [14, 15, 16]. Hence, twenty coefficients of water demand allocation between WTWs and water resources (Fig. 2), the tank capacity and the start year for both water reuse schemes related to each of the six subcatchments constitute decision variables of the optimisation model as shown in Fig. 1.

The real-world integrated UWS of Kerman was modelled in WaterMet^2 using the main UWS components including water supply, wastewater and stormwater collection as shown in Fig. 2. Six subcatchments were used to define water consumption and drainage basins. The water demand profiles in the UWS were split into domestic, industrial (commercial), garden watering and unregistered public use in the six subcatchments (Table 1). The domestic (indoor) water demand per capita was further split into six types of appliances and fittings including dishwasher, washing machine, hand basin, toilet, shower and kitchen sink. The Kerman UWS consists of ~160,000 household properties with the occupancy of approximately 4 inhabitants.

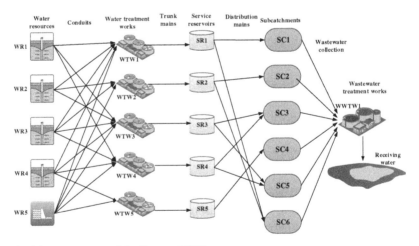

Fig. 2 Main components of the Kerman UWS

Table 1 Main features of water demand profiles in the case study

Subcatchment No	Population	Household demand (M³/month)	Industrial demand (M³/month)	Irrigation demand (M³/month)	Unregistered public demand (M³/month)
1	283,000	1,358,400	67,900	101,900	186,800
2	72,900	376,000	32,800	37,200	63,400
3	178,700	857,700	37,500	53,600	96,500
4	24,500	120,000	8,100	12,500	22,000
5	7,300	35,200	1,500	2,200	4,200
6	70,800	350,300	23,400	27,600	46,700
Sum	637,000	3,097,400	171,200	234,900	419,600

Note that the historical time series of rainfall and river inflow to the new dam reservoir over the past 30 years were used for this analysis. The distribution of monthly average values of these variables are shown in Fig. 3 with the annual average rainfall of 123 mm and 65 million cubic metres (MCM) water inflow to the dam reservoir.

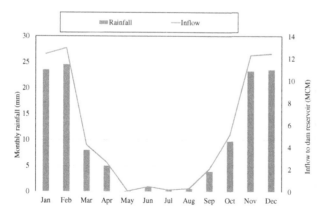

Fig. 3 Distribution of monthly average values of rainfall and inflow to the dam reservoir in Kerman UWS

4   Results and Discussion

The problem of optimising water supply management was analysed for determining water allocation coefficients in two distinct strategies: 1) only for existing water resources including groundwater and the new surface water reservoir and 2) for all existing and potential water resources including previous resources as well as water recycling schemes. Hence, the decision variables of Strategy 1 is limited to coefficients of water demand allocation (the first 20 genes in Fig. 1) while Strategy 2 considers the entire decision variables (i.e. 44 genes in Fig. 2). Furthermore, three scenarios related to different maximum allowable annual groundwater abstraction (i.e. 40, 50, 60 and 70 MCM per year) are included in the analysis to identify the impact of various regulations on the derived operating policies.

After a limited number of trial runs, the NSGA-II parameters were set as follows: population size of 52, tournament selection operator, one-point crossover with the probability of 0.75 and mutation probability of 0.05 for Strategy 1 and 0.025 for Strategy 2. These values were rigorously checked so that a fast convergence could be obtained, and the solutions reached were robust enough in different optimisation runs. The Pareto optimal front was obtained from running the optimisation model with 500 generations with no further progress afterwards.

Fig. 4 illustrates the Pareto optimal trade-offs between the two objective functions (reliability and operational cost) for the defined strategies and groundwater abstraction limits. Each optimal solution in this front represents a set of optimal operating policies for water demand allocation from different water resources over the long-term planning horizon. The figure show that the Pareto fronts obtained in different groundwater abstraction in Strategy 2 outperforms the counterpart fronts in Strategy 1, and thus the solutions in Strategy 2 can be recommended as the most cost-effective solutions. For instance, when the limitation of groundwater abstraction is 70 MCM per year, 100% reliability of water supply can be achieved with € 5.7 M per year in strategy 2, whereas strategy 1 in the same rate of groundwater limitation can achieve a maximum of 99.1% reliability by spending € 6.6 M per year as operational expenditure. As the regulation of groundwater becomes more stringent, the difference of operational costs between the two strategies will increase (compare other Pareto fronts in Fig. 4). For the most stringent regulation (i.e. groundwater abstraction of 40 MCM per year), none of the solutions in Strategy 1 can achieve a perfect value reliability of water supply (near 100%). This can be attributed to the limitation of surface water (i.e. dam reservoir) to meet the increased water demand over the later years of the planning horizon. Note that in such cases of not being able to fully provide water demands, prioritisation of water demands would be usually on the agenda and the water is supplied sequentially in order of importance (i.e. highest priority demands), which are usually sorted as domestic, industry and irrigation and unregistered public use. In the analysed case study, the domestic demands are accounted for 80% of the total demands and proportion of other water demands are 4% for industry, 6% irrigation and 11% for unregistered public use.

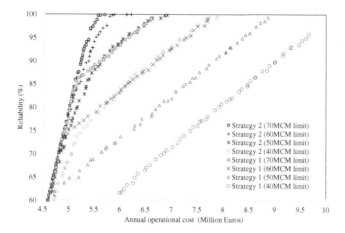

Fig. 4 Pareto trade-offs for the two water resource strategies and different groundwater abstraction limits

Further analysis of Pareto front solutions can be carried out by focusing on the result of the two extreme groundwater abstraction limits (i.e. 40 and 70 MCM per year). Fig. 5 shows the percentage contribution of all non-dominated solutions of the relevant Pareto front in ascending order of operational costs for the two strategies. As expected, the major source of water supply in all optimised strategies are groundwater (ranging averagely between 97% and 61% for different Pareto fronts) although its contribution reduces in different solutions for either providing higher reliability (i.e. larger solution number in the Pareto front) or including water recycling source. The main reason that groundwater is the primary water source in all solutions is due to its lower cost compared to other water resources in the case study. Therefore, this water source is replaced with other sources only when regulatory withdrawal limitations are imposed. Out of the other competitive sources (i.e. surface water, RWH and GWR schemes), The GWR option is accounted for the highest surrogate and thus the most cost-effective option with respect to the total operational cost in the UWS. More specifically, once the groundwater abstraction limits preclude the UWS from using further groundwater, the GWR source in strategy 2 indicates averagely 95% of surrogate water for solutions of the Pareto front with 70 MCM/year of groundwater abstraction and %73 for solutions of the Pareto fronts with 40 MCM/year of groundwater abstraction. This high proportion can be probably linked to the saving costs of GWR due to decreasing in both raw water abstraction and wastewater generation [2]. In addition, RWH option seems to be an ineffective option for the case study due mainly to small rate of average annual rainfall over the planning horizon (123 mm/year) [10].

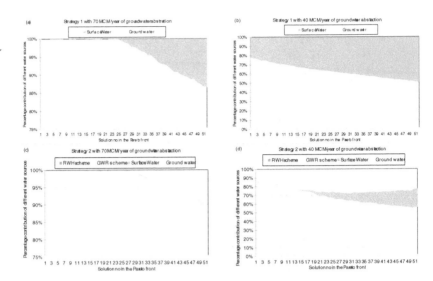

Fig. 5 Percentage contribution of different water resources in the Pareto optimal solutions for both strategies and two annual groundwater abstraction limits

5   Conclusions

Long-term planning of optimal water allocation was analysed in this paper for the real-world Kerman UWS, Iran, using existing water resources and potential ones including conventional water resources development and two types of water recycling schemes. Optimal monthly water allocation coefficients were determined by using the demand-oriented WaterMet[2] simulation linked to the NSGA-II multiobjective evolutionary algorithm (MOEA). The two water harvesting/recycling type schemes were also considered in conjunction with traditional sources. The two objectives used in the MOEA model were reliability of water supply and the total cost of the UWS operation.

The results obtained suggest that in the case of stringent regulation of groundwater abstraction, the optimised use of conventional water resources may not be able to fully supply the water demand required, especially for higher supply reliabilities. However, the inclusion of alternative water sources such as optimal water harvesting and recycling schemes could provide additional water required to achieve this and at reduced operational cost. In particular, grey water recycling scheme seems to be a good alternative to limited groundwater abstraction even under most stringent regulations. On the other hand, rainwater harvesting seems to represent a less attractive alternative to groundwater in the case study analysed given very little rainfall in summer months, i.e. when water is needed most.

Although the results obtained here present some promising conjunctive-use strategies, they should be further evaluated on other case studies with different structures and operational layout in order to derive more general conclusions for real-world applications of water reuse schemes in water supply management problems in the future.

References

1. Nair, S., George, B., Malano, H.M., Arora, M., Nawarathna, B.: Water–energy–greenhousegas nexus of urban water systems: review of concepts. state-of-art and methods. Resour. Conserv. Recycl. **89**, 1–10 (2014)
2. Behzadian, K., Kapelan, Z.: Advantages of integrated and sustainability based assessment for metabolism based strategic planning of urban water systems. Sci. Total Environ. **527–528**, 220–231 (2015). doi:10.1016/j.scitotenv.2015.04.097
3. Liu, S.C., Makropoulos, C.K., Memon, F.A., Butler, D.: Object based household water cycle model: Concept and construction, water practice and technology. IWA Online Journal **2**, 2 (2006)
4. Metcalf, L., Eddy, H.P., Tchobanoglous, G.: Wastewater engineering: treatment, disposal, and reuse. McGraw-Hill (2010)
5. Behzadian, K., Kapelan, Z., Morley, M.S.: Resilience-based performance assessment of water-recycling schemes in urban water systems. Procedia Eng. **89**, 719–726 (2014)
6. Najia, F., Lustig, T.: On-site water recycling—a total water cycle management approach. Desalination **188**(1), 195–202 (2006)
7. Rozos, E., Makropoulos, C., Butler, D.: Design robustness of local water-recycling schemes. J. Water Resour. Plann. Manage. **136**(5), 531–538 (2010)
8. Trembath, M.: Desal plant to close down. St George & Sutherland Shire Leader (Fairfax Media) (2012) (Access at http://www.theleader.com.au/story/267462/desal-plant-to-close-down/) (retrieved June 10, 2015)
9. Behzadian, K., Kapelan, Z.: Modelling Metabolism Based Performance of Urban Water System Using WaterMet2. Resources Conservation and Recycling **99**, 84–99 (2015). doi:10.1016/j.resconrec.2015.03.015
10. Behzadian, K., Kapelan, Z., Venkatesh, G., Brattebo, H., Saegrov, S.: WaterMet2: a tool for integrated analysis of sustainability-based performance of urban water systems. Drinking Water Eng. Sci. **7**, 63–72 (2014). doi:10.5194/dwes-7-63-2014
11. Rozos, E., Makropoulos, C.: Source to tap urban water cycle modelling. Environ. Modell. Software **41**, 139–150 (2013)
12. Deb, K., Pratap, A., Agarwal, S., Meyarivan, T.A.M.T.: A fast and elitist multiobjective genetic algorithm: NSGA-II. IEEE Trans. Evol. Comput. **6**(2), 182–197 (2002)
13. Behzadian, K., Kapelan, Z., Morley, M.S.: Resilience-based performance assessment of water-recycling schemes in urban water systems. Procedia Eng. **89**, 719–726 (2014)
14. Nazari, S., Mousavi, J., Behzadian, K.: Sustainable urban water management: a simulation optimization approach. In: HIC2014– 11th International Conference on Hydroinformatics, New York, USA (2014)
15. Nazari, S., Mousavi, J., Behzadian, K., Kapelan, Z.: Compromise programming based scenario analysis of urban water systems management options: Case study of Kerman City. In: HIC2014– 11th International Conference on Hydroinformatics, New York, USA (2014)
16. Behzadian, K., Kapelan, Z., Morley, M.S., Govindarajan, V., Brattebø, H., Sægrov, S., Jamshid Mousavi, S.: Quantitative assessment of future sustainability performance in urban water services using watermet$^{2.}$ In: IWA/TRUST Cities of the Future Conference, Mülheim, An Der Ruhr, Germany (2015)

Seismic Reliability-Based Design of Water Distribution Networks Using Multi-objective Harmony Search Algorithm

Do Guen Yoo, Donghwi Jung, Ho Min Lee, Young Hwan Choi
and Joong Hoon Kim

Abstract In the last four decades, many studies have been conducted for least-cost and maximum-reliability design of water supply systems. Most models employed multi-objective genetic algorithm (e.g., non-dominated sorting genetic algorithm-II, NSGA-II) in order to explore trade-off relationship between the two objectives. This study proposes a reliability-based design model that minimizes total cost and maximizes seismic reliability. Here, seismic reliability is defined as the ratio of available demand to required water demand under earthquakes. Multi-objective Harmony Search Algorithm (MoHSA) is developed to efficiently search for the Pareto optimal solutions in the two objectives solution space and incorporated in the proposed reliability-based design model. The developed model is applied to a well-known benchmark network and the results are analyzed.

Keywords Optimal design · Multi-objective · Harmony search algorithm · Earthquake hazard

1 Introduction

Earthquake is one of the natural disasters causing huge economic and social losses. It causes significant impact on critical lifelines such as water, electricity,

D.G. Yoo · D. Jung
Research Center for Disaster Prevention Science and Technology,
Korea University, Seoul 136-713, South Korea
e-mail: godqhr425@korea.ac.kr, donghwiku@gmail.com

H.M. Lee · Y.H. Choi · J.H. Kim(✉)
School of Civil, Environmental and Architectural Engineering,
Korea University, Seoul 36-713, South Korea
e-mail: {dlgh86,younghwan87,jaykim}@korea.ac.kr

© Springer-Verlag Berlin Heidelberg 2016
J.H. Kim and Z.W. Geem (eds.), *Harmony Search Algorithm*,
Advances in Intelligent Systems and Computing 382,
DOI: 10.1007/978-3-662-47926-1_43

459

bridges, and transportation systems [1]. There are two ways to reduce the losses of earthquake. One is to design system resistant to the event and the other is to restore the damaged system promptly after the event occurs. The former is achieved by strengthening the system so it can resist the outer interruption with small failure of system components. The latter focuses on the quick recovery of the system to minimize the consequences after the event occurs [2].

In this regard, related researches on the reliability assessment of water supply networks for seismic hazard have been carried out over the past 30 years. However, most recent researches have focused on the development of techniques enhancing system's ability to recover after seismic damage. The only few researches have studied methodologies for enhancing system reliability [3-5]. Li et al. [5] conducted a research to find the optimal topology of gas supply networks that are resistant to seismic hazards. The proposed model was to find the least-cost network topology while the seismic reliability between the sources and each terminal satisfies predetermined reliability constraints. For this optimization problem, a genetic algorithm (GA) was used to search for the optimal solutions. The proposed method was demonstrated in the design of a simple example network consisting of 10 nodes and an actual network with 391 nodes located in a large city in China [5].

In case of optimal design of water distribution networks, In the last four decades, Mmany studies have been proposed conducted for least-cost and maximum-reliability design of water supply systems for last four decades. Early studies took into account economic cost as a single objective [6-8]. In recent years, however, most design methods have considered more than two objective functions simultaneously to minimize economic cost and to maximize system reliability [9-10]. Most models employed multi-objective genetic algorithm (e.g., non-dominated sorting genetic algorithm-II, NSGA-II) in order to explore trade-off relationship between the two objectives.

This study proposes a reliability-based multi-objective design model that minimizes total cost and maximizes seismic reliability under earthquake hazards. Impacts of seismic wave to water mains such as pipe, pump, and tank can be considered to determine the resulting failure statuses, proposed by Yoo et al. [1]. Seismic reliability is defined as the ratio of available quantity of water to required water demand under earthquakes. Multi-objective Harmony Search Algorithm (MoHSA) is used to efficiently search for the Pareto optimal solutions in the two objectives solution space and incorporated in the proposed reliability-based design model. The developed model is applied to a benchmark network and the results are analysed.

2 Methodology

The following sections provide the detailed descriptions of the objective functions, constraints, and Multi-objective Harmony Search Algorithm (MoHSA).

2.1   Objective Functions and Constraints

The objective functions and constraints that were used in this study are shown in the following equations. The objective functions are to minimize the cost (C_c) and to maximize System serviceability (S_s). S_s in Eq. (1) is used as an surrogate measure of water supply network reliability. S_s is a factor for the evaluation of water supply ability against seismic hazards. It can be determined as the ratio of the required demand to the available demand of the entire system. This value shows the supply ability of water supply networks and is normally defined and used as availability or serviceability. To calculate S_s, the available demand of the system that is obtained as the result of hydraulic analysis of the water supply network after the occurrence of an earthquake should be calculated. Yoo et al. [1] performed repetitive probabilistic simulations to quantify hydraulic reliability of system components under possible seismic scenarios. In this model, probabilistic seismic hazards are produced in target area using Monte Carlo simulation (MCS) and hydraulic reliability is quantified by the proposed reliability index. In this paper, the same approach is adopted to calculate the available demand in the system.

The available demand is calculated according to the node pressure that appears after an earthquake as shown in Eq. (3). If the pressure of a node is zero, the node cannot supply water and if the nodal pressure is above a certain level (P_{min}), the node can deliver all water demanded. If the pressure is between zero and P_{min}, the amount of water that can be supplied by the node will decrease in proportion to the value of $\sqrt{P_i/P_{min}}$. The optimal designs for all pipes aim to satisfy the standards for minimum nodal pressure requirement, as shown in Eq. (2).

Objective Functions

$$\text{Maximize System Serviceability } (S_s) = \frac{\sum Q_{avl,i}}{\sum Q_{ini,i}} \tag{1}$$

$$\text{Minimize Construction Cost } (C_c) = \sum_{i=1}^{n}(uc(D_i) \times L_i)$$

Subject to,

$$P_{i,n} \geq P_{min} \tag{2}$$

Here, C_c = Pipe construction cost, $uc(D_i)$ = the unit cost of the pipe of diameter D_i (USD/ft) (i=1, …, n), n = number of pipes, L_i = the length of pipe.

$Q_{avl,i}$ = Available nodal demand at node i, which is expressed as a function of nodal pressure as follows:

$$Q_{avl,i} = \begin{cases} 0 & \text{when } P_i = 0 \\ Q_{new,i} \times \sqrt{\dfrac{P_i}{P_{min}}} & \text{when } P_i < P_{min} \\ Q_{new,i} & \text{when } P_i \geq P_{min} \end{cases} \tag{3}$$

Here, $Q_{new,i}$ = Updated nodal demand after pipe breakage at node i,

P_i = Nodal pressure at node i after earthquake,
P_{min} = Allowable minimum nodal pressure,
$Q_{in\,i,i}$ = Required nodal demand at node i,
$P_{i,n}$ = Nodal pressure under normal condition at node i

2.2 Multi-objective Harmony Search Algorithm

The Harmony Search Algorithm (HSA) is an algorithm inspired by the artificial musical phenomenon, which was initially proposed by Geem et al. [11] and Kim et al. [12]. Various sounds made by multiple musical instruments combine to form a chord and the chord may be a harmony, in which sounds are harmonized with each other, or a discord, in which sounds are not harmonized with each other. Discords disappear gradually with practice. Among those chords that are appropriate as harmonies (local optimum) are the most aesthetically beautiful chords (global optimum), which can be achieved through much practice. The HSA is a technique that regards the optimal chords found through practice processes as the optimum solutions to be found. As with other meta-heuristic techniques, the HSA uses several factors, such as harmony memory size (HMS), harmony memory considering rate (HMCR), and pitch adjusting rate (PAR). After the development of the HSA in 2001, studies [13-14] to revise or improve the algorithm, such as changing the parameters of the algorithm itself, have been conducted and successfully applied to engineering problems. In addition, the concept of multi-objective HSAs, which can solve two or more objectives, has been presented by some researchers [15].

A multi-objective optimization problem is defined as follows: give an n-dimensional decision variable vector x={x1,...,xn} in the solution space X, find a vector x* that minimizes a given set of N objective functions z(x*)={z1(x*),...,zN(x*)}. The solution space X is generally restricted by a series of constraints, such as Cj(x*)=bj for j=1, ...,m and bounds on the decision variables. A reasonable solution to a multi-objective optimization problem is to find a set of solutions, which satisfies the objectives without being dominated by any other feasible solutions. If all objective functions are for minimization, a feasible solution x is said to dominate another feasible solution y, if and only if, $z_i(x) \le z_i(y)$ for i=1,...,N and $z_j(x) < z_j(y)$ for least one objective function j (j=1,...,N) [16]. A solution is said to be Pareto optimal if it is not dominated by any other solution in the solution space. A Pareto optimal solution cannot be improved with respect to any objective without worsening at least one other objective. The set of all feasible non-dominated solutions in X is referred to as the Pareto optimal set, and for a given Pareto optimal set, the corresponding objective function values in the objective space are called the Pareto front. In MoHSA, the fitness assignment procedure is based on Pareto-ranking approaches. Various methods can be used for determining non-dominated solutions. In this study, the ranking approach proposed by Fonseca and Fleming [17] shown in Fig.1. Table 1 shows the Pseudo code of MoHSA.

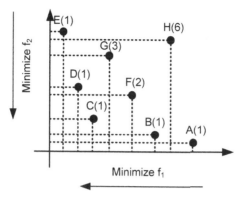

Fig. 1 Ranking method [17] used in the MoHSA

Table 1 Pseudo code of MoHSA

Begin
Objective function $f_1(x)$, $x=(x_1, x_2, ..., x_d)^T$
Objective function $f_2(x)$, $x=(x_1, x_2, ..., x_d)^T$
Generate initial harmonics (Define Harmony Memory and Size, HM & HMS)
Define pitch adjusting rate (PAR), pitch limits and bandwidth (BW)
Define harmony memory considering rate (HMCR)
Determining non-dominated solutions by Fonseca and Fleming [17]
while (*t<Max number of iterations*)
Generate new harmonics by accepting best harmonics
Adjust pitch to get new harmonics (solutions)
if *(rand<HMCR), choose an existing harmonic randomly*
else if *(rand<PAR), adjust the pitch randomly within limits*
else *generate new harmonics via randomization*
end if
Determining non-dominated solutions by Fonseca and Fleming [17]
Accept the new harmonics (solutions) if better
end while
Find the current best solutions (Pareto optimal solutions and Pareto fronts)
End

3   Application and Results

3.1   Study Network and Input Parameter

The proposed model is applied for the optimal design of a well-known benchmark WDS, Anytown network as shown in Fig. 2. Anytown network firstly published by Walski et al. [18] was modified by Jung et al. [10]. The unit costs of the commercial pipes are adopted from Walski et al. [18]. The pipe size can be selected from commercial pipe sizes (6, 8, 10, 12, 14, 16, 18, 20, 24, and 30 inch). In order to generate random locations of earthquake, 9 points were uniformly created and laid on the study network. Earthquake magnitude of 4 (M4) was generated at the points.

Total 900 stochastic earthquakes were generated by MCS (100 simulation on average at each point), considering the fact that the number of seismic points used is 9. The minimum pressure of the Anytown netowork is set to be 40 psi (pounds per square inch). In case of parameters of MoHSA, the HMS is set to 30, and the maximum number of times of repetition is determined to be 50,000. The values of HMCR and PAR were determined to be 0.95 and 0.05, respectively.

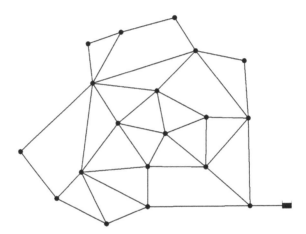

Fig. 2 Anytown network

3.2 Application Results

The obtained optimization results by MoHSA are shown in Fig. 3. The final Pareto optimal set was determined after 50,000 iterations. We found the representative solutions from the Pareto front sets shown as rectangular shape points. The optimal solutions show the better convergence and diversity ability of solution quality than the result until 1,000 iterations. The value of seismic reliability ranges from 0.318 to 0.416. In case of the construction cost, then minimum and maximum values were 13.3M and 14.6M USD, respectively. Therefore, one solution from the Pareto sets could be selected considering the budget and environmental conditions.

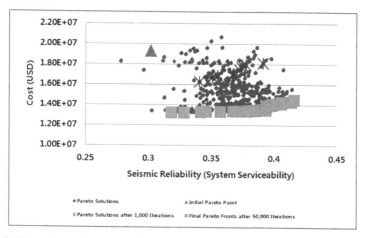

Fig. 3 MoHSA optimization results for Anytown network

Fig. 4. shows the optimal one candidate solution among optimal Pareto solutions. The seismic reliability is 0.406 and the construction cost is 14.6M USD. The corresponding optimal pipe layout is also presented in Fig. 4. The minimum size (6 inch) pipe should be installed even if it is the most vulnerable pipe to earthquakes. It means that the optimal solution shows the compromised results between the reliability and cost. It is also caused by network's layout. As we can see in Fig. 4, there are many diverse routes to supply the water at each node because the Anytown network is a looped network.

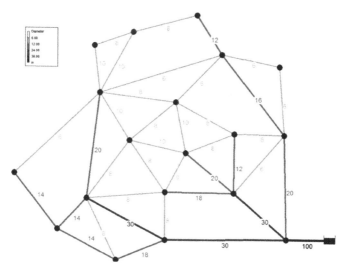

Fig. 4 The one candidate solution among optimal Pareto optimal soultions

4   Conclusions

This study developed a model to find optimal set of pipe diameter combinations that show minimum construction cost and maximum seismic reliability satisfying minimum require pressure requirement at the nodes. To solve this problem, MoHSA was used as optimization technique. The developed model was applied to a well-known benchmark water supply networks, and the results were analyzed. The results showed that MoHSA can be efficiently applied to search for the Pareto optimal solutions in the two objectives solution space and incorporated in the proposed reliability and cost based design model. Therefore, it was concluded that optimal pipe design of developed model can enhance seismic reliability and increase economic efficiency.

In future research, the proposed methods should be applied to real size networks for the purpose of additional verification.

Acknowledgment This work was supported by the National Research Foundation (NRF) of Korea under a grant funded by the Korean government (MSIP)(NRF-2013R1A2A1A01013886) and the Korea Ministry of Environment as "The Eco-Innovation project (GT-11-G-02-001-2)".

References

1. Yoo, D.G., Kang D.S, Kim. J.H.: Seismic reliability assessment model of water supply networks. In: Proceedings of the World Environmental and Water Resources Congress, Cincinnati, OH, USA (2013)
2. Kang, D.S., Lansey, K.: Post-earthquake Restoration of Water Supply Infrastructure. World Environmental and Water Resources Congress **2013**, 913–922 (2013)
3. Tan, R., Shinozuka, M.: Optimization of Underground Water Transmission Network Systems under Seismic Risk. Soil. Dyn. Earthq. Eng. 1(1), 30–38 (1982)
4. Hoshiya, M., Yamamotob, K., Ohno, H.: Redundancy Index of Lifelines for Mitigation Measures against Seismic Risk. Probabilist. Eng. Mech. **19**(3), 205–210 (2004)
5. Li, J., Liu, W., Bao, Y.F.: Genetic Algorithm for Seismic Topology Optimization of Lifeline Network Systems. Earthq. Eng. Struct. D. **37**(11), 1295–1312 (2008)
6. Alperovits, E., Shamir, U.: Design of optimal water distribution systems. Water. Resour. Res. **13**(6), 885–900 (1977)
7. Simpson, A., Dandy, G., Murphy, L.: Genetic algorithms compared to other techniques for pipe optimization. J. Water. Res. Pl. ASCE **120**(4), 423–443 (1977)
8. Savic, D., Walters, G.: Genetic algorithms for least-cost design of water distribution networks. Water. Res. Pl. ASCE **123**(2), 67–77 (1997)
9. Kapelan Z., Savic D., Walters G.: Multiobjective design of water distribution systems under uncertainty. Water. Resour. Res. **41**(11) (2005)
10. Jung, D., Kang, D., Kim, J.H., Lansey, K.: Robustness-based design of water distribution systems. J. Water. Res. Pl. ASCE **140**(11), 04014033 (2014)
11. Geem, Z.W., Kim, J.H., Loganathan, G.V.: A New Heuristic Optimization Algorithm: Harmony Search. Simulation **76**(2), 60–68 (2001)

12. Kim, J.H., Geem, Z.W., Kim, E.S.: Parameter Estimation of the Nonlinear Muskingum Model using Harmony Search. J. Am. Water. Res. As. **37**(5), 1131–1138 (2001)
13. Paik, K.R., Kim, J.H., Kim, H.S., Lee, D.R.: A Conceptual Rainfall Runoff Model Considering Seasonal Variation. Hydrol. Process. **19**(19), 3837–3850 (2005)
14. Geem, Z.W.: Improved harmony search from ensemble of music players. In: Gabrys, B., Howlett, R.J., Jain, L.C. (eds.) KES 2006. LNCS (LNAI), vol. 4251, pp. 86–93. Springer, Heidelberg (2006)
15. Sivasubramani, S., Swarup, K.S.: Multi-Objective Harmony Search Algorithm for Optimal Power Flow Problem. Int. J. Elec. Power **33**(3), 745–752 (2011)
16. Konak, A., Coit, D.W., Smith, A.E.: Multi-objective optimization using genetic algorithms: A tutorial. Reliab. Eng. Syst. Saf. **91**(9), 992–1007 (2006)
17. Fonseca, C.M., Fleming, P.J.: Genetic algorithms for multi-objective optimization: formulation, discussion and generalization. In: Proceeding of the 5th International Conference on Genetic Algorithms, pp. 416–423 (1993)
18. Walski, T., Brill Jr., E., Gessler, J., Goulter, I., Jeppson, R., Lansey, K., Lee, H., Liebman, J., Mays, L., Morgan, D., Ormsbee, L.: Battle of the network models: Epilogue. J. Water. Res. Pl. ASCE **113**(2), 191–203 (1987)

Retraction Note to: Optimization of Water Distribution Networks with Differential Evolution (DE)

Ramin Mansouri, Hasan Torabi and Hosein Morshedzadeh

Retraction to: J.H. Kim and Z.W. Geem (eds.),
Harmony Search Algorithm,
Advances in Intelligent Systems and Computing 382,
DOI: 10.1007/978-3-662-47926-1_39

DOI 10.1007/978-3-662-47926-1_44

After publication of the chapter "Optimization of Water Distribution Networks with Differential Evolution (DE)" (Pages 403–419) in the book Harmony Search Algorithm, it has come to the attention of the Editors that the authors were not registered at the conference to present their paper. As it is a condition of publication in the proceedings that all papers should have been presented at the conference, the Editors have decided to retract the chapter.

The original online version for this chapter can be found at
DOI: 10.1007/978-3-662-47926-1_39

R. Mansouri(✉) · H. Torabi
Water Engineering Department, Lorestan University, Khoramabad, Iran
e-mail: ramin_mansouri@yahoo.com

H. Morshedzadeh
Economics and Management Department, Tehran University, Tehran, Iran

© Springer-Verlag Berlin Heidelberg 2016   E1
J.H. Kim and Z.W. Geem (eds.), *Harmony Search Algorithm*,
Advances in Intelligent Systems and Computing 382,
DOI: 10.1007/978-3-662-47926-1_44

Author Index

Printed in the United States
By Bookmasters